P9-AGA-522

WITHDRAWN
UTSA LIBRARIES

RENEWALS 458-4574
DATE DUE

GAYLORD PRINTED IN U.S.A.

MICROELECTRONIC DESIGN OF FUZZY LOGIC–BASED SYSTEMS

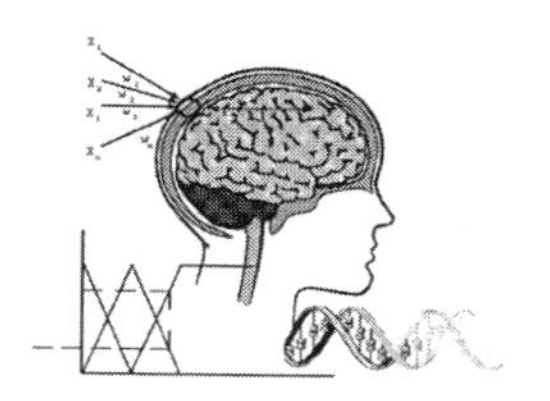

The CRC Press

International Series on Computational Intelligence

Series Editor

L.C. Jain, Ph.D., M.E., B.E. (Hons), Fellow I.E. (Australia)

L.C. Jain, R.P. Johnson, Y. Takefuji, and L.A. Zadeh
Knowledge-Based Intelligent Techniques in Industry

L.C. Jain and C.W. de Silva
Intelligent Adaptive Control: Industrial Applications in the Applied Computational Intelligence Set

L.C. Jain and N.M. Martin
Fusion of Neural Networks, Fuzzy Systems, and Genetic Algorithms: Industrial Applications

H.-N. Teodorescu, A. Kandel, and L.C. Jain
Fuzzy and Neuro-Fuzzy Systems in Medicine

C.L. Karr and L.M. Freeman
Industrial Applications of Genetic Algorithms

L.C. Jain and B. Lazzerini
Knowledge-Based Intelligent Techniques in Character Recognition

L.C. Jain and V. Vemuri
Industrial Applications of Neural Networks

H.-N. Teodorescu, A. Kandel, and L.C. Jain
Soft Computing in Human-Related Sciences

B. Lazzerini, D. Dumitrescu, L.C. Jain, and A. Dumitrescu
Evolutionary Computing and Applications

B. Lazzerini, D. Dumitrescu, and L.C. Jain
Fuzzy Sets and Their Application to Clustering and Training

L.C. Jain, U. Halici, I. Hayashi, S.B. Lee, and S. Tsutsui
Intelligent Biometric Techniques in Fingerprint and Face Recognition

Z. Chen
Computational Intelligence for Decision Support

L.C. Jain
Evolution of Engineering and Information Systems and Their Applications

H.-N. Teodorescu and A. Kandel
Dynamic Fuzzy Systems and Chaos Applications

L. Medsker and L.C. Jain
Recurrent Neural Networks: Design and Applications

L.C. Jain and A.M. Fanelli
Recent Advances in Artifical Neural Networks: Design and Applications

M. Russo and L.C. Jain
Fuzzy Learning and Applications

J. Liu
Multiagent Robotic Systems

M. Kennedy, R. Rovatti, and G. Setti
Chaotic Electronics in Telecommunications

H.-N. Teodorescu and L.C. Jain
Intelligent Systems and Techniques in Rehabilitation Engineering

I. Baturone, Á. Barriga, S. Sánchez-Solano, C.J. Jiménez-Fernández, and D.R. López
Microelectronic Design of Fuzzy Logic–Based Systems

T. Nishida
Dynamic Knowledge Interaction

C.L. Karr
Practical Applications of Computational Intelligence for Adaptive Control

MICROELECTRONIC DESIGN OF FUZZY LOGIC–BASED SYSTEMS

Library
University of Texas
at San Antonio

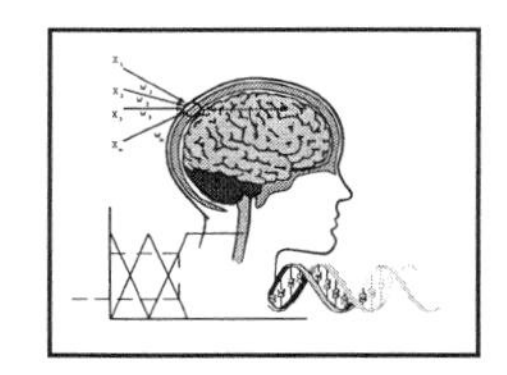

I. Baturone,
Á. Barriga,
S. Sánchez-Solano,
C.J. Jiménez-Fernández,
and
D.R. López

CRC Press
Boca Raton London New York Washington, D.C.

Library
University of Texas
at San Antonio

Library of Congress Cataloging-in-Publication Data

Catalog record is available from the Library of Congress.

This book contains information obtained from authentic and highly regarded sources. Reprinted material is quoted with permission, and sources are indicated. A wide variety of references are listed. Reasonable efforts have been made to publish reliable data and information, but the author and the publisher cannot assume responsibility for the validity of all materials or for the consequences of their use.

Neither this book nor any part may be reproduced or transmitted in any form or by any means, electronic or mechanical, including photocopying, microfilming, and recording, or by any information storage or retrieval system, without prior permission in writing from the publisher.

The consent of CRC Press LLC does not extend to copying for general distribution, for promotion, for creating new works, or for resale. Specific permission must be obtained in writing from CRC Press LLC for such copying.

Direct all inquiries to CRC Press LLC, 2000 N.W. Corporate Blvd., Boca Raton, Florida 33431.

Trademark Notice: Product or corporate names may be trademarks or registered trademarks, and are used only for identification and explanation, without intent to infringe.

© 2000 by CRC Press LLC

No claim to original U.S. Government works
International Standard Book Number 0-8493-0091-6
Printed in the United States of America 1 2 3 4 5 6 7 8 9 0
Printed on acid-free paper

Foreword

During the last 35 years fuzzy logic has evolved from a mathematical curiosity to a mature scientific body covering many different disciplines. It has attracted considerable attention because of its capability for representing complex phenomena and, particularly, for mimicking human reasoning mechanisms. By moving from classical mathematical models to logical models involving linguistic parameters, fuzzy sets paved the way for emulating human behavior by incorporating imprecision and vagueness.

The more the theoretical basis of the field has been developed the greater the number of applications have been foreseen for fuzzy logic. Traditionally, both numerical and algebraic methods have predominated in science and engineering, but many problems have been recognized to be more efficiently handled when we shift from differential-difference equations to fuzzy rules. In particular, the incorporation of these new concepts in control problems has been successfully explored, especially the so-called rule-based fuzzy paradigm.

Since the mid-1980s, there has been a growing activity in considering fuzzy models for building controllers. Combining logical operations for the connectives with fuzzy numeric values for the variables and parameters seems very appealing. Furthermore, this unique combination may lead to efficient results in practical situations where high numerical precision is neither attainable nor necessary, or where imprecision can provide a robustness that cannot be achieved by classical controllers.

Associated with this evolution, electronic implementations of fuzzy controllers have been reported, and some of them have even gone to production. However, although quite a few examples can be found in the technical literature covering a large application area, I personally believe that the practical incorporation of fuzzy techniques into the microelectronics industry is still in its infancy. With the advent of Systems-On-Chip (SOC), the potential usefulness of soft computing in general, and of fuzzy logic in particular, has broadened. There is room to benefit from the advantages offered by fuzzy systems to achieve significant savings in area, power, and cost, as well as improvements in speed, design complexity, and testing effort.

In terms of practical realizations, most of the examples given fall into one of two categories: (a) systems based on conventional digital hardware (general-purpose computers) but implementing (in one way or another) fuzzy rules by software and (b) special-purpose circuits (analog, digital, or mixed analog-digital) corresponding to a restricted view of the implementation space.

Since there is no universal solution for resolving the huge number of design needs a practicing engineer must face, the application of fuzzy logic to real-world problems can be seen as something that enables the fuzzy paradigm to be incorporated within a complex system in the most efficient manner. To fully exploit the potentialities of fuzzy logic, there are two requirements: first, popularizing the paradigm through a comprehensive, yet complete coverage of its main concepts in a language familiar to the electrical engineer; and second, filling in the gap between abstract fuzzy modeling and the techniques available for designing integrated circuits. The adoption of fuzzy technology requires:

(a) A comprehensive review of the main concepts and techniques forming the theoretical basis of fuzzy logic. This can be considered as an "engineering view" of those advanced mathematical principles that must be understood in order to take advantage of the potential usefulness of fuzzy logic.

(b) A solid background on microelectronic implementations of those components, which can be used as basic building blocks for constructing a fuzzy system. This task must be performed in a way that opens the door to different developments and allows the designer to select the most convenient implementation technique for his/her application and constraints (especially cost limitations).

(c) A firm support frame linking high-level fuzzy abstractions and the electrical circuits implementing them. In particular, VHDL descriptions, behavioral simulation and CAD tools integrated in the regular design flow are essential.

Tackling these issues is the main goal of this book, giving the reader the chance to build a system or subsystem based on fuzzy logic by resorting to common practices in the semiconductor industry. The book aims to provide methodologies, procedures, circuit techniques and CAD tools leading to practical, low-cost solutions, compatible with existing digital and analog processing technologies. In a few words, an essential feature of this book is its focus on how to map fuzzy concepts into working silicon. A design environment is presented in depth, allowing us to select the most convenient circuit technique or even share several design techniques. An example is the combination of analog and digital methods to obtain the best of both worlds.

This book was written by an enthusiastic group of young, yet mature researchers who have been exploring practical microelectronic applications of the fuzzy paradigm for the last 10 years. Undoubtedly, I am not impartial in my views because I supervised most of them when they began their careers as researchers, but declaring how proud I am of their final product, presented in the twelve chapters that follow, does not invalidate my technical appreciation of the quality of this material.

In summary, this book is a unique yet practical and readable text intended to cover the practical implementation of fuzzy systems from the point of view of the practicing circuit designer. Its aim is to broaden and popularize the use of fuzzy circuits in any electronics application field.

J. L. Huertas
Centro Nacional de Microelectrónica, CSIC–University of Sevilla

Preface

Emulating human performance by machines has been a constant challenge for scientists and engineers. Particularly, those human ways to solve problems that can be expressed in terms of a mathematical algorithm like complex numerical calculations are efficiently implemented on digital computers. However, activities that we perform easily every day, such as pattern recognition, prediction, control, or decision making, are very difficult to simulate by conventional computers. In these tasks, human reasoning mechanisms usually show an ability to extract correct conclusions from ambiguous, imprecise, or incomplete information.

Fuzzy logic offers inference mechanisms to cope with uncertainty and provides a mathematical framework into which we can translate the solutions that a human expert expresses linguistically. A fuzzy logic–based system combines the capability of representing the structured rule-based knowledge that a human expert usually employs with the numerical information processing for which hardware implementations are suitable. Software implementation of fuzzy systems as fuzzy expert systems has been the initial approach. However, a wide range of demanding areas requiring real-time operation and/or small size and low power consumption, such as control, robotics, automotive, signal processing, etc., call for hardware realizations tailored to the fuzzy paradigm.

The material presented in this book is intended to provide the reader with a comprehensive discussion on the theoretical and practical issues concerning the hardware design of fuzzy systems, with special emphasis on microelectronic realizations. On the one hand, this book is a self-contained reference that offers microelectronic solutions for scientists and practicing engineers involved in the area of knowledge-based computational techniques that employ fuzzy logic. On the other hand, this book can be used for educational purposes as a text for tutorials or advanced courses on subjects such as non-conventional processors, advanced integrated circuit design, VLSI microsystems design, applications of fuzzy logic, or fuzzy software design.

Our main goals with this book have been the following:

- To show clear and concise explanations and illustrations for understanding the translation of fuzzy logic theory into dedicated hardware.
- To write a comprehensive description of the fundamentals and details of VLSI design of fuzzy systems, from the architectural to the circuit level.

- To provide a reference that covers the most up-to-date microelectronic realizations of fuzzy systems: analog, digital, and mixed-signal approaches together with their cost and capability of solving demanding, real-world problems.
- To thoroughly explain the design and practice of CAD tools for implementing fuzzy systems.

The material is organized into 12 chapters.

The first three chapters present an overview of fuzzy logic theory and introduce the different concepts employed in the following chapters. **Chapter 1** introduces fuzzy logic, summarizing its history and application domain. **Chapter 2** reviews the basic mathematical foundations of fuzzy set theory. The mathematical formalisms of fuzzy inference mechanisms are presented in **Chapter 3**. The structure of a fuzzy logic–based system is analyzed in order to identify the various stages and building blocks that will be considered in their software and/or hardware implementation.

The development of a fuzzy system may be a complex task due to the large number of possibilities available to the designer. In this sense, computer-based tools are very helpful in assisting the designer at the different levels of the design process. **Chapter 4** discusses the features these CAD environments may offer, such as the use of a formal language for system specification; facilities for system definition like graphical user interfaces; capabilities for fuzzy operator extensions; integration with other development tools; facilities for tuning, verifying, and analyzing the system; and synthesis capabilities for obtaining the final implementation of the system. One of these design environments, called *Xfuzzy*, is presented in this chapter. It integrates a set of tools that will be used along the rest of the book to illustrate the whole design process of a microelectronic fuzzy system.

The first level in this design process is the algorithmic level, which consists of selecting the inference mechanism and defining the knowledge base. Two techniques that are basic at this level are learning techniques, which allow tuning the knowledge base automatically, and simulation techniques, which permit evaluating the fuzzy system behavior inside its operational environment. **Chapter 5** explains these verification techniques and, particularly, the verification tools provided by *Xfuzzy*.

Most development environments for designing fuzzy systems allow implementing them with a high-level programming language. Software implementations of fuzzy systems are easily obtained and provide a high flexibility, but they are not adequate for applications demanding fuzzy systems with small size, low power consumption, and high inference speed. To meet these requirements, hardware approaches must be adopted. **Chapter 6** reviews the different hardware approaches, also known as fuzzy hardware. One approach is to employ general-purpose microprocessors, occasionally expanded with new instructions or new circuitry. The other approach is to design dedicated hardware or fuzzy chips, as they have been named in the literature, that is, integrated circuits (ICs) optimized for fuzzy logic–based inference systems. The rest of the book focuses mainly on the last procedure, primarily considering CMOS technologies. The fundamental concepts concerning the design and fabrication of integrated circuits are discussed in this chapter. This material serves as an introduction to the following chapters where microelectronic realizations of fuzzy systems with different design techniques are described in more detail.

Analog techniques for designing fuzzy integrated circuits are explained in Chapters 7 and 8. **Chapter 7** describes continuous-time analog techniques, which offer a very good relation between area occupation and inference speed. These techniques were employed to implement the first analog fuzzy ICs. Three basic architectures that allow parallel rule processing are discussed. The rest of the chapter is devoted to describing the design of the basic building blocks. The design of a CMOS prototype and its experimental results illustrate the theoretical background provided. Circuits designed with continuous-time analog design styles work with voltages or currents that are continuous in the time domain and whose amplitudes can take a continuous value within a defined range. In the discrete-time analog design styles, the signals are not continuous in time but discrete, and their amplitudes maintain a continuous range of values. The use of the latter techniques for designing fuzzy circuits is addressed in **Chapter 8**, namely switched-capacitor (SC) and switched-current (SI) techniques. While continuous-time techniques are adequate for parallel signal processing, discrete-time techniques are very well suited to sequential signal processing because data can be stored in some blocks and transmitted to other blocks according to control signals. Therefore, sequential architectures of fuzzy ICs are described. Afterwards, the ways of designing the building blocks are discussed and an SC CMOS prototype is described.

Digital techniques for designing fuzzy integrated circuits are explained in **Chapter 9**. Digital circuitry provides greater precision than analog circuitry due to its higher robustness against noise, distortion, and fabrication process variations, although in fuzzy applications these requirements are not usually strict. The realization of the different building blocks with digital circuits is generally accomplished by using library cells and well-known design techniques in which CAD tools play a relevant role; hence, their cost development is lower than analog circuits although their area and power consumption is generally higher for achieving the same operation speed. Consequently, Chapter 9 does not describe the design of the different building blocks at a transistor level, as in the two previous chapters, but at a logic level that focuses more on architectural design. In particular, an efficient architecture is described and its different implementation options regarding programmability, selection of operators, etc., are analyzed. A wide range of analyses concerning temporal behavior and area occupation of integrated circuits designed with this architecture are provided and two CMOS prototypes are presented.

The two main factors that limit the realization of a microelectronic system are its complexity and its development time. In this sense, it is very valuable to have synthesis CAD tools that automatically translate the high level description of a fuzzy system into a physical description ready to be realized as an integrated circuit. This means translating the verified definition of the system, which can be obtained with CAD tools, into a final hardware realization, so as to automatize all the design process. **Chapter 10** describes synthesis tools for fuzzy integrated circuits, in particular, those available at the *Xfuzzy* environment. Two different approaches for hardware synthesis are considered. One approach implements the fuzzy system as a look-up table describing its input/output mapping. The second alternative employs a hardware description language, namely VHDL, and a specific architecture. Two implementation techniques are analyzed: field programmable gate arrays (FPGAs) and application specific integrated circuits (ASICs).

The rest of the book is dedicated to illustrating how all the previous issues are applied to solve particular problems. **Chapter 11** focuses on control applications since they have been the major force driving the popularity of fuzzy solutions. This chapter reviews some basic concepts of conventional control systems and strategies and briefly describes the features and realization of fuzzy controllers. An application example is studied in detail to illustrate both the automatic design methodology introduced in Chapter 10 and the application of some of the circuits described in the preceding chapters.

The successful applications of fuzzy systems to many fields find an explanation in their universal approximation capability, that is, their capability to approximate any input/output mapping to any degree of accuracy. This property is discussed in **Chapter 12**. The applications illustrated in this chapter employ fuzzy systems to reproduce a desired input-output mapping, such as identification and modeling of highly non-linear static and dynamic systems. In these cases, the required fuzzy systems are usually designed by using supervised learning algorithms taken from the neural network practice. This is why these systems are often called neuro-fuzzy systems. Concerning their microelectronic implementation, two situations can be distinguished. One situation is that the learning mechanism is implemented off chip. The other situation is that the learning mechanism is implemented on chip together with the inference mechanism. In the latter case, the design of adaptive fuzzy chips is addressed.

Much of this book is the result of research carried out by the authors at the Instituto de Microelectrónica de Sevilla, which is part of the Centro Nacional de Microelectrónica (IMSE-CNM); the Universidad de Sevilla; and the Centro Informático Científico de Andalucía (CICA). This research has been supported by different national and European research projects and by grants provided by the CSIC and the Andalusian Government. Our current research is partially funded by the Spanish CICYT Projects TXT98-1384 and TIC98-0869.

We are grateful to Prof. José L. Huertas, not only for writing the foreword to this book, but also for his valuable suggestions which have always helped and encouraged our research on integrated circuit design of fuzzy systems.

We would like to thank Francisco José Moreno Velo for developing the learning tool *xfbpa*, for his help in writing Chapter 5, and for proofreading the manuscript.

We would also like to thank other colleagues of our "Xfuzzy-team" who have contributed in a variety of ways: Javi Massa for beginning the development of *Xfuzzy*, and Raouf Senhadji Navarro, Enrique Lago García, Miguel Ángel Hinojosa Romero, Emilio Ramírez Ríos, and Diego Galán Fernández, for their contributions in developing the different *Xfuzzy* tools.

If you would like to send comments or questions on this book, you can access our Web page at:

`http://www.imse.cnm.es/Xfuzzy/`

Iluminada Baturone
Ángel Barriga
Santiago Sánchez-Solano
Carlos J. Jiménez-Fernández
Diego R. López

Contents

Chapter 1

INTRODUCTION

Since it was introduced by Zadeh in 1965, fuzzy logic has demonstrated to be an adequate tool for emulating the approximate reasoning mechanisms used by the human brain. Inference systems based on fuzzy logic share certain features with other artificial intelligence paradigms. A fuzzy system acquires and represents its knowledge base symbolically, while knowledge is processed numerically. The first feature permits the use of fast and simple mechanisms for representing and acquiring knowledge, structured in rules, and the second allows the employment of fast numerical algorithms and the direct implementation of the system by a microelectronic circuit. The ability of fuzzy logic to describe a complex system by means of a simple and intuitive set of behavioral rules has motivated an increasing interest for applying it in fields such as industrial control, non-linear systems modeling, and decision-making systems.

1.1 METHODS FOR INFORMATION REPRESENTATION AND PROCESSING

There are a great number of situations in real life where we have to choose between mutually exclusive alternatives. For example, we can only answer "YES" or "NO" to certain questions when completing a questionnaire, we can only turn on or off the lights in a room, we can only qualify a sentence as "TRUE" or "FALSE" when making a test, etc. However, things are usually a little more complex. Natural phenomena and our own sensations and responses to external events are associated with a continuous range of possible values. The strict yes or no responses, absolute trueness or falseness, are no more than extrapolations that admit a huge scale of shades. Two important characteristics can be found when analyzing human reasoning schemes. First, we tend to use a series of ambiguous or imprecise terms, even in the case that the information we are dealing with is multi-valued. We say that the volume of the radio is '*too low*' or that the grade obtained for an examination is '*quite high*'. We can correctly interpret without problems these kinds of terms using their context. The second characteristic is our ability to extract correct conclusions from this imprecise information by means of approximate reasoning, based on our own experience.

The nature of information and the method used to represent it have a direct influence on the procedures used to deal with this information. Bi-valued information is very easily stored and processed by electronic devices performing in accordance with the mathematical formalism known as binary logic, which constitutes the basis for digital computers. In fact, computers are extremely efficient in solving a great number of problems, particularly all those problems that can be expressed in terms of a mathematical algorithm describing a model of behavior. A computer can perform in a few seconds complex numerical calculations that would take several days for a human. However, information representation mechanisms and basic operating principles of computers differ greatly from the memory and reasoning mechanisms of the human brain. These differences make human activities that we perform easily every day, like pattern recognition or decision making, very difficult to implement even in the most sophisticated computers.

The attempts to build machines with computational models approximating human reasoning mechanisms gave birth, in the mid-1950s, to different *Artificial Intelligence* (AI) techniques [Winston, 1984; Charniak, 1985]. The first of these approaches, known as *Expert Systems*, tries to emulate our ability to structure information [Payne, 1990]. Expert systems store and process rules of the type "*if A then B*", associating actions (*B*) to conditions (*A*). A set of these rules makes up the knowledge base to which the inference process is applied. The inference is made by a *forward-chaining* process when it is directed to answer questions like "What?" or "If". A *backward-chaining* inference process is applied for questions like "Why?" or "What is the cause of?" [Hayes, 1983]. Traditional expert systems acquire, store, and process knowledge in the form of symbolic rules, without using numerical entities. This means a simple and fast way to represent knowledge, structured in rules, but which makes it very difficult to use mathematical tools or to implement this kind of system in integrated circuits [Bolc, 1988].

A completely different approach led to the introduction of *Neural Networks*. This technique is based on the expectation of (partially) reproducing the power and flexibility of the human brain by means of artificial procedures that emulate the brain organization [Grosberg, 1988]. Neural networks are made of a huge number of interconnected elements that perform simple calculations. They resemble the structure of the brain, where a great number of neurons are connected by more complex and structured links. Neural networks can be programmed or exercised to associatively store, recognize, or recover patterns or entries in a database. Unlike expert systems, neural networks permit a use of their numerical nature with efficient calculation algorithms and can be easily implemented on analog or digital integrated circuits; but, on the other hand, neural networks cannot directly represent symbolic information.

1.2 FUZZY SETS AND FUZZY LOGIC

Fuzzy Set Theory, developed by L. A. Zadeh in the mid-1960s, and the inference techniques based on it (*Fuzzy Logic*) provide a tool that is adequate for modeling the uncertainty present in natural language and for emulating the approximate reasoning mechanisms used by the human brain. This representation is based on the core concept of *fuzzy set*, a type of set in which elements (in the universe of discourse where it is defined) are its members to a certain degree. Zadeh himself expresses this concept in the following way: «*most of the basic tasks performed by humans do not require a high degree of precision in their execution. The human brain takes advantage of this tolerance for imprecision by encoding the "task-relevant" (or "decision-relevant") information into labels of fuzzy sets which bear an approximate relation to the primary data*» [Zadeh, 1973].

Fuzzy logic states that everything is a matter of degree, so traditional elements of binary logic ({0, 1}, {White, Black}, {True, False}) are no more than the extremes of a corresponding multivalued or continuous logic (the [0, 1] interval, shades of gray, or levels of certainty, respectively).

An inference system based on fuzzy logic, a *fuzzy system*, is constituted by a set of rules such as "*if x is A then y is B*", where x and y are system variables, and A and B are linguistic terms like '*high*', '*low*', '*medium*', '*positive*', '*negative*', etc. The reasoning techniques used by fuzzy logic are a generalization of the inference mechanisms employed traditionally by binary logic. One important difference with respect to a conventional expert system is that, in a fuzzy system, several rules can be simultaneously active with different activation degrees. The system evaluates all its rules to obtain a conclusion or an output whenever any input changes.

As in the cases of expert systems and neural networks, fuzzy logic–based systems obtain their output without the necessity of employing an analytical model. They share with the former their capability of representing knowledge in a structured manner (by means of rules) and with the latter the attributes derived from their numerical information processing characteristics [Kosko, 1992]. This last feature makes the hardware implementation of fuzzy logic–based systems especially appealing [Kandel, 1991].

1.3 A BRIEF HISTORY OF FUZZY LOGIC

The introduction of the fundamentals of fuzzy set theory, made by Zadeh in 1965 in his paper "*Fuzzy Sets*" [Zadeh, 1965], is commonly accepted as the starting point for fuzzy logic. The main concepts of fuzzy theory were developed by Zadeh in the following decade. Another important contribution established in 1973 the fundamentals of fuzzy control [Zadeh, 1973], which found its first practical application in the control of a laboratory model of a steam machine, realized by Mamdani and Assilian in 1975 [Mamdani, 1975]. A few years later the first industrial application of fuzzy logic was set up in Denmark: a system for controlling a concrete oven [Holmblad, 1982].

Although basic theoretical ideas in the area came from the U.S., practical applications of fuzzy logic were adopted in Japan much faster. Some authors justify this in the difference between Zen philosophy, predominant in Eastern society, and the Aristotelian logic that still prevails in Western science and engineering [Kosko, 1993]. Apart from these cultural considerations, the fundamental reason for the rapid development of fuzzy logic in Japan is the focus of Japanese industry on two main directions: the search for ad-hoc applications and, especially, the focus on the microelectronic implementation of basic ideas. At the beginning of the 1980s the first Japanese fuzzy applications were developed, a waste-water handling plant [Yagishita, 1985] and a fuzzy robot [Sugeno, 1985]. One of the more reputed applications of fuzzy logic was the automated control for the subway of the city of Sendai, developed by Hitachi, that started operating in 1987. At that time, the first implementations of microelectronic circuits for fuzzy logic were introduced. Togai and Watanabe presented in 1985 the first digital circuits [Togai, 1985], and Yamakawa presented analog chips in 1986 [Yamakawa, 1986]. The Japanese stake in fuzzy logic was realized in 1989 with the foundation of the *Laboratory for International Fuzzy Engineering Research* (LIFE) and the creation of the *Fuzzy Logic Systems Research Institute* (FLSI) in 1990. The results of these initiatives are more than 1000 patents in multiple fields, such as home appliances, video cameras, the automotive and aerospace industries, robotics, etc. [Munakata, 1994]. Many of these applications use the term "*fuzzy*" as a marketing motto. As a consequence, many Japanese companies such as Canon, Matsushita, Nissan, Minolta, Sharp, Mitsubishi, Toshiba, or Fujitsu sell products that use fuzzy logic.

The consequence of this boom in Japan was an increasing interest in fuzzy logic applications in the U.S. and Europe. Beginning in 1990, greater funds and efforts have been applied and new forums have been created for the growing community of researchers and practitioners, such as the *IEEE Conference on Fuzzy Systems* and the journal *Transactions on Fuzzy Systems* (1993). Several American (General Electric, Motorola, Boeing, General Motors, Chrysler, Whirlpool) and European (Hoechst, Siemens) companies and institutions such as NASA are working with fuzzy logic.

During its almost 35 years of history, fuzzy logic has also attracted controversy. The first objection to its theoretical basis came from mathematicians, since some authors argued that fuzzy logic was not more than some kind of fancy probability theory. The relationship between fuzzy logic and probability has been the subject of a great number of papers ([Kosko, 1990] and the special March 1994 number of *IEEE Transactions on Fuzzy Systems*). Other, more recent criticisms have come from the field of classical artificial intelligence: the special August 1994 issue of the journal *IEEE Expert* presents one of the most notorious discussions on this subject.

1.4 FUZZY LOGIC APPLICATION DOMAIN

Nevertheless, on the threshold of the 21st century, fuzzy logic is considered by most researchers as a useful tool for representing the human reasoning mechanisms. Its usefulness is substantiated by the great number of practical applications that take advantage of the characteristics of fuzzy systems [Terano, 1994].

As we have stated above, a fuzzy system is a system based on knowledge. This knowledge is structured in a set of symbolic rules of the type "*if...then*" that use natural language terms to represent imprecise or vague information. On the other hand, from a mathematical point of view, a fuzzy system is a deterministic system that provides a non-linear mapping from its inputs to its outputs (it has been demonstrated that certain types of fuzzy systems are universal approximators [Castro, 1995]). These two characteristics are the basis for *fuzzy control*, the field of engineering where fuzzy logic has been applied more often and has yielded better results.

The application of fuzzy logic to control problems is based on the so-called principle of incompatibility: «*As the complexity of a system increases, our ability to make precise yet significant statements about its behavior diminishes until a threshold is reached beyond which precision and significance (or relevance) become almost mutually exclusive characteristics*» [Zadeh, 1973]. The use of fuzzy logic makes it possible to replace the mathematical model that describes the behavior of a complex system with a linguistic model that defines the control strategies applied by an expert operator. In general terms, this approach implies a series of advantages that have been widely reported in the literature. By means of fuzzy systems it is possible to automate complex control tasks that could only be performed previously by human operators. Furthermore, these kinds of systems provide a robust non-linear control, able to deal with higher perturbations than conventional controllers, and produces soft control actions with a small number of rules. Finally, in many cases the characteristics of fuzzy systems allow us to reduce the development and maintenance costs of controllers, since the control strategies can be more easily expressed in terms of fuzzy control rules than by means of differential equations or conventional control algorithms.

Another fundamental feature of fuzzy systems is their ability to cope with ambiguity. A fuzzy system can extract conclusions from estimated values obtained using incomplete or uncertain information, and so they have been used in different fields of engineering and social sciences [Munakata, 1994; Schwartz, 1994]. Among the different application domains we can highlight *pattern recognition systems* (used in speech and image processing, classification, etc.), *fuzzy expert systems* (used in medical diagnosis, financial analysis, decision making, etc.), and *fuzzy databases* (that are able to process imprecise information and permit users to issue queries in natural language).

Regardless of the great development that fuzzy logic has undergone in recent years, there are several problems that still merit further research in the area of fuzzy systems. From the point of view of applications, the two main issues are related to the set up of systematic procedures for defining and tuning fuzzy systems, and the translation of the inference mechanisms into hardware in order to apply fuzzy logic to problems requiring real-time operation. The wide industrial acceptance of the fuzzy paradigm as a complement to other conventional techniques also imposes the need for the development of CAD environments to ease design tasks, both architectural ones (choice of the system structure, selection of rules, etc.) and those related to integrated circuit realization, by defining a design methodology that leads to the implementation once the knowledge base has been validated.

REFERENCES

1. **Bolc, L., Coombs, M. J.**, *Expert System Applications*, Springer-Verlag, 1988.
2. **Castro, J. L.**, Fuzzy logic controllers are universal approximators, *IEEE Transactions on Systems, Man, and Cybernetics*, Vol. 25, N. 4, pp. 629-635, 1995.
3. **Charniak, E., McDermott, D.**, *Introduction to Artificial Intelligence,* Addison-Wesley Pub., 1985.
4. **Grosberg, S.**, *Studies of Mind and Brain. Neural Principles of Learning, Perception, Development, Cognition and Motor Control*, D. Reidel Pub., 1988.
5. **Hayes-Roth, F., Waterman, D. A., Lenat, D. B.**, *Building Expert Systems*, Addison-Wesley Pub., 1983.
6. **Holmblad, L. P., Ostergaard, J.**, Control of cement kiln by fuzzy logic, in *Fuzzy Information and Decision Processes,* Gupta, M. M., Sánchez, E., Eds., North-Holland, 1982.
7. **Kandel, A.**, *Fuzzy Expert Systems*, CRC Press Inc., 1991.
8. **Kosko, B.**, Fuzziness versus probability, *Int. Journal of General Systems*, Vol. 17, pp. 211-240, 1990.
9. **Kosko, B.**, *Neural Networks and Fuzzy Systems, a Dynamical Systems Approach to Machine Intelligence*, Prentice-Hall Int., 1992.
10. **Kosko, B.**, *Fuzzy Thinking: The New Science of Fuzzy Logic,* Hyperion, 1993.
11. **Mamdani, E. H., Assilan, S.**, An experiment in linguistic synthesis with a fuzzy logic controller, *Int. Journal of Man-Machine Studies*, Vol. 7, N. 1, pp. 1-13, 1975.
12. **Munakata, T., Jani, Y.**, Fuzzy systems: an overview, *Communications of the ACM*, Vol. 37, N. 3, 1994.
13. **Payne, E. C., McArthur, R. C.**, *Developing Expert Systems*, Wiley, 1990.
14. **Schwartz, D. G., Klir, G. J., Lewis, H. W., Ezawa, Y.**, Applications of fuzzy sets and approximate reasoning, *Proceedings of the IEEE*, Vol. 82, N. 4, pp. 482-498, 1994.
15. **Sugeno, M., Nishida, M.**, Fuzzy control of model car, *Fuzzy Sets and Systems*, Vol. 16, pp. 103-113, 1985.
16. **Terano, T., Asai, K., Sugeno, M.**, Eds., *Applied Fuzzy Systems*, Academic Press, 1994.
17. **Togai, M., Watanabe, H.**, A VLSI implementation of a fuzzy inference engine: toward an expert system on a chip, *Information Science*, Vol. 38, N. 2, pp. 147-163, 1985.
18. **Winston, P. H.**, *Artificial Intelligence*, Addison-Wesley Pub., 1984.
19. **Yagishita, O., Itoh, O., Sugeno, M.**, Application of fuzzy reasoning to the water purification process, in *Industrial Application of Fuzzy Control*, Sugeno, M., Ed., North-Holland, pp. 19-40, 1985.
20. **Yamakawa, T., Miki, T.**, The current mode fuzzy logic integrated circuits fabricated by the standard CMOS process, *IEEE Transactions on Computer*, Vol. 35, N. 2, pp. 161-167, 1986.
21. **Zadeh, L. A.**, Fuzzy sets, *Information and Control*, Vol. 8, pp. 338-353, 1965.
22. **Zadeh, L. A.**, Outline of a new approach to the analysis of complex systems and decision processes, *IEEE Transactions on Systems, Man, and Cybernetics*, Vol. 3, N. 1, pp. 28-44, 1973.

Chapter 2

FUZZY SET THEORY

This chapter presents the basic mathematical foundations of fuzzy set theory. With this purpose, the concept of fuzzy set as a generalization of the classical concept of set will be introduced, and the terminology associated with this type of set will be defined. Later, the different operations that can be defined on fuzzy sets will be described. Finally, the concept of fuzzy relation will be introduced and the different kinds of fuzzy relation compositions will be analyzed, since these compositions are key elements in the application of approximate reasoning techniques.

2.1 FUZZY SETS

When we use mathematical language to define the value of certain magnitudes or to indicate the number of elements of a set, we employ exact expressions such as "*the temperature is 25°C*" or "*the number of individuals is 133*". However, in natural language we tend to employ ambiguous terms such as "*the water is hot*" or "*there are few individuals*". When using expressions such as {*cold, warm, hot*} or {*few, enough, many*} we are establishing a series of classes or categories with not well-defined limits and a semantic value that depends on the context (most of us will associate the term '*hot*' with a temperature near 40ºC when talking about a human body, or near 100ºC when talking about the cooling fluid in a motorcar). We also use imprecise terms when referring to uncountable magnitudes (*much sugar, a little water*) or characteristics that are difficult to measure (*sweet, bitter, sour*). The concept of fuzzy set was introduced by Zadeh as a mechanism for representing the vagueness and imprecision of natural language as humans use it [Zadeh, 1965]. Fuzzy sets have no precise limits: they are classes of objects where the transition from membership to non-membership is gradual instead of steep. In fact, Zadeh uses them to define classes of objects in the real world where it is difficult to establish precise membership criteria.

From a formal point of view, the concept of fuzzy set is a generalization of the classical concept of set. The fundamental difference lies in that, while in the classical set theory established by Georg Cantor, a certain element can be a member of a set or not, in the fuzzy set theory proposed by Zadeh, an element can be a member of several sets with different degrees of membership. The membership degree is a real number in the interval [0, 1] and indicates to which extent a certain element is a member of the set. This way, an element with a membership degree of '0' for a given set is not a member of that set. Again, an element with a membership degree of '1' is fully a member of the set.

Figure 2.1 graphically compares the concepts of classical set and fuzzy set. In the representation shown in Figure 2.1a the temperature can either be '*high*' or '*low*'; there are no intermediate cases. If the cross point is located at 36 °C, any value less or equal than this would be considered a low temperature, while a few tenths above would be considered high. Conversely, in the representation of Figure 2.1b temperatures below 35 °C are considered '*low*' and those above 37 °C are considered '*high*', but the temperatures between those values are considered to be in both sets. Specifically, a temperature of 36 °C has a membership degree of 0.5 in both high and low temperature sets, what is more in accordance with the way we would express this in natural language: "*the temperature is neither high nor low*".

2.1.1 THE CONCEPT OF MEMBERSHIP FUNCTION

A fuzzy set *F* defined on a *universe of discourse* U (containing all the possible elements or values for a certain variable) is characterized by a *membership function* μ_F (u) taking values in the interval [0, 1]:

$$\mu_F (u): U \rightarrow [0, 1] \tag{2.1}$$

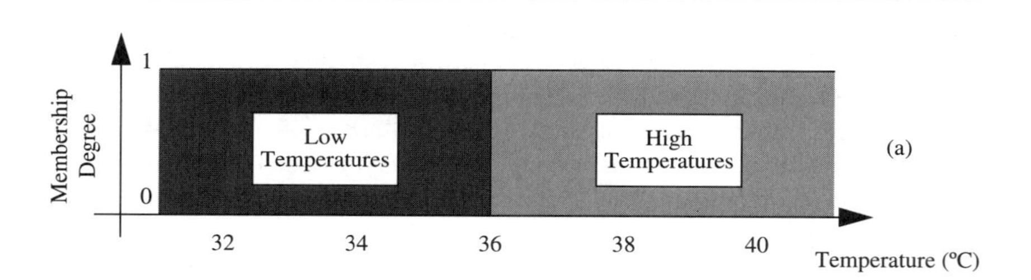

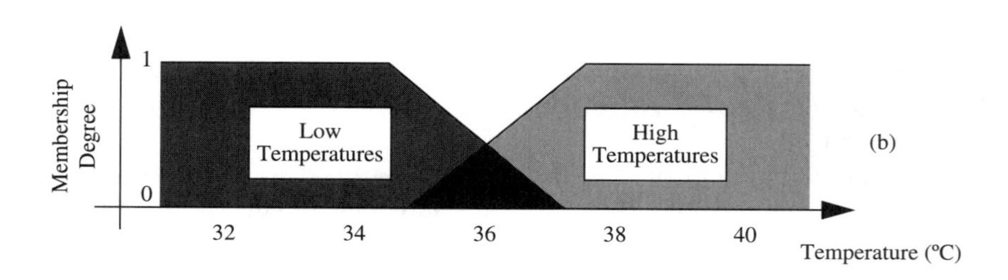

Figure 2.1. Graphical representation of two sets: (a) Classical; (b) Fuzzy.

According to this definition, a classical set C is a particular case of fuzzy set where the membership function (or characteristic function) is a bi-valued function taking only the values '0' or '1':

$$\mu_C(u) : \{U \rightarrow \{0,1\} \mid \mu_C(u) = 1 \text{ if } u \in C, 0 \text{ if } u \notin C\} \tag{2.2}$$

A fuzzy set can be represented by a set of ordered pairs that assign a membership degree for each element u in the universe of discourse U:

$$F = \{(u, \mu_F(u)) \mid u \in U\} \tag{2.3}$$

More concisely, depending on a continuous or discrete universe of discourse, (2.3) is often expressed as:

$$F = \int \mu_F(u) / u \tag{2.4}$$

$$F = \Sigma\, \mu_F(u) / u \tag{2.5}$$

where the integral and the sum symbols do not hold their usual mathematical meaning, but represent an infinite or a finite collection of elements in U.

Therefore, a membership function describes the membership degree in the fuzzy set for the different elements in the universe of discourse. The selection of a shape for a membership function is subjective and depends on the context. Nevertheless, for practical reasons most functions described in the literature are triangular, trapezoidal, or bell-shaped [Ross, 1995], as shown in Figure 2.2.

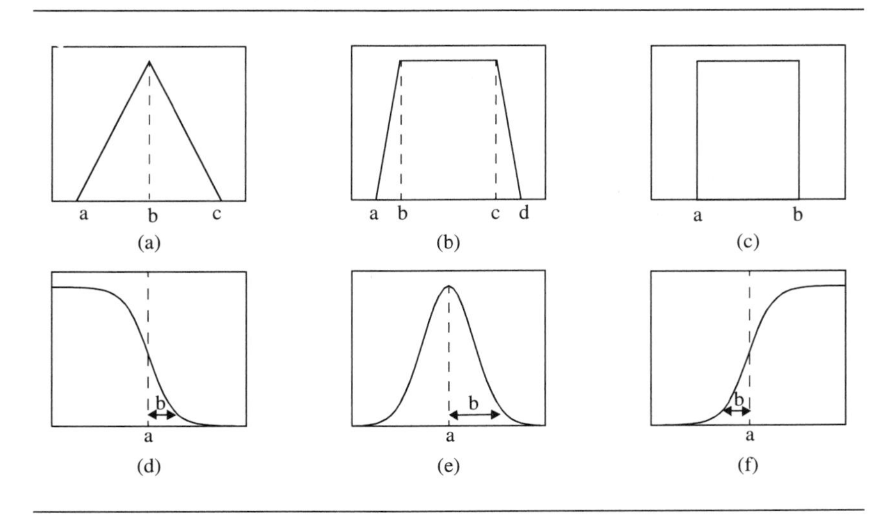

Figure 2.2. The most common membership function shapes: (a) Triangular; (b) Trapezoidal; (c) Rectangular; (d) Z-function; (e) Gaussian; (f) S-function.

2.1.2 TERMINOLOGY FOR FUZZY SETS

The basic concepts and terminology for fuzzy sets are depicted in Figure 2.3.

The *support* of a fuzzy set is the set of points in the universe of discourse where the membership function takes a value greater than zero:

$$support\ (F) = \{\ u \in U \mid \mu_F(u) > 0\ \} \tag{2.6}$$

A fuzzy set whose support has only one element, u_o, is called a *fuzzy singleton* (or fuzzy singularity):

$$\mu_F(u) = 0 \text{ if } u \neq u_o,\ \mu_F(u_o) = 1 \tag{2.7}$$

As we will see in the following chapters, the use of singletons considerably simplifies the inference process and allows for an efficient electronic implementation of fuzzy systems.

The *height* of a fuzzy set is the maximum value of the membership function. In this book we will consistently use fuzzy sets with height 1. Such a fuzzy set is called a *normal fuzzy set* [Cox, 1999].

The set of elements in the universe of discourse where the membership function takes values equal or greater than a threshold value $\alpha \in (0, 1]$ is called the α-*cut*, C_α, of the fuzzy set:

$$C_\alpha = \{u \in U \mid \mu_F(u) \geq \alpha\} \tag{2.8}$$

Likewise, the strong α-*cut,* $C_{\overline{\alpha}}$, is defined as the set of elements in the universe of discourse where the membership function takes values strictly greater than α:

$$C_{\overline{\alpha}} = \{u \in U \mid \mu_F(u) > \alpha\} \tag{2.9}$$

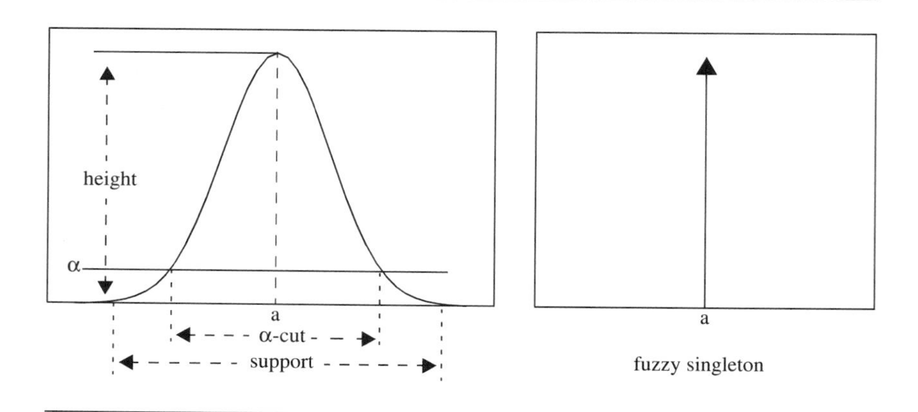

Figure 2.3. Some basic concepts for fuzzy sets.

2.2 OPERATIONS WITH FUZZY SETS

As we will see in this section, most of the operations in classical set theory can be extended for fuzzy sets [Ruspini, 1998].

• **Equality**: Given two fuzzy sets A and B, defined on the same universe of discourse U, set A is equal to set B if and only if both have the same membership function:

$$A = B \text{ iff } \mu_A(\mathrm{u}) = \mu_B(\mathrm{u}),\ \forall\ \mathrm{u} \in \mathrm{U} \tag{2.10}$$

• **Containment**: Similarly, set A is contained in set B if and only if, for each element in the universe of discourse, the membership function of A is less than or equal to the membership function of B:

$$(A \subseteq B) \text{ iff } \mu_A(\mathrm{u}) \leq \mu_B(\mathrm{u}),\ \forall\ \mathrm{u} \in \mathrm{U} \tag{2.11}$$

2.2.1 UNION, INTERSECTION, AND COMPLEMENT

As in the case of classical sets, selecting the "min" and "max" operators to represent the union and intersection of fuzzy sets agrees with the intuitive idea of these operators. Thus, given two fuzzy sets, A and B, defined on the same universe of discourse U, the operations of union, intersection, and complement can be defined, in terms of their membership functions, by:

• **Union**: The union of fuzzy sets A and B is a fuzzy set with a membership function that, for each element in U, takes the maximum value of the membership functions of A and B for that element:

$$\mu_{A \cup B}(\mathrm{u}) = \max\ [\mu_A(\mathrm{u}), \mu_B(\mathrm{u})\}],\ \forall\ \mathrm{u} \in \mathrm{U} \tag{2.12}$$

• **Intersection**: The intersection of fuzzy sets A and B is a fuzzy set with a membership function that, for each element in U, takes the minimum value of the membership functions of A and B for that element:

$$\mu_{A \cap B}(\mathrm{u}) = \min\ [\mu_A(\mathrm{u}), \mu_B(\mathrm{u})\}],\ \forall\ \mathrm{u} \in \mathrm{U} \tag{2.13}$$

• **Complement**: The complement of a fuzzy set A is a fuzzy set $\overline{A}$ with a membership function defined by:

$$\mu_{\overline{A}}(\mathrm{u}) = 1 - \mu_A(\mathrm{u}),\ \forall\ \mathrm{u} \in \mathrm{U} \tag{2.14}$$

A graphical representation of these operations is shown in Figure 2.4. Expressions (2.12) and (2.13) can be also used to define the union and intersection of classical sets with membership functions corresponding to (2.2). However, in the case of fuzzy sets this is not the only possible choice. Zadeh, in his first work about fuzzy sets, proposes algebraic sum and algebraic product as alternative operators for fuzzy set union and intersection, respectively. The next section introduces these operators, and some others that can be defined axiomatically, the so-called triangular norms: t-norms and s-norms [Weber, 1983].

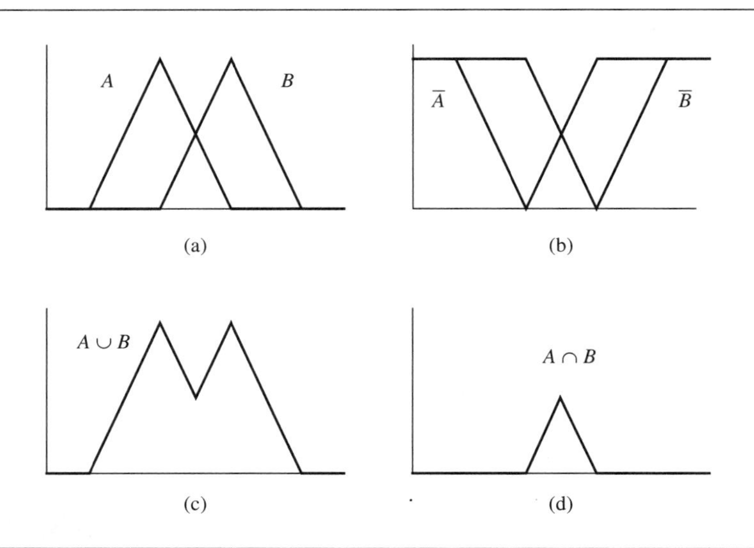

Figure 2.4. Basic operations with fuzzy sets: (a) Original sets; (b) Complement; (c) Union; (d) Intersection.

2.2.2 FUZZY INTERSECTION: T-NORMS

A triangular norm, or *t-norm*, is a function $T : [0, 1] \times [0, 1] \rightarrow [0, 1]$ that verifies the following properties:

- Commutativity: $T(x, y) = T(y, x), \forall\, x, y \in [0, 1]$
- Associativity: $T[x, (y, z)] = T[(x, y), z], \forall\, x, y, z \in [0, 1]$
- Monotonicity: If $x \leq y$ and $w \leq z \Rightarrow T(x, w) \leq T(y, z)$, $\forall\, x, y, w, z \in [0, 1]$
- Boundary conditions: $T(x, 0) = 0, \forall\, x \in [0, 1]$
  $T(x, 1) = x, \forall\, x \in [0, 1]$

In addition to the minimum operator, the following functions verify the properties of a t-norm:

- Algebraic Product:

$$T_p(x, y) = x \cdot y \tag{2.15}$$

- Bounded Product:

$$T_{bp}(x, y) = x \otimes y = \max\{0, x + y - 1\} \tag{2.16}$$

- Drastic Product:

$$T_{dp}(x, y) = x \bar{\cap} y = \{x, \text{ if } y = 1;\ y, \text{ if } x = 1;\ 0, \text{ if } x, y < 1\} \qquad (2.17)$$

The minimum operator is the biggest t-norm, while drastic product is the lower bound for the family of t-norms (Figure 2.5a). The following inequality can be demonstrated:

$$T_{dp}(x, y) \leq T(x, y) \leq \min(x, y) \qquad (2.18)$$

2.2.3 FUZZY UNION: S-NORMS

A triangular conorm or *s-norm* is a function $S : [0, 1] \times [0, 1] \rightarrow [0, 1]$ that verifies the following properties:

- Commutativity: $S(x, y) = S(y, x),\ \forall\ x, y \in [0, 1]$
- Associativity: $S[x, (y, z)] = S[(x, y), z],\ \forall\ x, y, z \in [0, 1]$
- Monotonicity: If $x \leq y$ and $w \leq z \Rightarrow S(x, w) \leq S(y, z)$, $\forall\ x, y, w, z \in [0, 1]$
- Boundary conditions: $S(x, 0) = x,\ \forall\ x \in [0, 1]$
  $S(x, 1) = 1,\ \forall\ x \in [0, 1]$

Beside the maximum operator, the following functions verify the properties of an s-norm:

- Algebraic Sum:

$$S_s(x, y) = x + y - xy \qquad (2.19)$$

- Bounded Sum:

$$S_{bs}(x, y) = x \oplus y = \min\{1, x + y\} \qquad (2.20)$$

- Drastic Sum:

$$S_{ds}(x, y) = x \bar{\cup} y = \{x \text{ if } y = 0;\ y \text{ if } x = 0;\ 1 \text{ if } x, y > 0\} \qquad (2.21)$$

The maximum operator is the lowest s-norm, while drastic sum is the upper bound of the family of s-norms (Figure 2.5b). The following inequality can be demonstrated:

$$\max(x, y) \leq S(x, y) \leq S_{ds}(x, y) \qquad (2.22)$$

Some other parametrical t-norms and s-norms have been proposed in the literature, as in the cases of Dombi, Dubois–Prade and Yager [Pedrycz, 1998]. All t-norms (s-norms) are equal when membership functions values are restricted to '0' and '1', that is, in all cases t-norms (s-norms) are extensions of the intersection (union) in classical set theory. From a practical point of view, some t-norms and s-norms can be more adequate than others in certain applications [Wang, 1997].

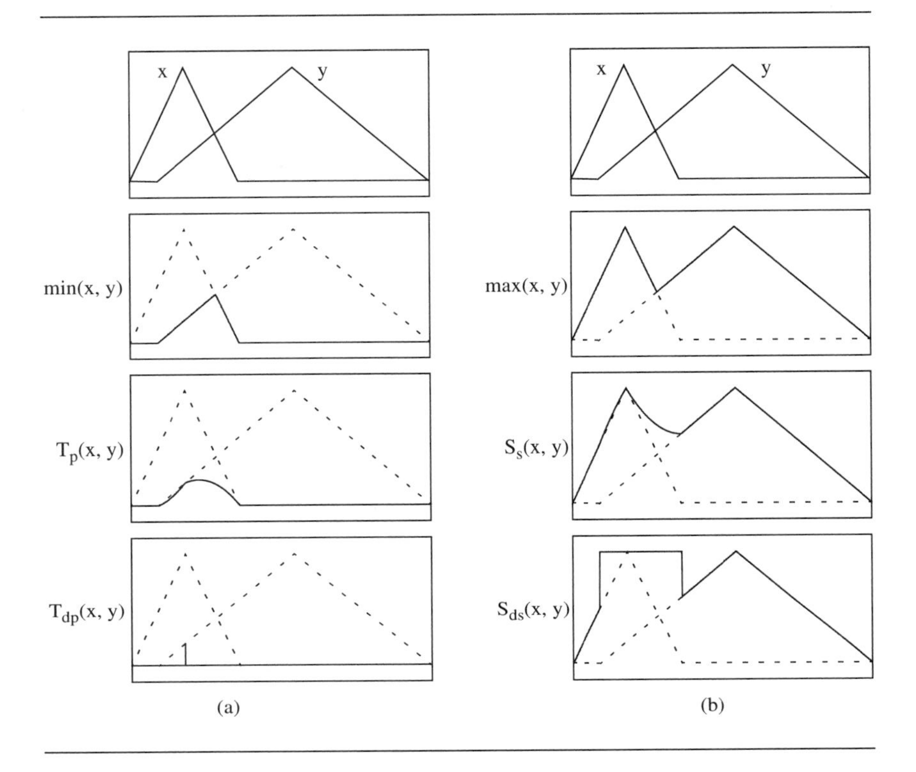

Figure 2.5. (a) t-norms; (b) s-norms.

2.2.4 FUZZY COMPLEMENT: C-NORMS

Although most practical applications use expression (2.14) to represent the complement of a fuzzy set, an axiomatic definition can be established as well. A fuzzy complement is a function C: [0, 1] → [0, 1] that verifies the following properties:

- Boundary conditions: $C(0) = 1, C(1) = 0$
- Nonincreasing condition: If $x < y \Rightarrow C(x) > C(y), \forall\, x, y \in [0, 1]$

Some proposed parametrical C-norms are:

- Sugeno:

$$C_\lambda(x) = (1 - x) / (1 + \lambda x),\ \lambda \in (-1, \text{infinity}) \tag{2.23}$$

- Yager:

$$C_w(x) = (1 - x^w)^{1/w},\ w \in (0, \text{infinity}) \tag{2.24}$$

Both classes of functions verify the above properties. The function defined by (2.23) collapses to the common fuzzy complement defined in (2.14) when $\lambda = 0$, as does the class of functions defined by (2.24) when $w = 1$.

2.2.5 PROPERTIES OF FUZZY SETS

Each of the t-norms T(x,y) defined in Section 2.2 can be associated with one of the s-norms S(x, y) defined in Section 2.3 so there is a fuzzy complement C(x) that verifies the generalized De Morgan's laws [Yager, 1994]:

$$C[T(x, y)] = S[C(x), C(y)] \tag{2.25}$$

$$C[S(x, y)] = T[C(x), C(y)] \tag{2.26}$$

These equations are verified by minimum-maximum and product-sum (algebraic, bounded and drastic) when used with the complement $C(x) = 1 - x$.

The minimum-maximum operators also verify distributivity:

$$T[x, S(y, z)] = S[T(x, y), T(x, z)] \tag{2.27}$$

$$S[x, T(y, z)] = T[S(x, y), S(x, z)] \tag{2.28}$$

Nevertheless, there are two fundamental laws in classical set theory that are not applicable to fuzzy set theory. They are the *law of the excluded middle* (which states that the union of a set and its complement is always the whole universe of discourse: $A \cup \overline{A} = U$) and the *law of the contradiction* (which states that an element cannot be simultaneously member of a set and its complement: $A \cap \overline{A} = \emptyset$). As the reader might deduce from the representation given in Figure 2.4, if A is a fuzzy set defined on the universe of discourse U, the following inequalities hold:

$$A \cup \overline{A} \neq U \; , \; A \cap \overline{A} \neq \emptyset \tag{2.29}$$

since A and $\overline{A}$ can overlap so that their union does not cover U completely. This comes from the essential nature of fuzzy sets, which have imprecise limits.

2.3 FUZZY RELATIONS

The concept of *fuzzy relation* is again a generalization of the concept of relation in classical set theory [Mendel, 1995]. While a relation between two classical sets describes the existence or not of an association between elements in both sets, a fuzzy relation describes the degree of association or interaction between the elements of two fuzzy sets.

Let us consider two universes of discourse, U and V. A fuzzy relation R is a subset of the Cartesian product $U \times V$, i.e., a fuzzy set that can be represented by means of an ordered sequence of pairs assigning a membership degree $\mu_R(u, v)$ for each element (u, v) in the product space $U \times V$:

$$R = \{[(u, v), \mu_R(u, v)] \mid (u, v) \in U \times V\} \tag{2.30}$$

According to the notation used in (2.4) and (2.5), and depending on the continuous or discrete nature of U and V, (2.30) can be written as follows:

$$R = \int \mu_R(\mathrm{u}, \mathrm{v}) / (\mathrm{u}, \mathrm{v}) \tag{2.31}$$

$$R = \Sigma\, \mu_R(\mathrm{u}, \mathrm{v}) / (\mathrm{u}, \mathrm{v}) \tag{2.32}$$

where the integral and the sum represent again the infinite or finite collection of elements in U × V.

In the discrete case, the fuzzy relation can be represented by a matrix, the *fuzzy relational matrix*, with elements taking values in the [0, 1] interval:

$$R = \begin{bmatrix} \mu_R(u_1,v_1) & \mu_R(u_1,v_2) & \dots & \mu_R(u_1,v_m) \\ \mu_R(u_2,v_1) & \mu_R(u_2,v_2) & \dots & \mu_R(u_2,v_m) \\ \dots & \dots & \dots & \dots \\ \mu_R(u_n,v_1) & \mu_R(u_n,v_2) & \dots & \mu_R(u_n,v_m) \end{bmatrix} \tag{2.33}$$

The fuzzy relations presented up to now, applied on two universes of discourse, are called binary fuzzy relations. Obviously, the concept of fuzzy relation can be generalized to n dimensions:

$$R = \{[(\mathrm{u}_1,\mathrm{u}_2, .. \mathrm{u}_\mathrm{n}), \mu_R(\mathrm{u}_1,\mathrm{u}_2, .. \mathrm{u}_\mathrm{n})] \mid (\mathrm{u}_1,\mathrm{u}_2, .. \mathrm{u}_\mathrm{n}) \in \mathrm{U}_1 \times \mathrm{U}_2, ... \times \mathrm{U}_\mathrm{n}\} \tag{2.34}$$

Chapter 3 will show the key role of the fuzzy relation concept in approximate reasoning mechanisms, since it will allow the interpretation of expressions such as "*x is practically equal to y*" or "*y is much greater than z*", where *x*, *y* and *z* are variables represented by means of fuzzy sets.

2.3.1 COMPOSITION OF FUZZY RELATIONS

A fuzzy relation is a fuzzy set defined on a product space. Therefore, it is possible to apply to it the union, intersection, and complement operators defined above. Given two fuzzy relations, *P* and *R* defined on U × V, the composition of these relations by the operators union and intersection are defined by:

$$\mu_{P \cap R}(\mathrm{u}, \mathrm{v}) = \mathrm{T}\, [\mu_P(\mathrm{u}, \mathrm{v}), \mu_R(\mathrm{u}, \mathrm{v})] \tag{2.35}$$

$$\mu_{P \cup R}(\mathrm{u}, \mathrm{v}) = \mathrm{S}\, [\mu_P(\mathrm{u}, \mathrm{v}), \mu_R(\mathrm{u}, \mathrm{v})] \tag{2.36}$$

where T and S can be any of the t-norms and s-norms defined in Section 2.

The composition of fuzzy relations on the same product space permits the interpretation of phrases like "*x is much greater than y or x is equal to y*". But the composition of relations defined in different product spaces is also possible, so as to interpret sentences like "*x is much greater than y and y is equal to z*". In the latter case, the resulting fuzzy relation will associate variables *x* and *z*.

The composition of fuzzy relations defined on different product spaces that share a universe of discourse is illustrated by the diagram in Figure 2.6. Given

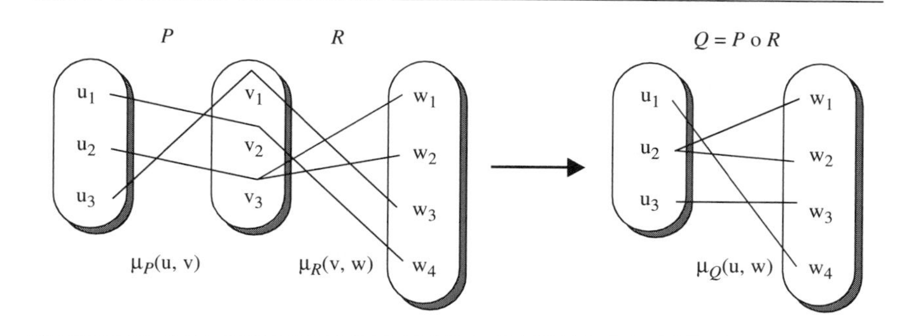

Figure 2.6. Composition of fuzzy relations.

two fuzzy relations, P and R defined on U × V and V × W, respectively, the composition of P and R is a new fuzzy relation ($Q = P \text{ o } R$) defined on U × W, where (u, w) will be a member of Q if and only if there is at least one element v in V so that (u, v) is member of P and (v, w) is member of R.

A formulation of fundamental importance in fuzzy inference processes is the so-called *sup-star* (supreme-star) composition [Patyra, 1996], which states that the membership function of the resulting fuzzy relation can be expressed as follows:

$$\mu_{PoR}(u, w) = \bigvee_{v} [\mu_P(u, v) * \mu_R(v, w)] \tag{2.37}$$

where $\bigvee$ represents the supremum operator (extended to all the elements in the universe of discourse V) and $*$ is a t-norm.

Using minimum and product as the t-norm in (2.37), two compositional rules widely referenced in the literature are obtained.

- **sup-min**: The *sup-min composition* of two fuzzy relations P(U, V) and R(V, W) is a fuzzy set Q(U, W) with the following membership function:

$$\mu_{PoR}(u, w) = \bigvee_{v} \{\min [\mu_P(u, v), \mu_R(v, w)]\} \tag{2.38}$$

- **sup-product**: The *sup-product composition* of two fuzzy relations P(U, V) and R(V, W) is a fuzzy set Q(U, W) with the following membership function:

$$\mu_{PoR}(u, w) = \bigvee_{v} [\mu_P(u, v) \cdot \mu_R(v, w)] \tag{2.39}$$

For discrete universes of discourse, the supremum operator becomes maximum and the above compositional rules are called *max-min composition* and *max-product composition*, respectively:

$$\mu_{PoR}(u, w) = \max_{v} \{\min [\mu_P(u, v), \mu_R(v, w)]\} \tag{2.40}$$

$$\mu_{PoR}(u, w) = \max_{v} [\mu_P(u, v) \cdot \mu_R(v, w)] \tag{2.41}$$

A very important particular case from the point of view of fuzzy inference techniques arises when one of the relations is a simple fuzzy set, i.e., when $\mu_P(u, v)$ becomes $\mu_P(v)$. Under these circumstances, (2.37) becomes:

$$\mu_{PoR}(w) = \bigvee_{v} [\mu_P(v) * \mu_R(v, w)] \tag{2.42}$$

Equation (2.42) shows how a fuzzy set can activate a fuzzy relation, thus providing a mathematical procedure to calculate the value of *z* from sentences such as "*x is big*" and "*z is greater than x*". This type of composition is the basis for approximate reasoning mechanisms, which will be studied in Chapter 3.

REFERENCES

1. **Cox, E.**, *The Fuzzy Systems Handbook*, 2nd edition, AP Professional, 1999.
2. **Mendel, J. M.**, Fuzzy logic systems for engineering: a tutorial, *Proceedings of the IEEE*, Vol. 83, N. 3, pp. 345-377, 1995.
3. **Patyra, M. J., Mlynek, D. M.**, Eds., *Fuzzy Logic. Implementation and Application*, Wiley & Teubner, 1996.
4. **Pedrycz, W., Gomide, F.**, *An Introduction to Fuzzy Sets. Analysis and Design*, MIT Press, 1998.
5. **Ross, T. J.**, *Fuzzy Logic with Engineering Applications*, McGraw-Hill Inc., 1995.
6. **Ruspini, E. H., Bonissone, P. P., Pedrycz, W.**, Eds., *Handbook of Fuzzy Computation*, Institute of Physics Pub., 1998.
7. **Wang, L. X.**, *A Course in Fuzzy Systems and Control*, Prentice-Hall Int., 1997.
8. **Weber, S.**, A general concept of fuzzy connectives, negations and implications based on t-norms, *Fuzzy Sets and Systems*, Vol. 11, pp. 115-134, 1983.
9. **Yager, R. R., Filev, D. P.**, *Essentials of Fuzzy Modelling and Control*, John Wiley & Sons, 1994.
10. **Zadeh, L. A.**, Fuzzy sets, *Information and Control*, Vol. 8, pp. 338-358, 1965.

Chapter 3

FUZZY INFERENCE SYSTEMS

A fuzzy inference system comprises a set of rules that employ linguistic terms similar to those used in natural language and an inference mechanism able to extract correct conclusions from approximate or incomplete data. This chapter elaborates the concepts of linguistic variables and rule bases, and introduces the mathematical formalisms for the numerical interpretation of fuzzy inference mechanisms. The different fuzzy operators will be considered in terms of their use as linguistic connectives, implication functions, aggregation operators and defuzzification methods. Finally, some of the inference mechanisms most widely used in practical applications of fuzzy logic will be analyzed.

3.1 LINGUISTIC VARIABLES

Going back to the example we used to introduce the concept of fuzzy set in Chapter 2, let us assume we use a thermometer to measure the body temperature of a person. We would obtain a numerical value such as 36°C or 37.2°C. However, if we ask this person, he/she would answer that his/her temperature is '*normal*', '*high*' or '*very high*'. A *linguistic variable* is a variable whose values can be expressed by means of natural language terms. The different terms or linguistic values are represented with fuzzy sets characterized by membership functions defined on the universe of discourse. Figure 3.1 shows the linguistic variable *Temperature* and the different linguistic values it can take.

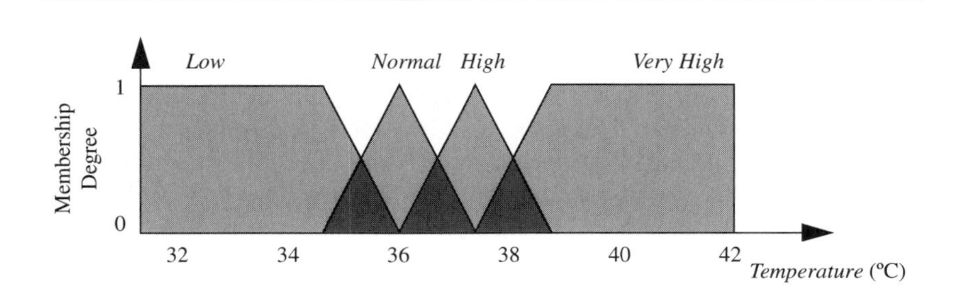

Figure 3.1. Definition of the linguistic variable *Temperature*.

Formally, a linguistic variable is characterized by a tuple of four elements (X, T, U, M), where X is the name assigned to the linguistic variable, T is the set of linguistic labels or values the variable can take, U is the application domain or universe of discourse where the variable is defined, and M is a semantic rule associating each linguistic label with a fuzzy set defined on the universe of discourse U. In the example we are considering, the name of the linguistic variable is '*Temperature*', the set of linguistic values is {'*Low*', '*Normal*', '*High*', and '*Very High*'}, the universe of discourse is the range 32–42°C and the meaning of the linguistic labels is defined by the membership functions shown in Figure 3.1.

3.2 FUZZY RULES

Vagueness or imprecision in natural language is not only a characteristic of the information representation mechanisms that humans use, but also the fundamental key of our approximate reasoning scheme. We can find in our daily activity many situations where we formulate rules like "*if the coffee is too bitter, add a lot of sugar*". Even in professional fields we employ fuzzy rules based on experience to control a complex system or to decide on an operation in the stock market. A *fuzzy rule* is a conditional sentence of the type:

IF <fuzzy proposition> THEN <fuzzy proposition> (3.1)

where, as in other kinds of rules, the proposition on the left side is called the *rule antecedent* or premise, and the one on the right side is called the *rule consequent* or conclusion.

An *atomic fuzzy proposition* is a simple phrase associating a linguistic label to a linguistic variable. A *compound fuzzy proposition* is any combination of atomic fuzzy propositions, employing linguistic connectives, either conjunctive ("and") or disjunctive ("or"), or negations ("not"). A pair of examples of compound fuzzy propositions could be "*temperature is high and pressure is low*" and "*error is big or error is very big and error variation is not small*".

3.2.1 LINGUISTIC CONNECTIVES

The meaning of an atomic fuzzy proposition such as "*x is A*" is defined by the membership function that represents fuzzy set *A*. Assigning a meaning for compound fuzzy propositions, such as "*x is A and y is B*" or "*(x is A or x is B) and y is C*", implies the calculation of the membership function that characterizes the fuzzy relation induced by the proposition. In order to do so, it is necessary to define the interpretation for the linguistic connectives "*and*" and "*or*" and for the operator "*not*". This interpretation, in any case, must be intuitive and correspond to the basic operations on fuzzy sets: intersection, union, and complement, respectively.

• **Connective "and"**: Given two linguistic variables, *x* and *y* defined on the universes of discourse X and Y, and two fuzzy sets, *A* and *B* on X and Y, respectively, fuzzy proposition "*x is A and y is B*" is interpreted as the fuzzy relation $A \cap B$ on X × Y with the following membership function:

$$\mu_{A \cap B}(x, y) = T\,[\mu_A(x), \mu_B(y)] \tag{3.2}$$

where T can be any t-norm.

• **Connective "or"**: Given two linguistic variables, *x* and *y* defined on the universes of discourse X and Y, and two fuzzy sets, *A* and *B* on X and Y, respectively, fuzzy proposition "*x is A or y is B*" is interpreted as the fuzzy relation $A \cup B$ on X × Y with the following membership function:

$$\mu_{A \cup B}(x, y) = S\,[\mu_A(x), \mu_B(y)] \tag{3.3}$$

where S can be any s-norm.

• **Operator "not"**: Given a linguistic variable, *x* defined on the universe of discourse X, and a fuzzy set, *A* on X, fuzzy proposition "*x is not A*" is interpreted as $\overline{A}$, with the following membership function:

$$\mu_{\overline{A}}(x) = C\,[\mu_A(x)] \tag{3.4}$$

where C can be any c-norm.

Thus linguistic connectives in a compound fuzzy proposition are interpreted in terms of fuzzy relations defined on the product space of the universes of discourse.

3.2.2 IMPLICATION FUNCTION

The preceding sections have allowed us to consider the meaning of atomic and compound fuzzy propositions where linguistic variables are used. We can now analyze the meaning of a fuzzy rule.

Given two linguistic variables, *x* and *y* defined on the universes of discourse X and Y, and two fuzzy sets, *A* and *B* on X and Y, respectively, fuzzy rule "*if x is A then y is B*" is interpreted as a fuzzy relation on X × Y with the following membership function:

$$\mu_{A \to B}(x, y) = \Phi\,[\mu_A(x), \mu_B(y)] \tag{3.5}$$

where Φ can be any of the so-called fuzzy implication operators.

Therefore, a fuzzy implication operator or *implication function* expresses the relation between the antecedent and the consequent of the rule. There are many fuzzy implication functions proposed in the existing literature. Most of them are extensions of the implications used in propositional or multivalued logic and use different options for performing the operations of set intersection, union, and complement [Lee, 1990; Driankov, 1993]. The existence of different implication functions resembles the different interpretations that humans can assign to the same sentence.

In traditional propositional logic a proposition can only take the values true 'T' or false 'F'. Figure 3.2 shows the truth tables for the five basic operations (conjunction, disjunction, negation, equivalence and implication). An expression of the form "*if p then q*" is represented by the implication ($p \rightarrow q$), which is true when any of the following conditions holds: (a) p is true, q is true; (b) p is false, q is false; and (c) p is false, q is true. The implication is false whenever (d) p is true, q is false. In other words, the only constraint is that a false conclusion cannot be derived from a true premise. It is easily demonstrated that $p \rightarrow q$ is equivalent to:

$$(\neg p) \vee q \tag{3.6}$$

$$(p \wedge q) \vee (\neg p) \tag{3.7}$$

By means of the isomorphism between propositional logic and set theory, the connectives $\vee$, $\wedge$, and $\neg$ can be substituted by the operators $\cup$, $\cap$, and $\overline{}$ (complement). Since in fuzzy rules p and q are fuzzy propositions, it is possible to interpret the fuzzy implication function by substituting the above mentioned operators by s-norms, t-norms, and c-norms, respectively. In formal terms, we can state that, given two linguistic variables, x and y defined on the universes of discourse X and Y, and two fuzzy sets, A and B on X and Y, respectively, fuzzy rule "*if x is A then y is B*" is interpreted as a fuzzy relation on $\mathrm{X} \times \mathrm{Y}$ whose membership function can be one of the following:

$$\mu_{A \rightarrow B}(\mathrm{x}, \mathrm{y}) = \mathrm{S}\ \{\mathrm{C}\ [\mu_A(\mathrm{x})], \mu_B(\mathrm{y})\} \tag{3.8}$$

$$\mu_{A \rightarrow B}(\mathrm{x}, \mathrm{y}) = \mathrm{S}\ \{\mathrm{T}\ [\mu_A(\mathrm{x}), \mu_B(\mathrm{y})], \mathrm{C}\ [\mu_A(\mathrm{x})]\} \tag{3.9}$$

depending on which equation, (3.6) or (3.7), is selected to define $A \rightarrow B$.

The concrete choice of the operators conforms the different implication functions.

p	q	$p \wedge q$	$p \vee q$	$\neg p$	$p \leftrightarrow q$	$p \rightarrow q$
T	T	T	T	F	T	T
T	F	F	T	F	F	F
F	T	F	T	T	F	T
F	F	F	F	T	T	T

Figure 3.2. Basic operations in propositional logic.

• **Dienes–Resher's implication**: Substituting C and S in (3.8) with the basic fuzzy complement (2.14) and the maximum operator (2.12), respectively, we obtain Dienes–Resher's implication or *Boolean implication*. The membership function of the fuzzy relation is:

$$\mu_{A \to B}(x, y) = \mu_{Rb}(x, y) = \max\,[1 - \mu_A(x), \mu_B(y)] \tag{3.10}$$

• **Mizumoto's implication**: Employing the algebraic sum (2.19) to implement the s-norm in (3.8) we get Mizumoto's implication, where the membership function of the fuzzy relation is:

$$\mu_{A \to B}(x, y) = \mu_{RM}(x, y) = 1 - \mu_A(x) + \mu_A(x)\,\mu_B(y) \tag{3.11}$$

• **Lukasiewicz's implication**: Using the bounded sum operator (2.20) to implement the s-norm in (3.8) we obtain Lukasiewicz's implication, also called *arithmetic implication*. The membership function of the fuzzy relation is:

$$\mu_{A \to B}(x, y) = \mu_{Ra}(x, y) = \min\,[1, 1 - \mu_A(x) + \mu_B(y)] \tag{3.12}$$

• **Dubois–Prade's implication**: Using the drastic sum operator (2.21) to implement the s-norm in (3.8) leads to Dubois–Prade's implication function. In this case, the membership function of the fuzzy relation is:

$$\mu_{A \to B}(x, y) = \mu_{RDP}(x, y) = \begin{cases} 1 - \mu_A(x), \text{ if } \mu_B(y) = 0; \\ \mu_B(y), \text{ if } \mu_A(x) = 0; \\ 1, \text{ if } \mu_A(x), \mu_B(y) > 0; \end{cases} \tag{3.13}$$

• **Zadeh's implication**: Taking the expression of propositional calculus (3.9) and the basic operators of fuzzy union (2.12), intersection (2.13), and complement (2.14), we obtain Zadeh's implication function, or *max-min implication*. The membership function of the fuzzy relation is:

$$\mu_{A \to B}(x, y) = \mu_{Rm}(x, y) = \max\,\{\min\,[\mu_A(x), \mu_B(y)], 1 - \mu_A(x)\} \tag{3.14}$$

The following implication functions come from multivalued logic:

• **Goguen's implication**: According to the constraints that implications must satisfy in multivalued logic, Goguen proposed a fuzzy relation for the "*if-then*" rules whose membership function is given by:

$$\mu_{A \to B}(x, y) = \mu_{R\Delta}(x, y) = \begin{cases} 1, \text{ if } \mu_A(x) \le \mu_B(y); \\ \mu_B(y) / \mu_A(x), \text{ if } \mu_A(x) > \mu_B(y); \end{cases} \tag{3.15}$$

• **Gödel's implication**: The implication function proposed by Gödel is one of the most used formulations in multivalued logic. The membership function of the fuzzy relation is given by:

$$\mu_{A \to B}(x, y) = \mu_{RG}(x, y) = \begin{cases} 1, \text{ if } \mu_A(x) \le \mu_B(y); \\ \mu_B(y), \text{ if } \mu_A(x) > \mu_B(y); \end{cases} \tag{3.16}$$

• **Sharp implication**: This implication function, also called *standard sequence*, is similar to the former, but more restrictive. The membership function of the fuzzy relation is defined by:

$$\mu_{A\to B}(x, y) = \mu_{Rs}(x, y) = \begin{cases} 1, & \text{if } \mu_A(x) \le \mu_B(y); \\ 0, & \text{if } \mu_A(x) > \mu_B(y); \end{cases} \quad (3.17)$$

The implication function of propositional logic is a global function, in the sense that truth tables in Figure 3.2 cover all the possible cases. When p and q are fuzzy propositions, a local interpretation of the implication function could be more sensible, so that the truth value of the function is high only when the truth values of p and q are high [Wang, 1997]. In formal terms, this means defining the implication function as a conjunction:

$$p \to q \equiv p \wedge q \quad (3.18)$$

Substituting the intersection in (3.18) by the operators minimum and product, we get the two most widely reported implication functions when applying fuzzy techniques to control.

• **Mamdani's implication**: According to this interpretation, an "*if-then*" rule induces a fuzzy relation R_c (c from conjunction) with a membership function given by:

$$\mu_{A\to B}(x, y) = \mu_{Rc}(x, y) = \min\,[\mu_A(x), \mu_B(y)] \quad (3.19)$$

• **Larsen's implication**: The membership function of the fuzzy relation is given in this case by:

$$\mu_{A\to B}(x, y) = \mu_{Rp}(x, y) = \mu_A(x) \cdot \mu_B(y) \quad (3.20)$$

The definition of new implication functions and the analysis of their properties have been the central point of interest of many researchers. The point of departure for the problem of the characterization of fuzzy implication functions is the establishment of a series of criteria that allow this characterization. In this sense, a set of desirable formal properties for implication functions has been defined [Baldwin, 1980], intuitive criteria for investigating the properties of implication functions have been proposed [Mizumoto, 1982], and the problem has been addressed from a pragmatical point of view, analyzing the influence of operators on the accuracy of fuzzy models [Kiszka, 1985]. The authors of these studies agree in concluding that the implication functions initially proposed by Zadeh, R_m (3.14) and R_a (3.12), despite the fact that they have a well-defined logical structure, are not suitable for the application of approximate reasoning techniques, since the results obtained from their use do not always correspond to those intuitively expected. Conversely, implication functions based on fuzzy conjunction, R_c (3.19) and R_p (3.20), are adequate for approximate reasoning techniques, especially in control applications, where inference is done in a single step, and inference chaining and contraposition do not play an important role. As we have previously stated, these characteristics, together with the ease of implementation, have made Mamdani's and Larsen's implication functions the most widely adopted in control applications. In applications where the inference mechanism becomes more complex, as in the case of decision-making systems, the use of the sharp implication, R_s (3.17), seems to be an adequate choice [Mizumoto, 1982; Lee, 1990].

3.3 APPROXIMATE REASONING TECHNIQUES

Reasoning techniques permit obtaining logical deductions, i.e., inferring conclusions from premises. The propositions of classical logic admit only two truth values {'0' and '1'}. However, in fuzzy logic the truth value of a proposition can be any number in the [0, 1] interval. This generalization is the basis for the approximate reasoning techniques that permit obtaining right conclusions from imprecise and vague premises.

The two most important inference rules in propositional logic are called *modus ponens* and *modus tollens*. Modus ponens states that, given the rule "*if x is A then y is B*" and the observation "*x is A*", the conclusion "*y is B*" is obtained. In symbolic form this can be written as:

$$(p \wedge (p \rightarrow q)) \rightarrow q \tag{3.21}$$

On the other hand, modus tollens states that, given the rule "*if x is A then y is B*" and the observation "*y is not B*", the conclusion "*x is not A*" is obtained. In symbolic form this can be written as:

$$(\bar{q} \wedge (p \rightarrow q)) \rightarrow \bar{p} \tag{3.22}$$

While modus ponens is used to obtain conclusions by means of *forward inference*, modus tollens is employed to deduce causes by means of *backward inference*. Another, more intuitive representation, of these inference rules is:

- Modus ponens -		**- Modus tollens -**	
rule:	*if x is A* $\rightarrow$ *y is B*	rule:	*if x is A* $\rightarrow$ *y is B*
premise:	*x is A*	premise:	*y is not B*
conclusion:	*y is B*	conclusion:	*x is not A*

3.3.1 GENERALIZED MODUS PONENS/TOLLENS

These two inference rules are extended in fuzzy logic to their generalized version. Fuzzy modus ponens (or *generalized modus ponens*) can be expressed as follows: given the two fuzzy propositions «*if x is A then y is B*» and «*x is A'*» then it is possible to obtain the conclusion «*y is B'*».

Rule:	*if x is A then y is B*
Premise:	*x is A'*
Conclusion:	*y is B'*

where A, B, A' and B' are linguistic terms represented by fuzzy sets, with the feature that B' will approximate more B as A' approximates more A.

In a similar way, fuzzy modus tollens (or *generalized modus tollens*) states that, given the two fuzzy propositions «*if x is A then y is B*» and «*y is B'*», then it is possible to obtain the conclusion «*x is A'*».

Rule:	*if x is A then y is B*
Premise:	*y is B'*
Conclusion:	*x is A'*

where A, B, A' and B' are linguistic terms represented by fuzzy sets, with the feature that A' will be more different from A as B' is more different from B.

3.3.2 COMPOSITIONAL RULE OF INFERENCE

Focusing on the case of generalized modus ponens, the problem is the numerical evaluation of B' from the similarity degree between A and A'. Zadeh gave a solution to this problem when he proposed the *compositional rule of inference* stating that, given a rule and an observation (A'), the conclusion (B') can be obtained as the sup-star composition ($\circ$) between the observation and the fuzzy relation induced by the rule:

$$B' = A' \circ R \tag{3.23}$$

Recalling the definition of sup-star composition in equation (2.37) of Section 2.3 in Chapter 2, the membership function for the fuzzy set obtained as conclusion is:

$$\mu_{B'}(y) = \bigvee_{x} \{\mu_{A'}(x) * \mu_{A \to B}(x, y)\} \tag{3.24}$$

Substituting in (3.24) the membership function of the fuzzy relation corresponding to the implication function defined by (3.5), the membership function of the fuzzy set B' can be expressed as:

$$\mu_{B'}(y) = \bigvee_{x} \{\mu_{A'}(x) * [\Phi[\mu_A(x), \mu_B(y)]]\} \tag{3.25}$$

Therefore, the result of the inference depends on the choices for the operator $*$ and the implication function Φ. With respect to the t-norm used in the compositional rule of inference, Zadeh initially proposed the use of a minimum (composition sup-min) [Zadeh, 1973]. Some other composition operators have been proposed, with other t-norms playing the role of $*$ (composition sup-product, sup-bounded product, or sup-drastic product). The sup-min and sup-product composition methods are, undoubtedly, the most widely employed, at least in control applications [Lee, 1990].

3.3.3 AGGREGATION OPERATOR

As we have seen in the previous section, the compositional rule of inference provides a mechanism for evaluating the result of a fuzzy rule. A fuzzy inference system normally contains a set of linguistic description rules of the form "*if - then*". In the most general case, antecedents and consequents in these rules include compound fuzzy propositions, combining multiple inputs and outputs. This type of system is called a MIMO (multiple input, multiple output) system. In any case, a MIMO system can always be considered a set of systems with multiple inputs and a single output (MISO). Rule antecedents can contain any combination of input variables by means of the connectives "*and*", "*or*" and "*not*", and rule consequents can contain one or more output variables by means of the connective "*and*". Yet, in order to avoid an excessive complexity in mathematical formulation, we will focus our discussion on fuzzy inference systems whose rule base has the following structure:

$$\text{Rule r: if } x_1 \text{ is } A_1{}^{r} \text{ and } x_2 \text{ is } A_2{}^{r} \text{ and} \ldots x_n \text{ is } A_n{}^{r} \text{ then } y \text{ is } B^{r} \quad (3.26)$$

where x_i and y correspond to the inputs and the output of the system and A_i^{r} and B^{r} are linguistic labels interpreted as fuzzy sets by means of their membership functions, $\mu_{A_i}{}^{r}(x_i)$ and $\mu_B{}^{r}(y)$, defined on the universes of discourse for the inputs and the output of the system.

According to the interpretation for the connective in the antecedents (3.2) and for the implication function (3.5), each rule in the knowledge base induces a fuzzy relation (R^{r}) in $X_1 \times X_2 \ldots \times X_n \times Y$ with a membership function defined by:

$$\mu_R{}^{r}(x_1,x_2 \ldots x_I,y) = \Phi \, \{T_c \, [\mu_{A_1}{}^{r}(x_1), \mu_{A_2}{}^{r}(x_2) \ldots \mu_{A_n}{}^{r}(x_n)] \, , \mu_B{}^{r}(y)\} \quad (3.27)$$

where T_c is the conjunction connective applied to the antecedents (with the properties of a t-norm) and Φ is one of the implication functions considered in Section 3.2.2.

In the most general case, the inputs to a fuzzy system are linguistic terms represented by fuzzy sets, i.e., an expression of the form "x_1 *is* A'_1 *and* x_2 *is* A'_2 ... *and* x_n *is* A'_n" which induces a fuzzy relation A' on $X_1 \times X_2 \ldots \times X_n$, whose membership function is defined by:

$$\mu_{A'}(x_1,x_2 \ldots x_n) = T_c \, [\mu_{A'_1}(x_1), \mu_{A'_2}(x_2) \ldots \mu_{A'_n}(x_n)] \quad (3.28)$$

Applying Zadeh's compositional rule of inference (3.23), the conclusion of each individual rule is given by:

$$B'^{r} = A' \circ R^{r} \quad (3.29)$$

In contrast to a conventional expert system, a fuzzy system can have several active rules simultaneously. The global conclusion must be calculated from the aggregation of the partial solutions supplied by each rule. Two questions arise at this point: first, how the aggregation is to be interpreted; second, what procedure is to be performed to accomplish aggregation. The answer to the first question is closely related to the intuitive meaning of the rule base. Considering that each one of the rules is independent from the others, a reasonable operator for interpreting aggregation is the union. The meaning of the rule base would be in this case "R^1 *or* R^2 ... *or* R^n" (this is why the aggregation operator is commonly referred to as the connective "also"). If, on the contrary, we consider that rules in the knowledge base are strongly coupled, so that conditions in all rules must match to infer a result ("R^1 *and* R^2 ... *and* R^n"), the obvious choice for the aggregation operator is intersection. When rules have local properties, as in the case of control applications or any other case where implication functions based on t-norms (like Mandani's or Larsen's) are used, the use of an s-norm as the aggregation operator is universally accepted. However, for other implication functions, such as Gödel's and those based on propositional calculus, interpreting aggregation by means of a t-norm makes more sense [Wang, 1997].

With respect to the way in which aggregation is performed, there are, again, two options [Driankov, 1993]. The first one, called *composition-based inference*, consists in combining the different rules in a single fuzzy relation and applying the compositional rule of inference on this global fuzzy relation. If we use the union as the aggregation operator, this mechanism yields the following result:

$$B' = A' \circ R_G \quad | \quad R_G = \bigcup_{r} (R^1, R^2, \ldots, R^R) \tag{3.30}$$

A second alternative, called *rule-based inference*, consists in calculating the contribution of each of the rules independently as stated in (3.29), and afterwards combining those contributions to obtain the global result from the rule base. If we again use the union as the aggregation operator, the result is of the form:

$$B' = \bigcup_{r} B'^r \quad | \quad B'^r = A' \circ R^r \tag{3.31}$$

The choice of one of these aggregation mechanisms is determined, in general, by a series of criteria that include the intuitive meaning of the rule base and the computational efficiency of the selected procedure. From the point of view of the hardware or software implementation of the inference system, the last condition is a key concern, since it can determine the viability of the implementation. Anyway, it can be demonstrated [Lee, 1990] that (3.30) and (3.31) produce the same results when using the sup-min or sup-product composition and the aggregation is interpreted by means of the union operator.

3.4 RULE-BASED INFERENCE MECHANISMS

According to the formulation presented up to now, a fuzzy logic–based inference system is determined by the options selected in the implementation of the connectives for antecedent conjunction (T_c) and rule aggregation ($\cup$ or $\cap$), the implication function (Φ), and the t-norm in the compositional rule of inference ($*$). This section will allow us to analyze in detail some of the inference mechanisms that appear when selecting these fuzzy operators. But before this, let us make some considerations about the type of the information provided to the input of the system and the impact that certain implication functions have on reducing the computational cost of the inference process.

Substituting expressions (3.27) and (3.28) in (3.29), the conclusion obtained when applying the premise "x_1 *is* A'_1 *and* x_2 *is* A'_2 ... *and* x_n *is* A'_n" to the r-th rule (3.26), can be calculated as:

$$\mu_{B'}{}^{r}(y) = \bigvee_{x} \{[\, T_c\,(\mu_{A'_1}(x_1) \ldots \mu_{A'_n}(x_n))] *$$

$$[\,\Phi\,[\,T_c\,(\,\mu_{A_1}{}^{r}(x_1) \ldots \mu_{A_n}{}^{r}(x_n))\,,\, \mu_{B}{}^{r}(y)\,]\,]\,\} \tag{3.32}$$

The computational complexity of this expression can be considerably reduced if the implication function is a t-norm (T_Φ), since in this case it is possible to apply the associativity and distributivity of t-norms with respect to the supremum operator $\bigvee$. Hence, (3.32) can be rewritten as follows:

$$\mu_{B'}{}^{r}(y) = T_\Phi\,\{T_c\,\{\,[\,\bigvee_{x_1}\,(\mu_{A'_1}(x_1) * \mu_{A_1}{}^{r}(x_1))\,]\ldots$$

$$\ldots[\,\bigvee_{x_n}\,(\mu_{A'_n}(x_n) * \mu_{A_n}{}^{r}(x_n))\,]\,\}\,,\, \mu_{B}{}^{r}(y)\,\} \tag{3.33}$$

Each of the terms between brackets in (3.33) defines the similarity degree between a fuzzy input and its corresponding antecedent in the rule. The composition of these similarity degrees by means of the T_c operator determines the *activation degree* (α) of the r-th rule. In this way, and assuming that the implication function is a t-norm, Equation (3.31) can be rewritten as follows for a set of rules:

$$B' = \bigcup_{r}\,[T_\Phi\;(\alpha^r,\; B^r)\,] \tag{3.34}$$

In many practical applications inputs to the inference engine are crisp (non-fuzzy) values. As an example, in control applications, inputs are usually provided by sensors that deliver numerical data. Each one of the terms in the compound proposition "x_1 *is* x_{o1} *and* x_2 *is* x_{o2} ... *and* x_n *is* x_{on}" can be interpreted as a fuzzy singleton:

$$\mu_{A'_i}(x_i) = 0 \text{ if } x_i \neq x_{oi},\ \mu_{A'_i}(x_{oi}) = 1 \tag{3.35}$$

This circumstance greatly simplifies the calculation of the activation degree used in expression (3.34), since this calculation becomes reduced to the evaluation of the membership degrees of each antecedent for its input value and their composition by means of the conjunction operator. The result of the inference can be expressed in this case as:

$$\mu_{B'}(y) = \bigcup_{r} \{ T_{\Phi} [\alpha^{r} , \mu_{B}{}^{r}(y)] \} \tag{3.36}$$

where

$$\alpha^{r} = T_{c} [\mu_{A_1}{}^{r}(x_{o1}), \mu_{A_2}{}^{r}(x_{o2}) \dots \mu_{A_n}{}^{r}(x_{on})] \tag{3.37}$$

In control applications the connective for antecedent conjunction (T_c) is usually the minimum function, the implication functions used (T_{Φ}) are either minimum (Mamdani's) or product (Larsen's), and the rule aggregation operators ($\cup$) maximum or sum. By combining these operators, a series of inference mechanisms widely reported in literature are obtained. The following sections will analyze these inference mechanisms, applying them, for the sake of simplicity, to a rule base with two fuzzy rules with two antecedents and one consequent:

$$\begin{aligned} &\text{Rule 1: if } x_1 \text{ is } A_1{}^1 \text{ and } x_2 \text{ is } A_2{}^1 \text{ then y is } B^1 \\ &\text{Rule 2: if } x_1 \text{ is } A_1{}^2 \text{ and } x_2 \text{ is } A_2{}^2 \text{ then y is } B^2 \end{aligned} \tag{3.38}$$

3.4.1 MIN-MAX INFERENCE

Min-Max (or Mamdani's) is, undoubtedly, the inference mechanism most widely reported in the literature. It uses the min t-norm as the implication function and the max s-norm as the aggregation operator. The membership function of the conclusion inferred from the application of the rule base is given by the following expression:

$$\mu_{B'}(y) = \max_{r} (\min (\alpha^{r}, \mu_{B}{}^{r}(y))) \tag{3.39}$$

where the activation degree of the r-th rule is:

$$\alpha^{r} = \min (\mu_{A_1}{}^{r}(x_{o1}), \mu_{A_2}{}^{r}(x_{o2}) \dots \mu_{A_n}{}^{r}(x_{on})) \tag{3.40}$$

Figure 3.3 shows the graphical interpretation of applying this inference mechanism to the rule base described by (3.38).

3.4.2 PRODUCT-SUM INFERENCE

This method uses the operators product and sum for implication and rule aggregation, respectively [Mizumoto, 1991]. The membership function of the resulting fuzzy set is given by:

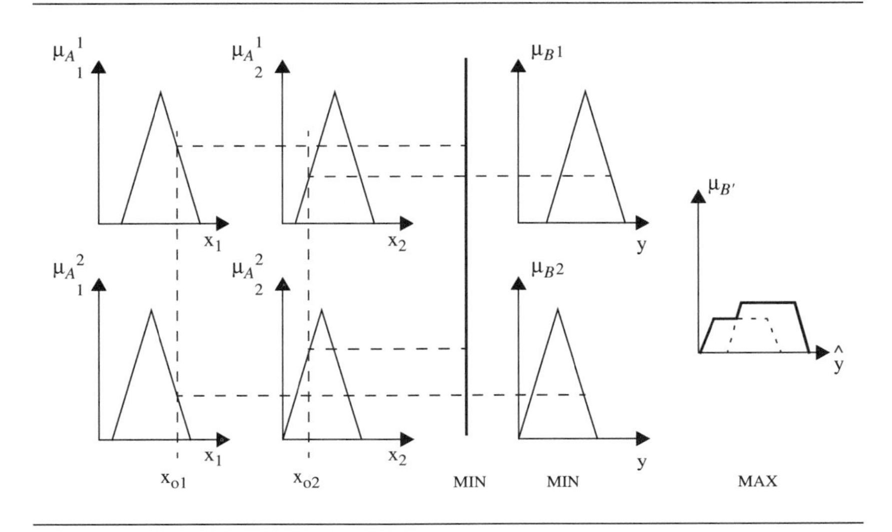

Figure 3.3. Min-Max inference mechanism.

$$\mu_{B'}(y) = \sum_r \alpha^r \cdot \mu_B{}^r(y) \tag{3.41}$$

where α^r is also expressed as in (3.40). The application of this inference mechanism to the rule base described in (3.38) is shown in Figure 3.4.

3.5 DEFUZZIFICATION METHODS

The result of any of the inference mechanisms we have just described is a fuzzy set. For this information to be used in certain applications, like control, it is necessary to procure a precise (crisp) value representing this set. The *defuzzification* process is expressed by means of a defuzzifying operator F^{-1} that transforms the membership function of a fuzzy set $\mu(y)$ in a definite element $\hat{y}$ of the universe of discourse Y:

$$F^{-1}: \mu(y) \rightarrow \hat{y} \tag{3.42}$$

A great number of defuzzification methods have been proposed in the literature. We will now describe some of these methods, analyzing them in the double perspective of their outcome efficiency and their computational simplicity (which, in conclusion, means facilitating their microelectronic implementation).

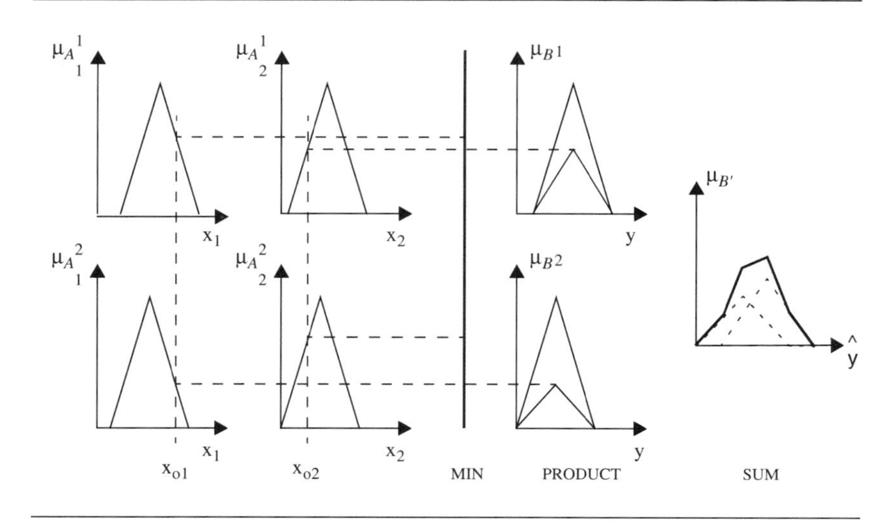

Figure 3.4. Product-Sum inference mechanism.

3.5.1 CONVENTIONAL DEFUZZIFICATION METHODS

• **Center of Area (CoA) / Center of Gravity (CoG) / Centroid**[1]: This is the defuzzification method commonly used in Mamdani-type systems. It determines the representative value for a fuzzy set as the center of the area delimited by the fuzzy set *B'* resulting from applying the different rules. Practical applications employ the discrete version of this method, that can be interpreted as a Riemann's sum [Jager, 1992]:

$$\text{CoA} \rightarrow \hat{y} = \frac{\sum_{i=1}^{n} y_i \cdot \mu_{B'}(y_i)}{\sum_{i=1}^{n} \mu_{B'}(y_i)} \tag{3.43}$$

where n is the number of quantization levels for the output (the number of elements in its universe of discourse).

1. The concept of gravity is commonly associated with multidimensional situations, while area is assumed to suit one-dimensional schemes. Since it is a mere semantic cast, we will use both terms without distinction.

Among the positive characteristics of this method it is worth pointing out that it is *continuous* (a small change in the inputs does not imply an abrupt change in the output) and *not-ambiguous* (it always supplies a single defuzzification value). However, the method has some disadvantages. First, it is not *plausible* (since the output can be a point in the support of the fuzzy set with a low membership degree). Secondly, since it starts from the global conclusion, when the inference mechanism employed is Min-Max it does not consider the overlapping zones of different partial conclusions (the method is not *weight counting*) [Hellendoorn, 1993a]. Furthermore, this method introduces non-linearities in the system even when rules are linear. Having to sweep the whole universe of discourse in order to evaluate the centroid is the main drawback for its microelectronic implementation: it requires a high area consumption if the process is parallelized or a slow inference cycle if a serial processing strategy is applied.

- **Center of Sums (CoS)**: In contrast to the above method, the objective of the Center of Sums method is to account for the contribution of each of the partial conclusions B'^{r}. For this purpose it uses sum instead of maximum for rule aggregation, so overlapping areas are reflected in its final result. In the discrete case it is formally expressed by:

$$\mathrm{CoS} \rightarrow \hat{y} = \frac{\sum_{i=1}^{n} y_i \cdot \sum_{r=1}^{R} \mu_{B'}^{r}(y_i)}{\sum_{i=1}^{n}\sum_{r=1}^{R} \mu_{B'}^{r}(y_i)} \tag{3.44}$$

Most software implementations of fuzzy systems use this method instead of CoA because it is faster. Furthermore, if the implication function is the product (Product-Sum inference (3.41)), expression (3.44) can be transformed into:

$$\hat{y} = \frac{\sum_{i=1}^{n} y_i \cdot \sum_{r=1}^{R} \alpha^{r} \cdot \mu_{B}^{r}(y_i)}{\sum_{j=1}^{r}\sum_{r=1}^{R} \alpha^{r} \cdot \mu_{B}^{r}(y_i)} = \frac{\sum_{r=1}^{R} \alpha^{r} \cdot \sum_{i=1}^{n} y_i \cdot \mu_{B}^{r}(y_i)}{\sum_{r=1}^{R} \alpha^{r} \cdot \sum_{i=1}^{n} \mu_{B}^{r}(y_i)} \tag{3.45}$$

The sum on i in the denominator of (3.45) corresponds to the area of the fuzzy set of the rule consequent, while the sum in the numerator is the product of this area by the centroid of the fuzzy set. The inference process can be accelerated precalculating the values for centroids $\hat{c}^{r}$ and areas S^{r}.

$$\text{CoS} \rightarrow \hat{y} = \frac{\sum_{r=1}^{R} \alpha^r \cdot S^r \cdot \hat{c}^r}{\sum_{r=1}^{R} \alpha^r \cdot S^r} \tag{3.46}$$

• **Center of Largest Area (CLA)**: When the union of the output fuzzy sets is not convex, for example, it consists of two convex subsets, this method determines the convex subset with the largest area and defines the defuzzified output to be the center of area of that subset. The result is more plausible than the center of area of the original set, but it can be ambiguous when both subsets have the same area.

• **First of Maxima (FoM) / Last of Maxima (LoM)**: Using the fuzzy set obtained from the aggregation of the rules, these methods provide as a result the lowest/highest value in the output domain for the maximum membership degree:

$$\text{FoM} \rightarrow \hat{y} = \inf \{y_j \in Y \mid \mu_{B'}(y_j) = \max_i \mu_{B'}(y_i)\} \tag{3.47}$$

$$\text{LoM} \rightarrow \hat{y} = \sup \{y_j \in Y \mid \mu_{B'}(y_j) = \max_i \mu_{B'}(y_i)\} \tag{3.48}$$

• **Mean of Maxima (MoM)**: This method yields the mean of all values in the output domain that have a maximum value for the membership function of the conclusion:

$$\text{MoM} \rightarrow \hat{y} = \sum_{j=1}^{J} y_j / J \mid \mu_{B'}(y_j) = \max_i \mu_{B'}(y_i) \tag{3.49}$$

The application of this defuzzification method can reduce the fuzzy characteristics of the system, since the membership functions that describe inputs may not have any effect on the value obtained (in fact, if classical sets instead of fuzzy sets are used, the result is not modified). Other drawbacks are that this method is not weight counting, and may present discontinuities since only the dominant rule influences the output. Moreover, it has a high computational complexity as it is necessary to sweep the whole universe of discourse in order to calculate the maxima.

3.5.2 SIMPLIFIED DEFUZZIFICATION METHODS

The use of inference and defuzzification mechanisms which are not costly in terms of area and are able to provide fast inference cycles is an essential requirement for an efficient implementation of fuzzy systems on silicon. This requirement leads us to discard the handling of fuzzy information in rule consequents and to substitute this information by a set of parameters representing it (as we have seen above in the case of the Product-Sum inference with CoS defuzzification). This leads to a series of methods that we will overall designate as *simplified defuzzification methods*. Each of these methods differs from the others in the number and meaning of the parameters it uses [Baturone, 1995].

• **Fuzzy Mean (FM) / Height Method (HM):** In this method, the output is obtained as:

$$\text{FM/HM} \rightarrow \hat{y} = \sum_r \alpha^r \cdot c^r / \sum_r \alpha^r \tag{3.50}$$

were c^r is the point of maximal membership of the r-th rule consequent and α^r is the activation degree of the r-th rule.

This expression can be obtained from CoS, assuming that the consequent membership functions are symmetrical and with the same shape (same area), so that c^r coincides with the centroid of the consequent fuzzy set. It can also be obtained from CoA/CoG by reducing gradually the number of quantization intervals for the output universe of discourse until it coincides with the number of distinct consequents [Jager, 1992].

Although this simplified method does not consider membership function support or area (so it is not possible to emphasize certain conclusions by choosing different sizes or shapes for consequents), this is currently the most used method, for its simplicity of implementation. Nevertheless, to alleviate its drawbacks, some other alternative methods have been proposed.

• **Weighted Fuzzy Mean (WFM) / Quality Method (QM) / ξ-Quality Method (ξ-QM):** These three methods obtain the defuzzified output with a expression similar to (3.50) but introducing a second weighting parameter (γ^r) for each rule:

$$\hat{y} = \sum_r \alpha^r \cdot \gamma^r \cdot c^r / \sum_r \alpha^r \cdot \gamma^r \tag{3.51}$$

In the case of Weighted Fuzzy Mean [Jager, 1992], γ^r is proportional to the basis of the consequent fuzzy set (d_r), so the conclusions provided by those consequents represented by a membership function with a wider basis (fuzzier information) contribute in a higher degree to the output. The results are very similar to CoA/CoG:

$$\text{WFM} \rightarrow \hat{y} = \frac{\sum_{r=1}^{R} \alpha^r \cdot c^r \cdot d_r}{\sum_{r=1}^{R} \alpha^r \cdot d_r} \quad (3.52)$$

Conversely, for the Quality Method [Hellendoorn, 1993a] the point is that those consequents represented by membership functions with a narrower basis correspond to more precise rules that, therefore, should make a stronger contribution to the final result. For this reason, this method uses γ^r parameters inversely proportional to the basis of the consequent sets:

$$\text{QM} \rightarrow \hat{y} = \frac{\sum_{r=1}^{R} \alpha^r \cdot c^r / d_r}{\sum_{r=1}^{R} \alpha^r / d_r} \quad (3.53)$$

Finally, the ξ-Quality Method [Hellendoorn, 1993b] corresponds to a modification of the previous one, where the output is given by:

$$\xi\text{-QM} \rightarrow \hat{y} = \frac{\sum_{r=1}^{R} \alpha^r \cdot c^r / d_r^{\xi}}{\sum_{r=1}^{R} \alpha^r / d_r^{\xi}} \quad (3.54)$$

For $\xi = 1$ the result is equivalent to QM, and for $\xi = 0$ it coincides with FM/HM. For values of ξ between 0 and 1 the result is somewhere between that provided by QM and FM/HM. Finally, for $\xi > 1$ the output is still more influenced by the less fuzzy rules. These last two defuzzification methods present an important property called by their authors *"quality regarding"*.

• **Center of Sums with minimun (CoSmin):** As we have seen above, CoS can be considered as a simplified method when using the product as the implication function. In this case, the output can be interpreted as the sum of conclusions ($\hat{c}^r$) weighted by the activation degrees (α^r) and the areas of the consequent membership functions (S^r). Both sets of parameters, c^r and S^r only depend on rule consequents, so they can be precalculated. For triangular membership functions $S^r = d_r \cdot 1/2$ and expression (3.46) becomes (3.52). That is, CoS is equivalent to WFM.

Instead of considering the product as the implication function, the proposal in [Pammu, 1995] is to consider the minimum and to use triangular membership functions with the same basis. The outcome of this method is provided by the following expression:

$$CoS_{min} \rightarrow \hat{y} = \frac{\sum_{r=1}^{R} c^r \cdot \alpha^r \cdot (2 - \alpha^r)}{\sum_{r=1}^{R} \alpha^r \cdot (2 - \alpha^r)} = \frac{\sum_{r=1}^{R} c^r \cdot (1 - (\beta^r)^2)}{\sum_{r=1}^{R} (1 - (\beta^r)^2)} \tag{3.55}$$

where $\beta = 1 - \alpha$ is the complement of the activation degree.

• **Level Grading Method (LGM):** With similar arguments to those that justify QM and ξ-QM, LGM intends to avoid the problem of the dominance of wider consequent membership functions in the output. But, instead of using a weighting parameter inversely proportional to the basis of the membership function, it employs a "certainty measure" which is inversely proportional to the length of the α-cut of the r-th consequent membership function, $\omega_B{}^r(\alpha)$ [Jung, 1994]. The output is given by:

$$LGM \rightarrow \hat{y} = \frac{\sum_{r=1}^{R} \alpha^r \cdot c^r / \omega_{B^r}(\alpha)}{\sum_{r=1}^{R} \alpha^r / \omega_{B^r}(\alpha)} \tag{3.56}$$

This method takes into account not only the support of consequent membership functions, as WFM, QM, and ξ-QM do, but also their shapes. In particular, it can be considered as an extension of QM, since both provide the same results for rectangular membership functions. If triangular membership functions are considered, the relationship between the basis of the membership functions (d) and the length of its α-cut $\omega_B(\alpha)$ is $\omega_B(\alpha) = d \cdot (1-\alpha)$. Hence, the expression (3.56) is transformed in:

$$LGM \rightarrow \hat{y} = \frac{\sum_{r=1}^{R} \alpha^r \cdot c^r / [d_r \cdot (1 - \alpha^r)]}{\sum_{r=1}^{R} \alpha^r / [d_r \cdot (1 - \alpha^r)]} \tag{3.57}$$

And, if all triangles are equal (d_r is constant):

$$\text{LGM} \rightarrow \hat{y} = \frac{\sum_{r=1}^{R} \alpha^r \cdot c^r/(1-\alpha^r)}{\sum_{r=1}^{R} \alpha^r/(1-\alpha^r)} \tag{3.58}$$

• **Yager's Method (YM):** In all the simplified methods described so far the representative values for the rule consequents (c^r) are assigned a priori and later aggregated using a weighted average, where the rule activation degrees and/or the supports or areas of the fuzzy sets are employed. The method proposed in [Yager, 1991] and the similar method described in [Berenji, 1992] dynamically calculate the values for each c^r. For this purpose, they use the arithmetic mean of the projections on the output universe of the intersections of the activation degree of each rule and the membership function of its consequent (Figure 3.5).

When using triangular or trapezoidal fuzzy sets, c^r can be expressed in terms of two parameters that depend on the shape of the membership function, and the result of this defuzzification method is given by:

$$\text{YM} \rightarrow \hat{y} = \frac{\sum_{r=1}^{R} \alpha^r \cdot (p_1^r + p_2^r \cdot \alpha^r)}{\sum_{r=1}^{R} \alpha^r} \tag{3.59}$$

This method is similar to the Level Grading Method in that it also takes into account the shapes of the consequent membership functions.

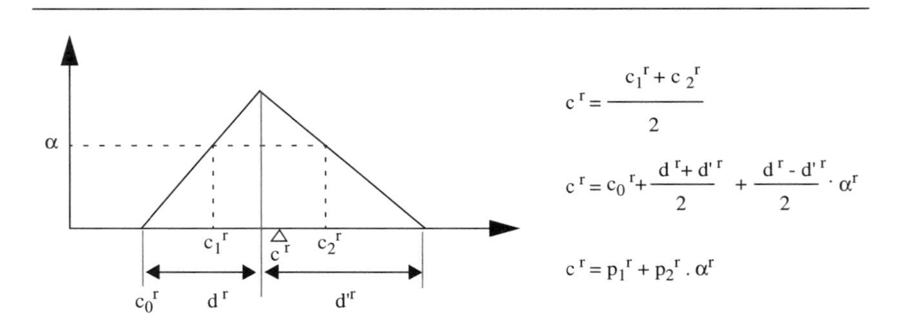

Figure 3.5. Yager's defuzzification method.

3.6 TYPES OF FUZZY SYSTEMS

The inference systems we have analyzed until now employ consequents described by fuzzy sets. These types of fuzzy systems are usually referred to as *"Mamdani-type systems"*. We will briefly describe other types of fuzzy systems that have been proposed in the literature.

• **Tsukamoto-type systems**: In this group are included the systems with antecedents and consequents described by monotonic functions such as the ones shown in Figure 3.6 [Tsukamoto, 1979]. The inference method applied, called Tsukamoto's method, calculates the system conclusion as a weighted average extended to the set of rules, employing the activation degree α^r as the weight for each rule, and the point where $\mu_B{}^r$ takes the value α^r as the representative point for each consequent. If we denote this point as $c_B{}^r$, the output of these systems is given by:

$$y_o = \sum_r \alpha^r \cdot c_B{}^r / \sum_r \alpha^r \qquad (3.60)$$

• **Takagi–Sugeno type systems**: Inference systems of this type are characterized by the use of functions of their input variables as consequents [Takagi, 1983]:

$$F_B{}^r (x_1, x_2, \ldots x_u) = \sum_k a_k{}^r \cdot x_k + a_o{}^r \qquad (3.61)$$

where $a_k{}^r$ (k = 0, 1, ... u) are constants.

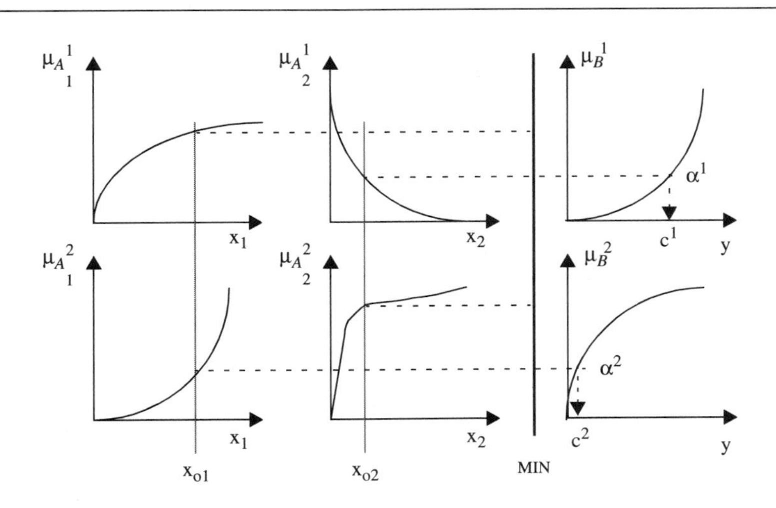

Figure 3.6. Inference mechanism for Tsukamoto-type systems.

The conclusions provided by the different rules coincide with the values that those functions take for a certain value of the inputs. To obtain the global output, each rule is weighted by its activation degree α^r:

$$y_o = \sum_r \alpha^r \cdot F_B{}^r(\mathbf{x}) \, / \sum_r \alpha^r \tag{3.62}$$

Zero-order Takagi–Sugeno type systems, in which $F_B{}^r(\mathbf{x}) = a_o{}^r$, have been widely employed because they allow very simple software as well as hardware implementations. Since in these systems the consequents are represented by singleton values, they are also known as *singleton fuzzy systems*. A Mamdani-type system which employs the Fuzzy Mean / Height Method as the defuzzification method is equivalent to a zero-order Takagi–Sugeno type or a singleton fuzzy system that employs the centroids of the consequent fuzzy sets as singleton values for the consequents.

3.7 STRUCTURE OF A FUZZY SYSTEM

From an implementation point of view, a fuzzy system includes the blocks shown in Figure 3.7. The kernel of the system is a knowledge base that contains the definition of membership functions for the antecedents and consequents used in rules, and an inference engine able to process this information according to any of the mechanisms described in the preceding sections. Apart from these basic elements, in most cases it is necessary to include two interface blocks connecting the inference engine with the inputs and outputs of the system. These blocks are termed fuzzifier and defuzzifier.

The fuzzifier is in charge of accepting the inputs to the inference system and evaluating the similarity degree between these inputs and the linguistic labels used in the rule antecedents. The practical realization of this block fundamentally depends on the type of inputs accepted by the system. In decision-

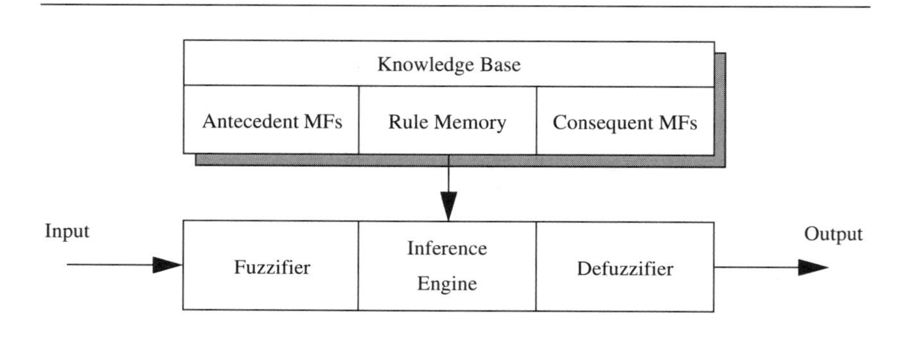

Figure 3.7. Structure of a fuzzy system.

making applications and in complex control problems, the inputs to the system may be described by means of possibility distributions represented by a fuzzy set. The similarity degree between this fuzzy set A' and the one representing an antecedent A is usually calculated as the maximum of the minimum (max-min composition) of both sets (Figure 3.8a). Conversely, in the vast majority of control applications the inputs come from sensors providing concrete values that can be interpreted as fuzzy singletons. The similarity degree between the input and an antecedent is in this case the value that the membership function of the antecedent takes for the value u_o of the input (Figure 3.8b).

The inference engine of the fuzzy system evaluates the different rules in the knowledge base. The activation degree of each rule is calculated from the activation degree of its antecedents and according to the interpretation of the different connectives in use. From this point, the output of each rule is calculated applying the activation degree to the consequent by means of the implication function.

Finally, the conclusions of the different rules are combined by the aggregation operator and, if the application requires it so, the defuzzifier is used to provide the output of the inference system. The operation of the inference and defuzzification processes depends on the defuzzification method in use. When conventional methods, such as those described in Section 3.5.1, are applied, it is necessary to sweep the universes of discourse of the output variables to obtain the fuzzy set that will be passed to the defuzzifier. However, if simplified methods, such as those analyzed in Section 3.5.2, are in use, the inference and defuzzification processes are combined by the weighted average calculation mechanism, extended only on the set of rules. As stated above, the choice between both defuzzification schemes greatly influences the efficiency of an inference system implementation.

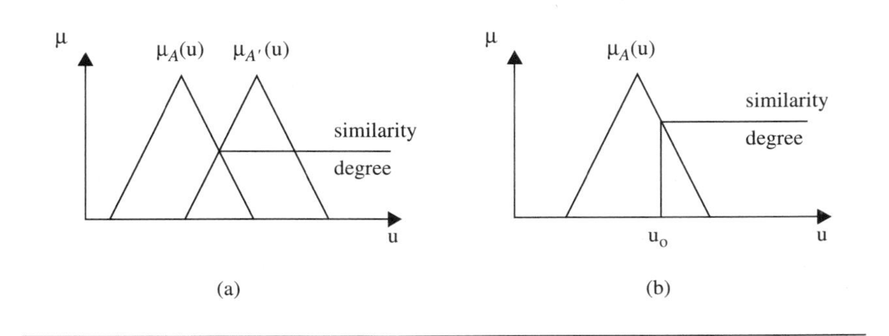

Figure 3.8. Functionality of the fuzzifier when the input is: (a) a fuzzy set; (b) a fuzzy singleton.

REFERENCES

1. **Baturone, I., Sánchez-Solano, S., Barriga, A., Huertas, J. L.**, Implementation of inference/defuzzification methods via continuous-time analog circuits, in *Proc. 6th IFSA World Congress*, pp. 623-626, São Paulo, 1995.
2. **Baldwin, J. F., Pilsworth, B. W.**, Axiomatic approach to implication for approximate reasoning with fuzzy logic, *Fuzzy Sets and Systems*, Vol. 3, N. 2, pp. 193-219, 1980.
3. **Berenji, H. R., Khedkar, P.**, Learning and tuning fuzzy logic controllers through reinforcements, *IEEE Transactions on Neural Networks,* Vol. 3, N. 5, pp. 724-740, 1992.
4. **Driankov, D., Hellendoorn, H., Reinfrank, M.**, *An Introduction to Fuzzy Control*, Springer-Verlag, 1993.
5. **Hellendoorn, H., Thomas, C.**, Defuzzification in fuzzy controllers, *Journal of Intelligent and Fuzzy Systems*, Vol. 1, pp. 109-123, 1993a.
6. **Hellendoorn, H., Thomas, C.**, The ξ-quality defuzzification method, in *Proc. 5th IFSA World Congress*, pp. 1159-1162, Seoul, 1993b.
7. **Jager, R., Verbruggen, H. B., Bruhx, P. M.**, The role of defuzzification methods in the application of fuzzy logic, in *Proc. Symposium on Intelligent Components and Instruments for Control Applications*, pp. 111-116, Málaga, 1992.
8. **Jung, S. H., Cho, K. H., Kim, T. G., Park, K. H.**, Defuzzification method for multishaped output fuzzy sets, *Electronics Letters*, Vol. 30, pp. 740-742, 1994.
9. **Kiszka, J. B., Kochanska, M. E., Sliwinska, D. S.**, The influence of some fuzzy implication operators on the accuracy of a fuzzy model, Parts I & II, *Fuzzy Sets and Systems*, Vol. 15, pp. 111-128, 223-240, 1985.
10. **Lee, C. C.**, Fuzzy logic in control systems: fuzzy logic controller, Parts I & II, *IEEE Transactions on Systems, Man, and Cybernetics*, Vol. 20, N. 2, pp. 404-432, 1990.
11. **Mizumoto, M., Zimmermann, H.**, Comparison of fuzzy reasoning methods, *Fuzzy Sets and Systems*, Vol. 8, pp. 253-283, 1982.
12. **Mizumoto, M.**, Min-max-gravity method versus product-sum-gravity method for fuzzy controls, in *Proc. 4th IFSA World Congress*, pp. 127-130, Brussels, 1991.
13. **Pammu, S., Quigley, S. F.**, Novel analogue CMOS defuzzification circuit, *IEE Proc. Circuits Devices and Systems*, Vol. 142, N. 3, pp. 495-498, 1995.
14. **Takagi, T., Sugeno, M.**, Derivation of fuzzy control rules for human operator's control actions, in *Proc. IFAC Symposium on Fuzzy Information, Knowledge Representation and Decision Analysis*, pp. 55-60, Marseille, 1983.
15. **Tsukamoto, Y.**, An approach to fuzzy reasoning method, in *Advances in Fuzzy Set Theory and Applications*, North-Holland, 1979.
16. **Wang, L. X.**, *A Course in Fuzzy Systems and Control*, Prentice-Hall Int., 1997.
17. **Yager, R. R.**, An alternative procedure for the calculation of fuzzy logic controller values, *Journal of Japanese Society of Fuzzy Technology,* Vol. 3, N. 4, pp. 736-746, 1991.
18. **Zadeh, L. A.**, Outline of a new approach to the analysis of complex systems and decision processes, *IEEE Transactions on Systems, Man, and Cybernetics*, Vol. 3, N. 1, pp. 28-44, 1973.

Chapter 4

FUZZY SYSTEM DEVELOPMENT

Once the theoretical foundations for fuzzy inference systems have been presented, we will focus on the study of the practical aspects pertaining to the development of these systems. With this aim, the discussion will be structured according to the three typical stages in the design of any system: description, verification, and synthesis. Concretely, this chapter will try to answer two key questions related to the description of fuzzy systems: (1) What are the different tasks a fuzzy system designer has to undertake? and (2) What techniques or tools can ease the fulfillment of these tasks? To answer the first question we will analyze the general structure of the inference system introduced in Section 3.7 of Chapter 3, identifying the activities associated with the definition of each of the parts of the system. This analysis will show the existence of a design space with considerable complexity that requires for its exploration the use of formal specification languages managed by computer-based tools. These tools shall assist the designer in the generation and graphical visualization of the system structure and properties. The last part of this chapter will introduce the fuzzy system specification language XFL and the development environment integrating the different design tools that will be used throughout the rest of the book.

4.1 DEFINITION OF A FUZZY SYSTEM

According to the structure and operational principles described in the previous chapter, the development of a fuzzy system requires the selection of the system variables, the definition of the rule base, and the choice of the different fuzzy operators from a series of implementation options. It is important to mention that there is no ideal methodical procedure providing an optimal solution. Nevertheless, the following sections will offer some general considerations that can be helpful when designing a fuzzy inference system.

4.1.1 SELECTION OF SYSTEM VARIABLES

The first step in the development of a fuzzy inference system is the assignment of linguistic variables for each of the inputs and outputs of the system. If we recall the definition given in Section 3.1 of Chapter 3, each linguistic variable requires a name or identifier and the definition of: (a) the linguistic labels or terms that will be used inside the rules; (b) the universe of discourse or application range for the variable; and (c) the membership functions describing or representing each of the linguistic labels. The system designer must answer three basic questions: How many membership functions must be used? What kind of membership functions must be used? How must the membership functions be distributed along the universe of discourse?

The number of labels, and therefore the number of membership functions, associated to a linguistic variable is directly related to the convenience or ability to distinguish different categories in the universe of discourse for the variable. In general, this number will depend on the application context of the inference system. Most of the practical applications reported in the literature use between three and seven labels per linguistic variable. This choice has a double motivation. First, increasing the number of antecedent labels implies increasing the number of rules, complicating system implementation and slowing down its operation. Second, when the number of labels and rules exceeds a certain threshold, the linguistic significance of the rules is lost. Let us simply consider that in natural language the number of terms used to qualify the range of variation for a certain magnitude is relatively reduced (we refer to a temperature as '*frozen*', '*cold*', '*warm*', '*hot*', '*burning*' or to the size of an object as '*tiny*', '*small*', '*fair*', '*big*', '*huge*'). Anyway, the selection of the number of linguistic terms always means a compromise between implementation simplicity and detail of description.

The types of membership function most often used in practical applications of fuzzy logic were presented when introducing the concept of fuzzy set in Chapter 2 (Figure 2.2). At this point, it is sufficient to note that, when considering the implementation of a fuzzy system by means of a program or a microelectronic circuit, the use of piecewise linear functions (triangular, trapezoidal) generally simplifies the mechanisms employed for storing or generating those functions. The use of differentiable functions (Gaussian or sigmoid), on the contrary, commonly provides softer output modifications and eases certain procedures employed for tuning the system knowledge base.

Once the number and shape of the membership functions for a linguistic variable have been selected, the remaining issue is the distribution of the different fuzzy sets along the universe of discourse the variable is defined on. Two important considerations can be made at this point. First, there must be a complete coverage of the universe of discourse. Otherwise, the inference system will not be able to generate a coherent output for some possible input values. Second, a certain overlapping degree among different membership functions is highly desirable, so several rules can be activated at the same time and the interpolation capabilities of fuzzy logic are conveniently exploited. Several authors have proposed that the crossing point for two overlapping membership functions must be 0.5 for control applications and a little lower for classifiers [Driankov, 1993]. Finally, some of the microelectronic implementation techniques for fuzzy systems described in this book base their efficiency in limiting the overlapping degree of the membership functions used in rule antecedents, so the number of rules that can be simultaneously activated in the system is also limited.

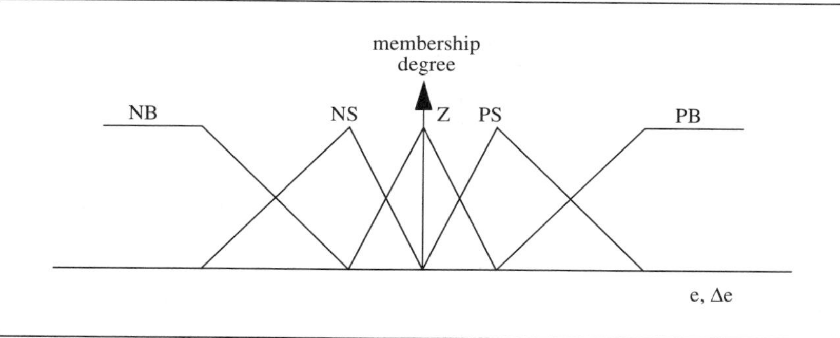

Figure 4.1. Membership functions typically used in control applications.

Figure 4.1 shows a set of membership functions typically used in control applications when representing the variables *error* (e) and *error_change* (Δe) in a fuzzy controller. The labels assigned to the five membership functions, 'NB', 'NS', 'Z', 'PS' and 'PB' are the initials of the terms '*negative big*', '*negative small*', '*zero*', '*positive small*' and '*positive big*', respectively. Those functions covering the extremes of the universe of discourse have saturations that take into account that when the system input values are in these ranges, the system is very far from equilibrium and the controller must apply the maximum corrective action permitted by the output range. The rest of the membership functions are triangles, whose basis decrease as they are nearer the steady position in the center of the universe of discourse. We must recall here that a fuzzy set with a larger support represents a fuzzier situation. As the steady position is approached, the controller output must become more precise, as suggested by the use of the triangle with the narrowest basis to represent the label 'Z'. Finally, the set of membership functions in Figure 4.1 has another important feature from the point of view of its implementation. The highest overlapping degree is two and, given a certain input value, the sum of the activation degrees is always equal to 1. In Chapter 9 we will discuss how these characteristics can simplify not only the circuits for generating membership functions but also, when using certain fuzzy operators, the realization of the inference mechanisms.

4.1.2 RULE BASE DEFINITION

The rule base of a fuzzy system describes the behavior of the inference system based on the linguistic terms associated with the input and output variables. For the example introduced in the section above, a typical control rule could be expressed with a sentence like "*if error is negative_big and error_change is positive_small then control_action must be negative_small*". Computer-based tools for the definition and development of fuzzy systems use different techniques for the representation of the rule base. The first consists of employing a formal language inspired by natural language. Something like:

if e is NB & Δe is NB -> out is NB
if e is NB & Δe is NS -> out is NB
...
if e is PB & Δe is PB -> out is PB

Figure 4.2 shows two alternative representation techniques that permit a more condensed notation. The tabular or relational format (Figure 4.2a) allows the description of rule bases with an arbitrary number of inputs and outputs, provided that a single connective is used in all antecedents. For systems with two inputs and one output the matrix format shown in Figure 4.2b is universally accepted.

There are certain properties of a rule base that determine the behavioral characteristics of the fuzzy system. A rule base is *complete* when for any combination of inputs there is at least one active rule. A rule base is said to be *consistent* if it has not any pair of rules with the same antecedents and different consequents. Finally, a rule base is *continuous* if there are no adjacent rules whose consequents have an empty intersection. This last property is highly related to the smoothness of the input-output characteristic of a fuzzy system.

The matrix representation of a fuzzy system allows a fast and simple verification of these properties for a certain rule base. This way, we can assert that, when using the membership functions in Figure 4.1, the rule base represented by Figure 4.2b is complete, consistent, and continuous. When a rule base is represented as a matrix, and provided that there is overlapping in the membership functions for the antecedents, having a definition for all the elements in the matrix is sufficient to guarantee that the rule base is complete. The matrix repre-

e	Δe	out
NB	NB	NB
NB	NS	NB
	...	
Z	Z	Z
	...	
PS	PB	PB
PB	PB	PB

(a)

Δe \ e	NB	NS	Z	PS	PB
NB	NB	NB	NB	NS	Z
NS	NB	NB	NS	Z	PS
Z	NB	NS	Z	PS	PB
PS	NS	Z	PS	PB	PB
PB	Z	PS	PB	PB	PB

out

(b)

Figure 4.2. Different representations for the rule base of a fuzzy controller. (a) Tabular format; (b) Matrix format.

sentation leads to rule bases that are "consistent by construction", since any element in the matrix represents a different combination of antecedents.

The number of rules in a fuzzy system depends on the number of inputs (that is, the number of antecedents) and on the number of labels for each antecedent. Assuming that the same number of labels, L, is used for each input variable, the number of possible rules for a fuzzy system with u inputs is L^u. One of the main problems that arises when the complexity of a fuzzy system increases is the so-called "curse of dimensionality", caused by the exponential growth in the number of possible rules as the number of input variables becomes larger. This problem implies, first, a loss of semantic clarity in the knowledge base and, second, an excessive growth of the resources required for implementing the systems. Both drawbacks can be alleviated by using a specification formalism that allows the partition of the knowledge base in terms of hierarchical structures, providing a flexible description framework for complex systems. In other cases, the way to alleviate the exponential growth is the use of a non-complete rule base.

Some important aspects not yet discussed are those related to the process of deriving the rules for a fuzzy inference system. We have repeatedly pointed out in this text that a fuzzy system is a knowledge-based system and that its objective is the emulation of the reasoning and acting scheme of a human expert faced with a specific situation. In these terms, fuzzy rules are basically "common sense rules". Nevertheless, if we analyze our own acting mechanisms we can conclude that there are two kind of situations: (1) those in which the rules can be directly formulated in an explicit form, like "*if the temperature is low, raise the thermostat of the heater*", and (2) situations where we know what to do, but a rationalization effort is necessary to express this "what to do" in terms of rules. Most of us find parking a car or choosing a fruit from a basket of apples easier than precisely formulating the rules that explain our behavior. In other words, the extreme cases of these two situations correspond, respectively, to conscious and unconscious mechanisms [Wang, 1997]. The procedures for obtaining the rules of a fuzzy system should be based on either of these mechanisms:

- From the knowledge of experts, trying to express their knowledge about the problem in terms of common sense rules that can be further refined by some tuning procedure.

- From data corresponding to a correct behavior, analyzing the actions of experts to extract general rules from their responses to particular cases.

In the latter case, when there is a set of data conforming to the input-output behavior to be emulated, it is very important to use techniques for automating the processes of extracting rules and adjusting the parameters of the knowledge base of the fuzzy system. Since these *learning techniques* play a very important role in the development of fuzzy systems, a significative part of the next chapter will describe the theoretical foundations and some implementation aspects of these methods.

4.1.3 SELECTION OF FUZZY OPERATORS

As we have seen in the previous chapters, the whole formalism of fuzzy logic is based upon a set of mathematical operators that define membership functions, the different logical connectives and operations, like implication functions and defuzzification methods. There is a large number of these functions and, in fact, new proposals for them are still being introduced.

There is not yet a well-established corpus about the appropriateness of each of these functions for the different applications of fuzzy systems, whereas the use of a certain set of operators can be steered by the nature of the intended final implementation. In any case, the implementation platform for a fuzzy system (computer algorithms, circuits) will necessarily impose constraints on the selection of the different fuzzy operators, as will be shown in the next chapters.

4.2 CAD TOOLS FOR FUZZY SYSTEMS

The use of computer-aided design (CAD) techniques and tools is a common practice in many fields of science and engineering. CAD tools ease the conception and further development of complex systems. The user is relieved from repetitive calculation and tuning tasks, so the effort can be focused on the more relevant aspects of system design. In the field of fuzzy logic–based systems, the need for CAD tools is even more justified by the nature of the technology and its application areas.

Fuzzy logic employs complex mathematical models aiming to approximate qualitative reasoning, where imprecision and uncertainty must be taken into account. Any inference performed by a fuzzy system, no matter how simple it is, implies a great number of calculations based on a set of parameters whose variation may imply a dramatic change in the behavior of the system. Therefore, it is more than desirable for the designer to have the means that allow him/her to concentrate on the system characteristics in terms of the fuzzy technology itself, without worrying about all the calculation framework that supports its realization.

On the other hand, the application of fuzzy logic in areas such as control, non-linear system modeling, or artificial intelligence implies that the system under development is being conceived to work in complex environments, so it is necessary to exercise its behavior before the final implementation is available. This way, it will be possible to guarantee that, at least as a prototype, the (or the consecutive) implementation(s) of the system has been thoroughly validated prior to using it in its actual working environment, thus saving time and effort.

There is a number of CAD tools for fuzzy systems, either commercial or in the public domain (Table 4.1 shows the Internet locations of the companies and institutions that offer some of these tools). To establish a useful reference framework for analyzing this kind of tools, we will introduce a series of criteria that permit the evaluation of their suitability to the designer's purposes. These criteria, which will be discussed in the following sections, are:

- Use of a formal language for system specification
- Facilities for system definition
- Capabilities for fuzzy operator extensions
- Integration with other development tools
- Facilities for tuning, verifying, and analyzing the system
- Synthesis capabilities for obtaining the final implementation of the system

4.2.1 SPECIFICATION LANGUAGE

Any CAD tool must store the definition of the systems under development in a certain format, so the designer can stop or resume working on it at any moment. In principle, there are many options for this format, from the use of plain text files containing lines with a predefined meaning for each field, to the employment of databases with specific tables. In any case, the use of a formal specification language is the most convenient for several reasons.

First, a formal specification language is human-readable and this can help the designer in understanding what the tool is doing with the system under development and how it is doing it, while it allows the designer to make modifications on the system definition in a fast and simple way, without having to execute the development environment. Second, the use of a formal language with well-defined syntax and semantics allows the tool to verify some system properties from its specification. Furthermore, the modifications applied to the system by the tool itself (for example, as a result of applying a learning method) can be validated in a natural way.

It is also worth noting that the use of a formal language eases the integration of the development environment with other tools, since the interfaces to these other tools can be accomplished by translators able to transform one representation into the other. This point is also applicable to synthesis tools, thus it is possible to get an implementation of the system for a set of specific conditions by means of programs implementing the semantics defined by the specification language on the target architecture. Lastly, the use of this kind of language assures the extensibility of the CAD tools and backward compatibility among its different versions, so systems designed with a former version can be used with the current version without complicated (and therefore, error prone) conversion processes.

Table 4.1. CAD tools for fuzzy systems.

Company/Institution	Product	Web address
Aptronix	FIDE [Yen, 1995]	http://www.aptronix.com/fide/fide.htm
Darmstadt University of Technology	FuNeGen	ftp://obelix.microelectronic.e-technik.th-darmstadt.de/pub/neurofuzzy/
Flexible Intelligence Software	FlexTool	http://www.flextool.com/
HyperLogic	CubiCalc	http://www.hyperlogic.com/products.html
Indigo Software	Fuzzy Expert	http://www.indigo.co.uk/fzy.htm
Inform	FuzzyTECH	http://www.fuzzytech.com/
Instituto de Microelectrónica de Sevilla	Xfuzzy [López, 1998]	http://www.imse.cnm.es/Xfuzzy/
Intelligent Machines	O'INCA Design Framework	http://www.meridian-marketing.com/O_INCA/
Logic Programming Associates	FLINT	http://www.lpa.co.uk/fln.html
Mathworks	MATLAB Fuzzy Logic Toolbox	http://www.mathworks.com/products/fuzzylogic/
MODiCO	Fuzzle	http://www.modico.com/downfuz.htm
SGS-Thomson Microelectronics	Fuzzy Studio	http://eu.st.com/stonline/products/support/fuzzy/
Siemens	SieFuzzy	http://www.atd.siemens.de/td_electronic/produkte/software/siefuzzy.htm
Togai Infralogic	TILShell, TILGen	http://www.ortech-engr.com/fuzzy/TilShell.html
University of Magdeburg	NEFCON, NEFCLASS [Nauck, 1994; 1995]	http://fuzzy.cs.uni-magdeburg.de/
University of Hannover	VSP Decision Program [Jeschke, 1998]	http://www.vspdecision.uni-hannover.de/
University of Missouri -St. Louis	FID [Janikow, 1998]	http://www2.cs.umsl.edu/~janikow/fid/
Universidad Nacional de Colombia	UNFUZZY	http://ohm.ing.unal.edu.co/ogduarte/Software.htm
University of New Mexico	Fuzzy Logic Software	http://ace.unm.edu/fuzzy/fuzzy.html
University of Oldenburg	FOOL & FOX	http://condor.informatik.uni-oldenburg.de/FOOL.html
University of Otago	FuzzyCOPE	http://divcom.otago.ac.nz:800/COM/INFOSCI/KEL/fuzzycop.htm
Vienna University of Technology	StarFLIP++ [Bonner, 1997]	http://www.dbai.tuwien.ac.at/proj/StarFLIP/

4.2.2 FACILITIES FOR SYSTEM DEFINITION

A common characteristic of virtually any current CAD tool is the use of a graphical user interface. The ability for understanding, handling, and analyzing the definition of a system is greatly enhanced by techniques that provide a graphical representation of its structure and behavior. Furthermore, the time needed to learn the use of a tool is remarkably reduced if it offers intuitive access mechanisms.

It is important to distinguish three different areas relating to graphical facilities for system definition:

- The graphical definition of the system structure, in terms of input and output variables, rule bases, and any other elements considered in the tool. An interface with these features permits a fast and simple visualization of the architecture of the system.
- The graphical definition of system variables, universes of discourse, and membership functions. All CAD tools for fuzzy systems offer more or less sophisticated interfaces for accessing graphical representations of membership functions and variables. These representations can usually be manipulated by drag-and-drop mechanisms.
- The (more or less) graphical definition of rule bases. The nature of rules in a fuzzy system makes it more difficult, if not impractical, to provide a completely graphical interface for rule definition. When working with a pair of inputs and one output, a tabular definition may be viewed as natural, but the representation becomes more and more complicated and unmanageable as the number of inputs and outputs grows even minimally. Most rule editors offer facilities for the construction of syntactically correct rules in the specification language used by the tool.

It must be taken into account that many CAD tools impose certain constraints on the characteristics of the system that can be defined by their use, such as a maximum number of variables or a maximum number of rules. These limits usually come from the fact that the tools are oriented to the definition of fuzzy systems with a predetermined architecture, which dictates restrictions on the system structure.

4.2.3 FUZZY OPERATORS

As stated in Section 4.1.3, any CAD tool for fuzzy systems must provide mechanisms for selecting the set of fuzzy operators to be applied in the realization of the system. Common facilities in this aspect offer the possibility of selecting, at least, fuzzy connectives, implication functions, and defuzzification methods from a list of predefined operators installed into the tool.

With respect to membership functions, the availability of a set of predefined classes of functions into the specification language is desirable, besides the possibility of representing any other type of function by means of a piece-wise linear approximation.

An additional advantage, unfortunately not so common in many current tools, is the availability of mechanisms for the direct definition of new operators into the tool. These facilities allow the evaluation of new techniques and formalisms, both in terms of algorithmic efficiency and of simplicity in the final system implementation.

4.2.4 INTEGRATION WITH OTHER TOOLS

When developing any minimally complex system, different methods must be used to obtain a complete implementation for it. For example, for a software-based system it is unlikely that the same elements will be used for the calculation modules and for the user interface modules. In the same way, a hardware-based system may need analog and/or digital modules to fit its specifications. This is especially true in our case, where we are dealing with elements specialized on the system behavior which do not cover the full range of requirements the system must fulfill.

Development tools tend to be delivered as an integrated whole entity in which, following the adequate steps, a final implementation of the system can be obtained from a high-level definition of it. But we cannot forget that the final implementation of the system must very often include other elements that have to be integrated to conform it. In this sense, the integration of a CAD tool into a (more or less) normalized design flow becomes very important, since it permits the use of the tool facilities in parallel with other similar ones. As far as possible, this integration should be available at any level inside the design flow, from the algorithmic level (for system verification and simulation) to the final implementation.

Notwithstanding that in other aspects is an advantage, having only a graphical user interface for accessing a CAD tool can be a drawback with respect to its integration capabilities. Essentially, a graphical user interface only accepts user interaction through the usual input elements (keyboard and mouse) and it is difficult to connect with other tools. Furthermore, certain phases in the development process can have a high cost in terms of time and computer resources, so it is convenient to provide mechanisms that do not require direct user attention at those costly stages.

Therefore, CAD tools should have the means for using a non-graphical interface, easy to integrate with other tools into the whole system design flow. This way, the designer can choose between using the graphical features of the development environment to specify the system, following the steps the environment suggests, or specifying the system in the same way that a program is written, using the tools offered by the environment together with any other needed to obtain a complete final implementation. Of course, this scheme should allow any intermediate situation between these two extremes.

4.2.5 FACILITIES FOR TUNING, VERIFICATION, AND ANALYSIS

Three groups of CAD tool features can be included in this category:

- Learning mechanisms
- Tuning and simulation mechanisms
- Analysis mechanisms

The use of learning methods for easing the building and tuning of a system is a fundamental feature of a development tool that intends to be employed for minimally complex systems. Simulation mechanisms provide the means for evaluating the algorithmic behavior of the system prior to its final implementation. Using these mechanisms it is possible to reduce costs and development time, since once the system is going to be tested in its real working environment, the designer has a prototype whose behavior has been tested in situations as close as possible to its actual operational conditions.

This is potentially one of the most important aspects when evaluating the characteristics of a learning or simulation tool. It should be possible to establish a simulation context closely resembling real conditions, while the system behavior must also approximate its intended final implementation (for example, in terms of quantization errors in a digital implementation).

What we refer to here as analysis mechanisms includes a set of procedures that allow us to obtain information about relevant parameters in the system, either by means of any kind of representation (graphical, numerical, or even real-time animations) or by real-time emulations of the system. The main objective of these mechanisms is to provide the designer with a better and more direct understanding of the system's internal behavior. This way, the designer may know more in-depth the effects of changes on the system and gain experience for future work, thus establishing design patterns.

One of the most usual representations is a surface, in which one of the system outputs is represented as a function of two of its inputs. Other common ways to picture system behavior are visualizations of rule activation degrees, usually connected in real time with the simulation procedures. Those tools more specifically oriented to fuzzy control often provide facilities to show the trajectory of the system inside the phase plane.

Due to the key role that these mechanisms play in the development of a fuzzy system, the next chapter will analyze in detail both their theoretical foundations and their practical application inside CAD tools.

4.2.6 SYNTHESIS CAPABILITIES

It is obvious that the final objective of a CAD tool is to produce an operating implementation, that is, the synthesis of a realization of the system from the specifications that the designer has defined and tuned. In this sense, we must recall that fuzzy logic defines a model independent from any particular realization that employs it and, therefore, a fuzzy system can be implemented according to different paradigms and architectures.

The most direct method for performing the synthesis of a specification by a CAD tool consists of translating the internal representation the tool holds into a software-based implementation. Software synthesis tools cover two fundamental goals in a CAD environment. First, they constitute an alternative for the final realization of the system. Second, they provide the fundamentals for the elements that allow the verification and tuning of the system by simulating its behavior.

The software synthesis of a fuzzy system can be performed by employing two different strategies: either by code that will be further interpreted by a core that implements fuzzy functionalities, or by making a direct translation into an executable program.

In the first case, an internal representation of the system (usually, the same representation used by the tool to store and verify it) runs on a run-time library. This library defines some kind of fuzzy virtual machine that executes a specific program, corresponding to the behavior of the system. This approach presents some drawbacks, since its results run slower than in the other case and there may be problems if the system is intended to run on a computer with a different architecture from the one running the development tool. Its main advantage is that, wherever the "fuzzy virtual machine" is available, information exchange between different modules and computers is faster and more efficient.

To perform a direct translation of the system specification into an executable program it is necessary to obtain a final implementation in machine language for the architecture and operating system of the computer that the system will be running on. The obvious choice is to use, as an intermediate element, a general purpose programming language. This language is the target of the translation process from the system specification the CAD tool holds. Once the system has been expressed in terms of this programming language, the compilation utilities commonly available in any operating system are used to produce the executable program that implements the fuzzy system.

Virtually all the currently available CAD tools for fuzzy systems include options for translating the definition of the system under development into a C program. The vast availability and the standardization of this programming language make it the obvious choice for this kind of task. In addition to producing standard C code suitable for standard computer architectures, some tools offer facilities for obtaining direct translations into assembler code for a series of microprocessors, microcontrollers, and specific fuzzy coprocessors, so it is possible to have a more efficient implementation of the inference system for the selected target architecture.

However, it is not so easy to find CAD tools for fuzzy systems providing mechanisms to perform a synthesis into dedicated hardware of the system under development. This lack of tools for hardware synthesis cannot be attributed to the absence of well-defined design methodologies for dedicated hardware. As will be shown in this text, these design methodologies, structured enough to support the automatic synthesis of fuzzy systems, have been already proposed.

The development environment that will be introduced in the next section was conceived with the objective, among others, of providing capabilities for synthesizing microelectronic circuit implementations for fuzzy systems.

4.3 THE XFUZZY DEVELOPMENT ENVIRONMENT

The *Xfuzzy* development environment for fuzzy systems is composed of a set of tools covering the different stages of description, verification, and synthesis of inference systems based on fuzzy logic. The different tools that constitute *Xfuzzy* share a common formal specification language. They are integrated into a graphical user interface, and employ an internal representation scheme that allows them to work in parallel on a fuzzy system definition [López, 1998].

Figure 4.3 shows the general design flow using *Xfuzzy*. The description of the system is performed using the XFL specification language [López, 1997]. An XFL specification holds information about the knowledge base (membership functions for antecedents and consequents, and the structure of the rule base) and the operators (connectives, the implication function, the defuzzification method) that define the fuzzy system. Each of the tools in *Xfuzzy* corresponds to a certain task in the development of a fuzzy system. In the description stage, the specification of the system can be inserted either using any text editor or by means of the graphical facilities of the environment. The XFL description of a fuzzy system is the starting point for the different tools of the environment.

Xfuzzy provides specific mechanisms to allow the user to define and experiment with new fuzzy operators. The definition of these new operators is made depending on the target of the final implementation, so it is possible to adapt the behavior of the operators to the precise aims of the system under development.

Since the environment integrates a set of tools that can be independently employed, *Xfuzzy* has no problem when working in combination with other development tools. Furthermore, those design stages that are expensive in terms of time or computer resources can be performed non-interactively, even taking advantage of systems with execution queues.

The use of the specification language XFL and the internal representation employed by the different components of *Xfuzzy* ease the setup of different tools suited to the verification stage. The simulation tool included in *Xfuzzy* allows for a very flexible definition of the system and its operational environment, since fuzzy logic-based elements and other external modules can be arbitrarily combined inside the tool, and the simulation can be directly connected to other programs in order to (for example) obtain more sophisticated output formats than those directly offered by the tool itself. Moreover, the environment includes a supervised learning tool based on different backpropagation algorithms, able to handle all the classes of membership functions supported by XFL, and a large number of fuzzy operators.

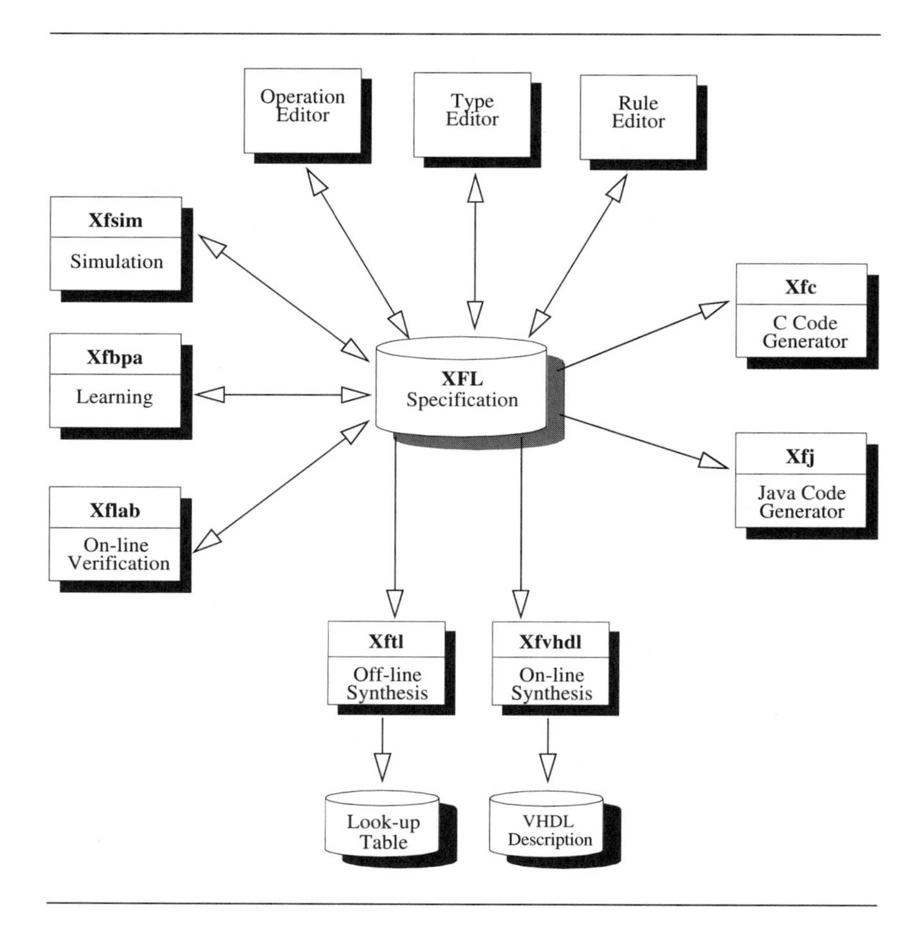

Figure 4.3. Design flow using *Xfuzzy*.

However, the synthesis features of *Xfuzzy* really make the difference with other currently available CAD tools for fuzzy systems. In addition to software synthesis capabilities, based on translators into C or Java, *Xfuzzy* includes two tools for the hardware synthesis of the systems described using XFL. These tools respond to different implementation strategies for fuzzy systems and consider the use of different techniques for the realization of microelectronic circuits.

Xfuzzy has been written in C using the Unix operating system. Its graphical user interface employs the X-Window system. *Xfuzzy* can be used on any Unix-like operating system with a C compiler and X-Window.

In this and other chapters the different components of *Xfuzzy* will be described to some extent. Our goal is not to provide a detailed description of the syntax and options of each of the components of the environment (the reader can refer to the reference manuals of the most recent version of *Xfuzzy*), but rather to give an overview of its features to aid in the understanding of the examples included in the text.

4.3.1 THE XFL LANGUAGE

XFL is a formal language specifically oriented to the definition of fuzzy systems. It has been designed according to a set of objectives that permit the use of XFL in any of the stages of the development of a fuzzy system, from its conception to its final implementation. These objectives are:

• **Simplicity**: Syntactical constructions in XFL are intentionally simple, so it is easy to implement programs able to read, modify, or generate a specification expressed in terms of the language, such as graphical editors, learning programs, or interpreters.

• **Flexibility**: XFL offers mechanisms that ease the reuse of already written (or automatically produced) definitions, so previous work can be employed to simplify the design effort.

• **Extensibility**: The model that XFL is based on allows for an easy incorporation of new theoretical contributions, since the only constraint it imposes is the set of predefined membership function classes. This set can be extended without affecting the compatibility with definitions made prior to the extension.

• **Neutrality**: An XFL definition is neutral as long as it does not impose any assumption on its final implementation. All aspects regarding the particular implementations of fuzzy operators, variable representations, or the characteristics of the arithmetic applied on the implementation are external to the XFL definition.

• **Expressiveness**: XFL does not impose any constraint on the complexity of the system being specified, either in the structure of its rule bases or in the definition of the rules themselves. This way, it is possible to define a system composed of multiple arbitrarily connected rule bases, allowing the hierarchical aggregation of rule bases [Berenji, 1990]. Furthermore, the antecedents of a rule can be freely combined by fuzzy connectives in structures as complex as the designer needs.

4.3.2 STRUCTURE OF AN XFL DEFINITION

The underlying model of a specification written in XFL responds to the three typical stages of fuzzification, inference, and defuzzification that were described for the general structure of a fuzzy system at the end of Chapter 3, although the model supports the use of fuzzy inputs and outputs as well. XFL offers specific constructions for facilitating the use of conventional and simplified defuzzification methods, as described in Sections 3.5.1 and 3.5.2 of Chapter 3, respectively.

The XFL definition of a fuzzy system consists of three fundamental parts: the definition of fuzzy operators, the declaration of types, and the specification of the module(s) conforming the system. In the first part the operators defining the inference mechanisms are selected. The second part contains the definition of the classes for the variables the system is going to operate with, including their universes of discourse and the membership functions associated with each of the fuzzy sets applied on them. The last part defines the structure and contents of the rule base that controls the system behavior.

• **Definition of Fuzzy Operators**

The mechanism for the definition of fuzzy operations plays a key role when using XFL as the core of a fuzzy system development tool. In essence, each tool that performs a translation from XFL into a target implementation format, either hardware or software, must support a configuration method (typically, a file) for defining, in terms of the target language, the fuzzy operations suited for the particular implementation method.

XFL-based tools must support mechanisms for the identification of the different operators. The selection of the fuzzy operations to be used for a certain implementation is made by including the directives shown in Table 4.2. into the XFL specification.

The configuration file consists of a series of sections delimited by a start symbol (*%Tnorm*, *%Tconorm*, *%Negation*, *%Implication* and *%Defuzzification*) and an end symbol (*%%*). Inside each of these sections, the definition of an operation has two parts: its identifier and the body of its definition. The body of operation definitions (except in the case of defuzzification methods, which

Table 4.2. XFL directives for the selection of fuzzy operators.

#and	T-norm to be used as operator and
#or	S-norm to be used as operator or
#not	Implementation of the operator not
#implication	Implication function
#composition	Preaggregation of rules with the same consequent
#also	Aggregation (operator also)
#defuzzification	Defuzzification method

will be described later) must contain the body of a native construction of the target language (function, macro, procedure, method, etc.) returning the result of the operation. For binary operators, the first operand is identified as *_a* and the second one as *_b*. For unary operators (actually, for the negation) the only operator is identified as *_a*.

Any definition can include lines containing sentences of the target language, using the following constructions: *%proc*, to define procedures or functions more complex than those permitted by the mechanisms described so far, and invoked by them, and *%data*, to define data structures employed either by the standard mechanisms or by those specified using the *%proc* construction. The end of these constructions is denoted by the symbol *%end*.

The definition of defuzzification methods is a little more complex: the definition lines are preceded by symbols that identify the requisites of the method in terms of the membership function classes the method is applicable for, and the point in the defuzzification process where the sentences are applied. These symbols are: *%requires* (to list the classes of membership function compatible with the method); *%init* (includes the sentences used for initializing the method); *%exit* (includes the sentences to be executed when the process is finished to assign the final value to the output); *%numloop* (denotes that the sentences should be executed inside a loop for the universe of discourse of the variable); *%linloop* (denotes that the sentences should be executed inside a loop for the linguistic labels of the variable); and *%end* (marks the end of any of the previous constructions). Figure 4.4 shows the aspect of some of the fuzzy operator definitions included in the configuration file used by the XFL to C compiler.

- **Type Definition**

As shown in Table 4.3, XFL supports a set of predefined membership function classes that include the vast majority of the functions described in the literature [Mizumoto, 1982; Lee, 1990], and provides a definition method for those functions not included in the predefined classes. This method is based on the definition of a series of points where the value of the membership function is defined and implies a linear approximation of the function for the interval defined by two consecutive points.

The most remarkable characteristic of the way universes of discourse and membership functions (termed type definition) are defined in XFL is the inclusion of inheritance mechanisms, with the aim of simplifying system specification and allowing the reuse of previously written modules. An XFL type also includes an attribute called *cardinality*, defining the number of different values to be taken into account for a variable of this type inside its universe of discourse. For integer types, it takes by default the value of the number of integers in the universe of discourse, while for real variables it is not defined (the default cardinality is infinite). Cardinality is suited for applications such as digital hardware synthesis (defining the number of bits needed for representing a vari-

```
%Tnorm
min
        ((_a)>(_b)?(_b):(_a))
prod
        ((_a)*(_b))
bounded_prod
        (((_a)+(_b)-1)>0?((_a)+(_b)-1):0)
%%

%Tconorm
max
        ((_a)>(_b)?(_a):(_b))
sum
        ((_a)+(_b)-(_a)*(_b))
bounded_sum
        (((_a)+(_b))>1?1:(_a)+(_b))
%%

%Negation
not
        (1-(_a))
%%

%Implication
min
        ((_a)>(_b)?(_b):(_a))
prod
        ((_a)*(_b))
%%
```

```
%Defuzzification
CenterOfArea
%init
        _in[0]=0.0;
        _in[1]=0.0;
%end
%numloop
        _in[0]+=_mx*_x;
        _in[1]+=_mx;
%end
%exit
        _out=_in[0]/_in[1];
%end


FuzzyMean
%requires delta bell triangle
%init
        _in[0]=0.0;
        _in[1]=0.0;
%end
%linloop
        _in[0]+=_a;
        if (_mft==_TRIANGLE) _in[1]+=_a*_p2;
        else _in[1]+=_a*_p1;
%end
%exit
        _out=_in[1]/_in[0];
%end
```

Figure 4.4. Fuzzy operator definition included in the configuration file for the XFL to C compiler.

able) and the implementation of defuzzification methods (setting the step the universe of discourse is swept with). Inheritance mechanisms allow the definition of *derived types* from other previously defined types. The derived type obtains the universe of discourse from the intersection of the universes of discourse of its ancestor types and the minimal cardinality among them. A derived type inherits automatically all the membership functions defined for its ancestor(s).

The general form of type definition in XFL is shown in Figure 4.5. To illustrate it with an example, let us consider the development of a controller to back up a truck to a loading dock in a planar parking lot [Kosko, 1992] (Figure 4.6). The input variables to the controller are the x-position of the rear center of the truck in the plane (x), and the angle the truck axis defines with the horizontal (ang). The output control variable is the angle that the truck wheels must

Table 4.3. Membership function classes supported by XFL.

Identifier	Defined by	Format
triangle	Three points conforming a triangle	
rectangle	Two points conforming a rectangle	
trapezoid	Four points conforming a trapezoid	
bell	Two points, a and b, conforming a bell-shaped function: $e^{-\left(\frac{(x-a)}{b}\right)^2}$	
sigma	Two points, a and b, conforming the function: $\frac{1}{1+e^{(x-a)/b}}$	
delta	A single point where the function takes the value 1	
points	A set of points that determine a piecewise linear approximation to an arbitrary function	

form with the horizontal (wheel). The goal of the control problem is twofold. On the one hand, the position of the truck has to be aligned with the desired loading dock. On the other hand, the truck has to arrive at the loading dock at a right angle (ang=90°). Figure 4.7 shows the XFL-type definitions of the system variables. These definitions correspond to the universes of discourse and membership functions illustrated on the right of Figure 4.7. The meanings of the membership functions labels are: LE (*LEft*), LC (*Left Center*), CE (*CEnter*), RC (*Right Center*), and RI (*RIght*); RB (*Right Below*), RU (*Right Upper*), RV (*Right Vertical*), VE (*VErtical*), LV (*Left Vertical*), LU (*Left Upper*), and LB (*Left Below*); NB (*Negative Big*), NM (*Negative Medium*), NS (*Negative Small*), ZE (*ZEro*), PS (*Positive Small*), PM (*Positive Medium*), and PB (*Positive Big*).

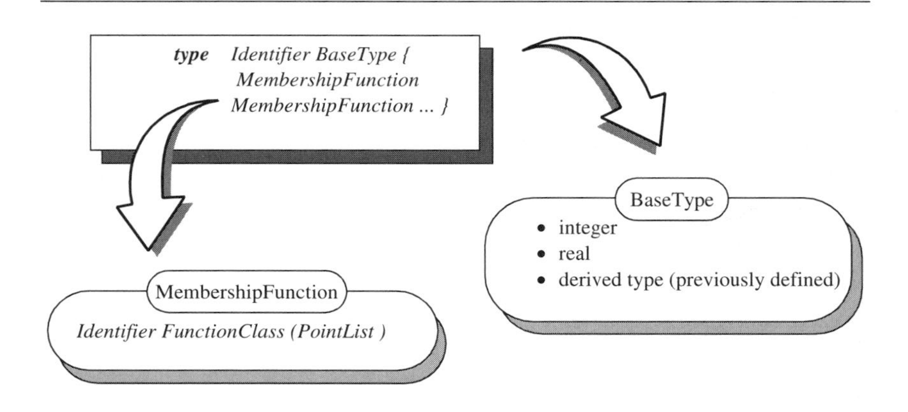

Figure 4.5. Type definition in XFL.

• Module Definition

The basic element in the XFL definition of the rule base that determines the behavior of a fuzzy system is called a module. A module consists of a declaration of the (input and output) variables for it and a definition of its structure. Each module has an associated identifier that permits referencing it in the structure of a more complex module. XFL requires any specification to have a module called "*system*", defining the global behavior of the system, whose input/ output variables are the inputs and outputs of the system.

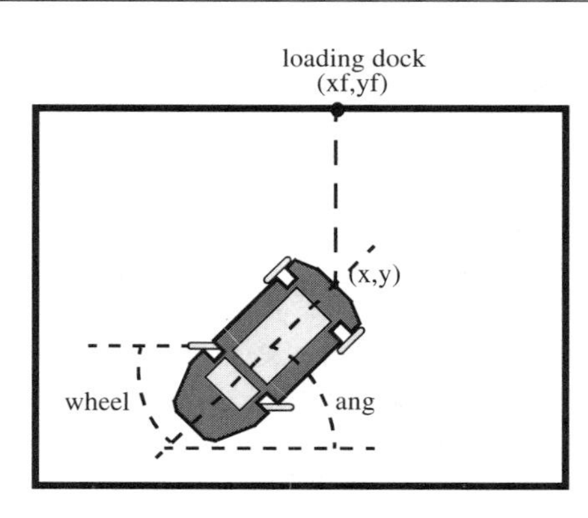

Figure 4.6. Example of the truck-dock problem.

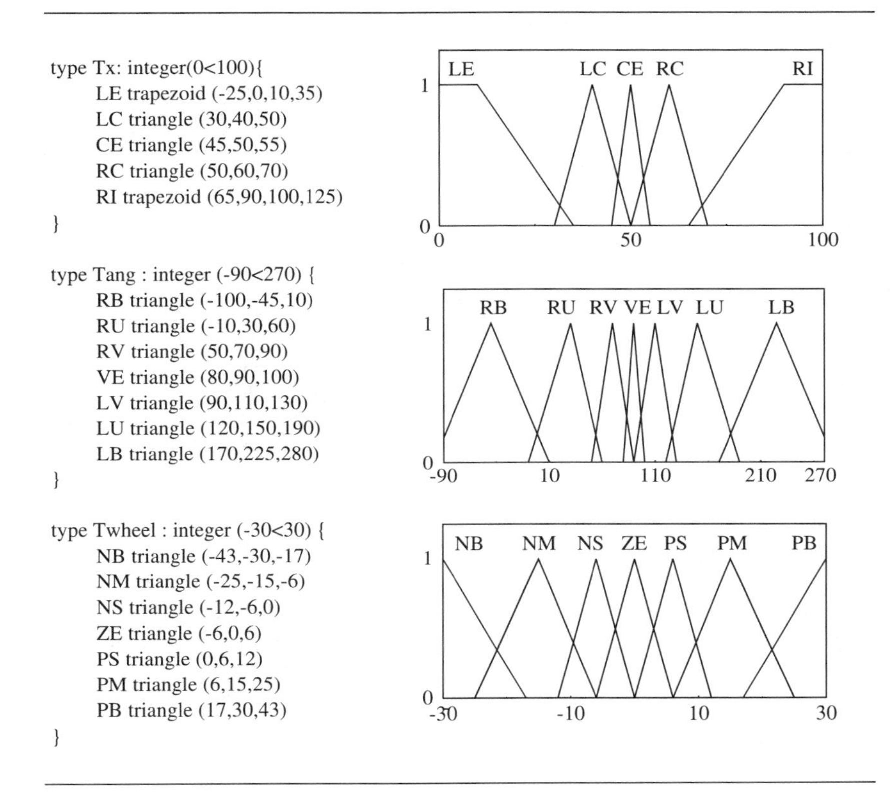

Figure 4.7. Examples of type definitions in XFL.

To illustrate this definition, the right part of Figure 4.8 shows the rule base of the truck-dock controller previously discussed. The XFL definition of this rule base with the system module is shown on the left of Figure 4.8.

XFL supports the definition of rule bases with an arbitrarily complex structure. To support this feature, the structure of an XFL module can be specified through two different (and exclusive) alternatives. In one of them, the rules defining the module are directly specified. In the other, the module is refined into the composition of some other modules (Figure 4.9).

The antecedent of a rule is an arbitrary combination of expressions binding one input variable to one of the membership functions defined for its type. The usual rules for associativity, distributiveness, and precedence are applicable. Of course, expressions can be grouped by means of parentheses. The consequent of a rule is a (comma-separated) list of expressions binding one of the output variables to one of the membership functions defined for its type.

```
system (Tx ? x, Tang ? ang, Twheel ! wheel)
  rulebase {

    if (x is RC & (ang is RB | ang is RU)) |
       x is RI & (ang is RB | ang is RU | ang is RV)
       -> wheel is PB

                 . . .

    if (x is LE & ang is LV) |
       ((x is LE | x is LC) & (ang is LU | ang is LB))
       -> wheel is NB
  }
```

ang \ x	LE	LC	CE	RC	RI
RB	PS	PM	PM	PB	PB
RU	NS	PS	PM	PB	PB
RV	NM	NS	PS	PM	PB
VE	NM	NM	ZE	PM	PM
LV	NB	NM	NS	PS	PM
LU	NB	NB	NM	NS	PS
LB	NB	NB	NM	NM	NS

wheel

Figure 4.8. Example of module definition in XFL: the truck-dock controller rulebase.

If a module is refined in terms of the composition of a set of other modules, the definition of its structure consists of an arbitrary combination of module references connected by means of the two compositional operators provided by XFL: parallel and serial. A module reference is made by including its identifier, and instantiating its input/output variables through references to the variables of the parent module or through "dummy" references in the case of serial compositions.

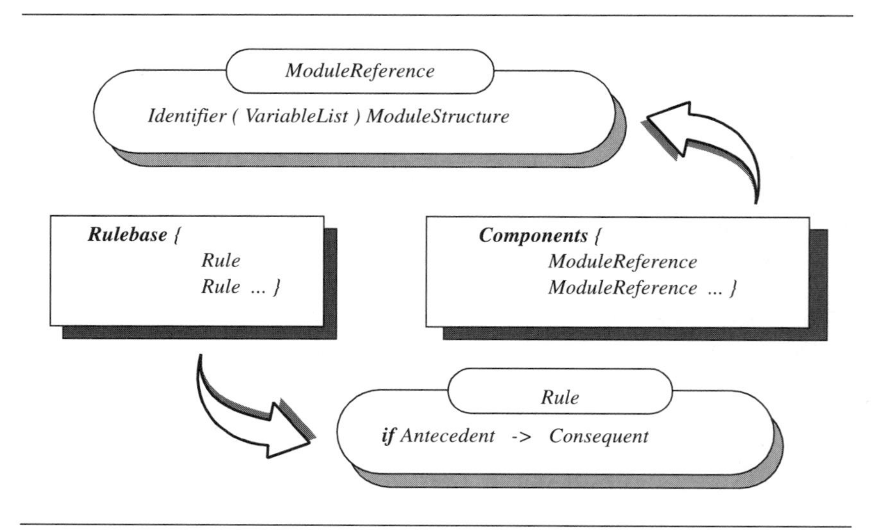

Figure 4.9. Module definition in XFL.

Parallel composition is limited to splitting a certain rule base into several simpler bases that can share common inputs with independent outputs. It does not imply any relationship between any of the composed modules. This composition can be used to integrate different already existing rule bases into another one with a higher number of inputs and outputs or to simplify the rule base format of a complex system.

Serial composition assumes that at least one of the outputs from the first rule base acts as input for the second. The outputs of the first rule base determine the support degree for each membership function in the corresponding inputs of the second rule base, thus allowing the definition of hierarchical rule bases.

Both compositions may be, as in the case of individual rules, freely combined to define the structure of the system global rule base: for example, a rule base serially composed with other two rules bases combined in parallel corresponds to a model where a primary rule base controls the activity of the secondary rule bases that eventually provide the system output, as shown in Figure 4.10.

4.3.3 DESCRIBING FUZZY SYSTEMS WITH XFUZZY

Xfuzzy integrates different tools for the design of the fuzzy systems under a common graphical user interface, so the designer can take advantage of the full power of XFL as a specification language and of the features offered by the tools based on it in an easy-to-use and intuitive manner. Furthermore, *Xfuzzy* offers additional facilities both for the definition of XFL specifications and for visualizing the results of some of the tools integrated within it.

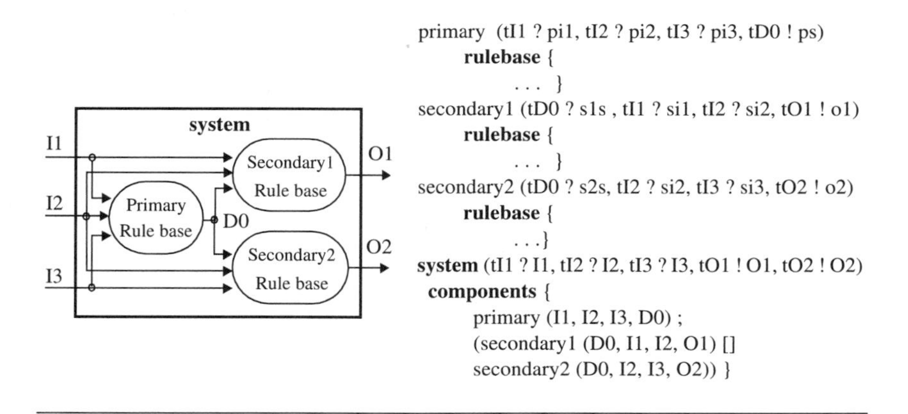


```
primary (tI1 ? pi1, tI2 ? pi2, tI3 ? pi3, tDO ! ps)
      rulebase {
            . . . }
secondary1 (tDO ? s1s , tI1 ? si1, tI2 ? si2, tO1 ! o1)
      rulebase {
            . . . }
secondary2 (tDO ? s2s, tI2 ? si2, tI3 ? si3, tO2 ! o2)
      rulebase {
            . . .}
system (tI1 ? I1, tI2 ? I2, tI3 ? I3, tO1 ! O1, tO2 ! O2)
  components {
        primary (I1, I2, I3, DO) ;
        (secondary1 (DO, I1, I2, O1) []
        secondary2 (DO, I2, I3, O2)) }
```

Figure 4.10. Examples of serial and parallel rule base compositions.

As we have stated previously, all the elements that compose *Xfuzzy* use the same internal representations for XFL specifications, so the designer can always work (independently of the design stage) with a unique and consistent representation of the system. In fact, it is possible (though not advisable) to define, verify, and get the implementation of a fuzzy system without using a single intermediate file to store its specification. This feature makes *Xfuzzy* an essentially open environment, since it is quite easy to integrate new XFL-based tools into the environment as they become available.

When *Xfuzzy* starts, it shows the main window of the environment (Figure 4.11). This window is divided into three areas corresponding to: (1) a set of buttons that allow the invocation of the different menus and options; (2) a selectable list that contains the XFL specifications available in the environment; and (3) a text area containing a log of the generated messages. This text area can be moved using the scrollbar at the left.

The button at the right end of the button area (labeled *Help*) accesses the help files for the environment. All *Xfuzzy* help is stored in HTML, so the *Help* button will run a Web browser showing the main help page.

• **Selection of Fuzzy Operators**

The option *Fuzzy Operations* in the *Set Up* menu of the main window gives access to a graphical interface for handling the definitions of fuzzy operators used by the different tools within the environment. As Figure 4.12 shows, this interface contains a set of selectable lists (one for each type of fuzzy operator) with the identifiers of the available operators, and a text area where the definition of the selected operator is shown. The buttons in the bottom part of the window permit loading a new operator definition file (*Load*), opening a window for editing the contents of the definition file currently in use (*Edit*), remov-

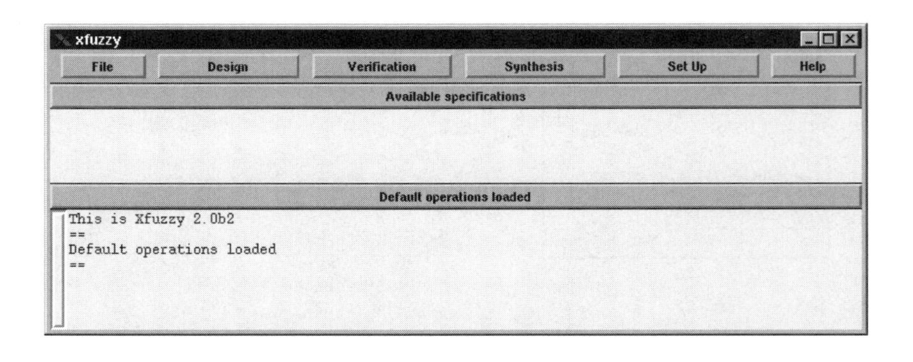

Figure 4.11. *Xfuzzy* main window.

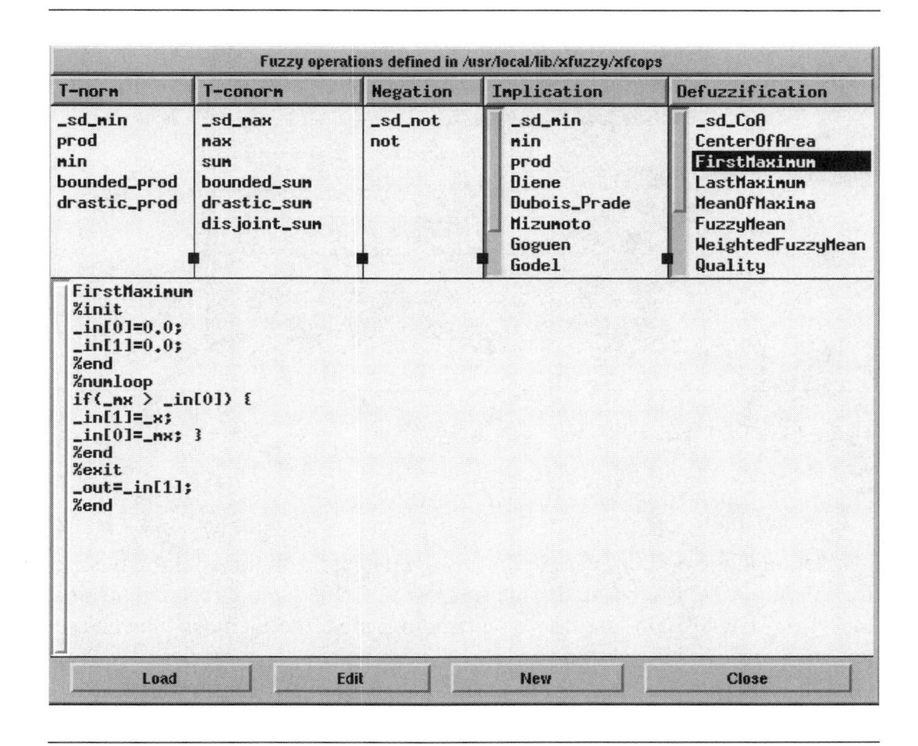

Figure 4.12. Operator definitions window.

ing the currently loaded operations to exclusively use the operators defined by default (*New*), and closing the window (*Close*).

When writing an XFL specification, the operations editor that is invoked by the option *Operations* in the menu *Design* of *Xfuzzy* is used for selecting the implementation of the different fuzzy operators to be employed. The editor presents the window shown in Figure 4.13, listing the values for each of the fuzzy operator directives accepted by XFL. The identifier of each directive is a button that pops up a list with the identifiers of the operators currently available to the environment. This way, the designer can select the most adequate operation from those loaded into the environment. The bottom part of the window is a button area with options for changing the operator set (*Select*), storing the changes made to the specification while closing (*OK*) or not (*Apply*) the window, reading again the current specification discarding the changes made so far (*Reload*), and leaving the editor without saving any changes (*Cancel*).

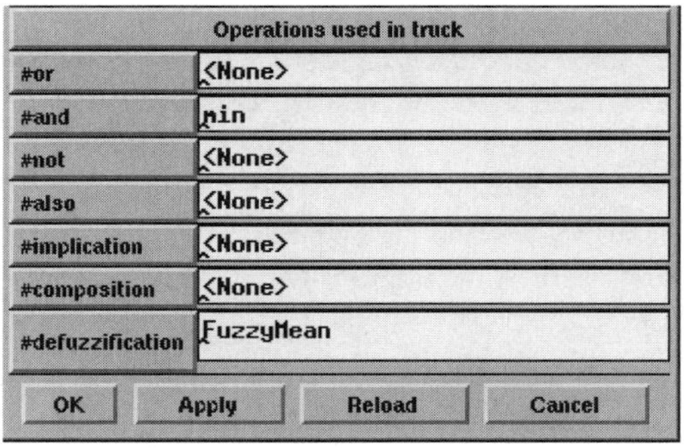

Operations used in truck	
#or	<None>
#and	min
#not	<None>
#also	<None>
#implication	<None>
#composition	<None>
#defuzzification	FuzzyMean

OK Apply Reload Cancel

Figure 4.13. Fuzzy operations editor.

• Type editing

Through the option *Types* in the *Design* menu, the window of the type editor (Figure 4.14) is presented. This window is divided into three vertical areas that, from left to right, are: the type handling area, the membership function handling area, and the graphics area.

The type handling area holds a selectable list of the XFL types defined inside the specification, where the active type (the type the editor is currently working with) is highlighted. Below this list, there are several buttons for accessing different type handling functions. In the same way, the membership function handling area contains a selectable list of the membership functions defined for the active type. The active function (the one the editor is currently accessing) is highlighted. The buttons for function handling include options for creating, copying, renaming, and removing membership functions. Finally, the graphics area contains a representation of the active function and (optionally) of the rest of the membership functions in the active type. Depending on the function class, there is a set of active points that allow modification of the membership function. By clicking on one of these points with the left button of the mouse, the point can be moved, according to the constraints associated with the function class (for example, the point defining the vertex of a triangular function cannot go beyond the points defining its basis). By clicking with the right button of the mouse it is possible to move the whole function without changing its shape. The graphics area also contains two buttons for adding or removing points in functions of the class "*points*".

The bottom part of the type editor window is an area of control buttons with the typical options for storing or discarding the changes made on the XFL specification.

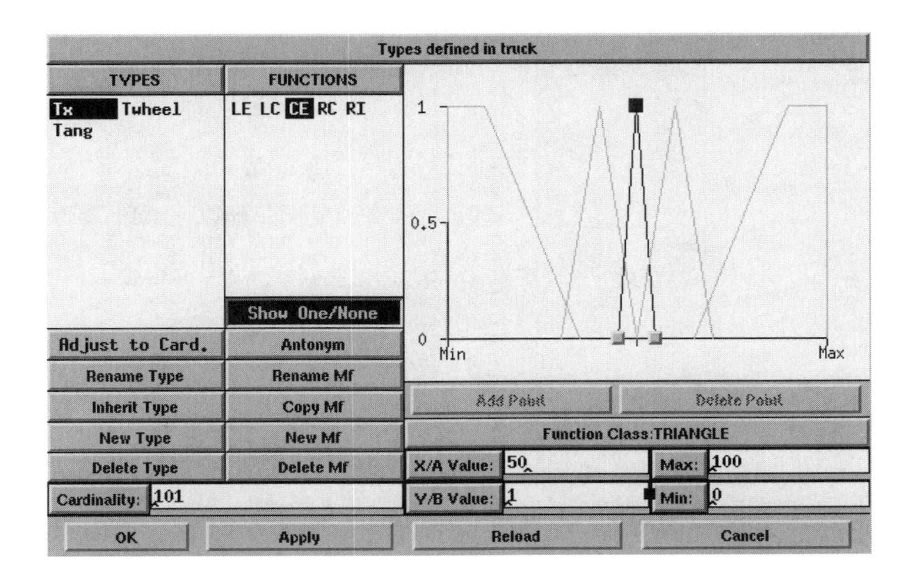

Figure 4.14. Type editor window.

• Module editing

The option *Modules* in the *Design* menu presents the rule editor window (Figure 4.15). The current version of the editor does not allow working with multimodule specifications; this is why we refer to it as a "rule editor" and not a "module editor". The editor window shows three areas, containing selectable lists and buttons, for defining the inputs and outputs of the system and its rule base.

The areas for inputs and outputs have a similar structure: a selectable list with the currently defined elements and two buttons for adding a new element (*New*) or removing the selected element (*Delete*). When a new element is added, the editor shows a dialog for defining its identifier and type. The area corresponding to the rule base contains a selectable list of the already specified rules and three buttons for defining a new rule (*New Rule*), and removing (*Delete Rule*) or modifying (*Edit Rule*) the selected rule.

The options *New Rule* and *Edit Rule* open a rule editing window (Figure 4.16), that assists in the definition of syntactically correct XFL rules. The window contains a list of the inputs (to be selected as antecedents) and the outputs (for consequents), together with the applicable membership functions. A set of

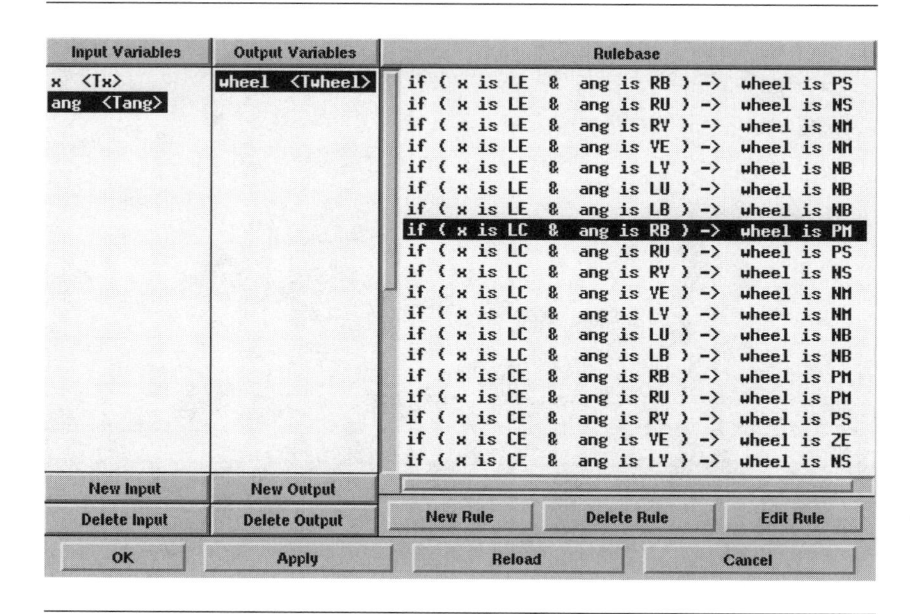

Figure 4.15. Module editor window.

buttons contain the operators that can be used in an XFL rule, while the text area shows the contents of the rule as it is built. In its bottom part, the window has three control buttons: for inserting the rule into the rule base (*OK*), for removing the definition (*Clear*), and for closing the window without modifying the rule base (*Cancel*).

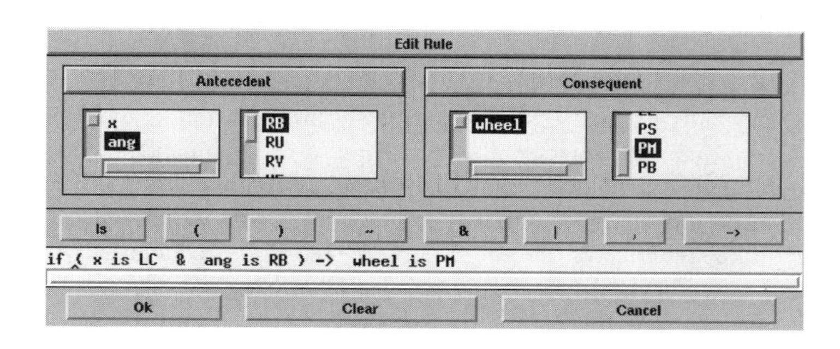

Figure 4.16. Rule editor window.

The bottom part of the rule editor window contains a button area (*OK*, *Apply*, *Reload* and *Cancel*) with the same functionality described for the type editor.

4.3.4 SOFTWARE SYNTHESIS OF XFL-BASED SYSTEMS

Although the description of software synthesis mechanisms is out of the scope of this text, it is worth mentioning here the main characteristics of these kinds of tools in *Xfuzzy*, since they make up the core of the tools that will be presented in the next chapter, and they are directly employed for the implementation of certain hardware synthesis strategies, as will be shown in Chapter 10.

The software synthesis tools of *Xfuzzy* translate an XFL specification into a programming language, particularizing the abstract definition of the system on constructions native to the target language. In essence, they are "compilers" of XFL for the target language. The current version of *Xfuzzy* offers two of these compilers, which use C and Java as target languages.

In the first case, C is the language used in an extremely large range of applications: from very high level programs to real-time systems running in small microprocessors. Therefore, it is an obvious choice for the software implementation of a fuzzy system. The characteristics of the XFL to C compiler make it possible to incorporate fuzzy logic–based elements in virtually any architecture, from a supercomputer to a microcontroller. Using Java as the target language for software synthesis it is possible to take advantage of object-oriented technology, and of the important momentum of this programming language, due to its binary portability across any platform implementing the language. A fuzzy system transformed into a Java program can be used, without changing the executable code, on practically any computer or operating system. Finally, due to its object-oriented nature, Java-based implementations of XFL-defined systems are especially suited to applications where a higher emphasis on the "fuzziness" of inputs and outputs is required.

REFERENCES

1. **Berenji, H. R., Chen, Y. Y., Lee, C. C., Jang, J. S., Murugesan, S.**, A hierarchical approach to designing approximate reasoning-based controllers for dynamic physical systems, in *Proc. 6th Conference on Uncertainty in Artificial Intelligence*, pp. 362-369, 1990.
2. **Bonner, M., Mayer, S., Raggl, A., Slany, W.**, FLIP++: A fuzzy logic inference processor library, in *Fuzzy Logic in Artificial Intelligence: Towards Intelligent Systems*, Martin, T. P., Ralescu, A. L., Eds., Springer-Verlag, 1997.
3. **Driankov, D., Hellendoorn, H., Reinfrank, M.**, *An Introduction to Fuzzy Control*, Springer-Verlag, 1993.
4. **Janikow, C. Z.**, Fuzzy decision trees: issues and methods, *IEEE Transactions on Systems, Man, and Cybernetics*, Vol. 28, N. 1, pp. 1-14, 1998.

5. **Jeschke, H., Wahle, M., Gotz, A.**, Fuzzy multiobjective decision making on modeled VLSI architecture concepts, in *Proc. IEEE Int. Symposium on Circuits and Systems*, Monterey, 1998.
6. **Kosko, B.**, *Neural Networks and Fuzzy Systems, a Dynamical Systems Approach to Machine Intelligence*, Prentice-Hall Int., 1992.
7. **Lee, C. C.**, Fuzzy logic in control systems: fuzzy logic controller, Parts I & II, *IEEE Transactions on Systems, Man, and Cybernetics*, Vol. 20, N. 2, pp. 404-432, 1990.
8. **López, D. R., Moreno, F. J., Barriga, A., Sánchez-Solano, S.**, XFL: A language for the definition of fuzzy systems, in *Proc. IEEE Int. Conference on Fuzzy Systems*, Vol. 3, pp. 1585-1591, Barcelona, 1997.
9. **López, D. R., Jiménez, C. J., Baturone, I., Barriga, A., Sánchez-Solano, S.**, Xfuzzy: A design environment for fuzzy systems, in *Proc. IEEE Int. Conference on Fuzzy Systems*, pp. 1060-1065, Anchorage, 1998.
10. **Mizumoto, M., Zimmermann, H.**, Comparison of fuzzy reasoning methods, *Fuzzy Sets and Systems*, Vol. 8, pp. 253-283, 1982.
11. **Nauck, D., Kruse, R.**, NEFCON-I: An X-Window based simulator for neural fuzzy controllers, in *Proc. IEEE Int. Conference on Neural Networks*, pp. 1638-1643, Orlando, 1994.
12. **Nauck, D.**, NEFCLASS - a neuro-fuzzy approach for the classification of data, in *Applied Computing*, George, K. M., Carrol, J. H., Deaton, E., Oppenheim, D., Hightower, J. Eds., pp. 461-465, ACM Press, 1995.
13. **Wang, L. X.**, *A Course in Fuzzy Systems and Control*, Prentice-Hall Int., 1997.
14. **Yen, J., Langari, R., Zadeh, L. A.**, Eds., *Industrial Applications of Fuzzy Logic and Intelligent Systems*, IEEE Press, 1995.

Chapter 5

FUZZY SYSTEM VERIFICATION

The process of fuzzy system verification involves different techniques for analyzing the behavior of the system under development and for tailoring this behavior to its appropriate targets. These techniques (and the design tools that employ them) can be roughly classified into two groups: (1) Learning techniques, which facilitate the definition of the inference system and provide mechanisms for automatically tuning its parameters; and (2) Simulation techniques, to analyze the system responses under different operational conditions.

The distinct learning techniques differ in the way the desired system behavior is described. When this behavior is expressed in terms of a set of expected outputs with respect to the applied inputs, these techniques are known as *supervised learning*. Conversely, when correct input/output data are not available, but some way of evaluating system behavior can be employed, the so-called *reinforcement learning* is applicable.

Simulation techniques are based on the use of approximate models for describing system behavior. These models usually correspond to an algorithmic abstraction level, so they do not take into account the constraints imposed by a particular implementation of the system. For example, it is possible to perform a simulation using double precision arithmetic even in the case that the final (hardware or software) implementation only allows the use of single precision arithmetic. Regardless of this limitation, the use of high level models in the first design stages is particularly important, since it allows a fast exploration of the design space, for selecting the options to be considered in further stages of the system implementation.

Besides this, in certain applications of fuzzy logic, such as control applications, it is especially useful to consider the verification of the fuzzy system under development (the controller) in conjunction with the system to be controlled (the plant). There are two general strategies for achieving this task. The first one involves the interaction of the fuzzy system model with a simplified model of the plant (this strategy is what is normally associated with the term *simulation*). A second alternative consists of including the actual plant in the simulation loop for the fuzzy system. This strategy is referred to in the literature as "*hardware in the loop*", and we will refer to it in this text as *on-line verification*.

This chapter analyzes the different tools for fuzzy system verification, according to the three categories described above: learning, simulation, and on-line verification. In the case of supervised learning, the error function is introduced, and the computational models that serve as ground for the different learning algorithms proposed in the scientific literature are reviewed, with a higher emphasis on gradient descent algorithms. As examples of the different types of tools, the verification tools provided by the development environment described in the previous chapter are described.

5.1 LEARNING TECHNIQUES FOR FUZZY SYSTEMS

The application of learning techniques to a fuzzy system allows the automation of the knowledge acquisition process, easing the definition or tuning of the inference system in accordance to some validity criteria that must be defined. As stated in the previous chapters, the functionality of a fuzzy system basically depends on its rule base, on the membership functions used by the rules, and on the selected fuzzy operators. Generally speaking, fuzzy operators are not usually modified during the learning process, since they correspond to fixed design parameters and their modification is problematic during the identification or tuning stages (this is especially true for hardware implementations). For membership functions, either their shapes (triangular, trapezoidal, etc.) or the parameters defining them (position, slopes, etc.) could be modified, but the shape of membership functions is usually subject to design criteria as well, and therefore is not modified by learning. With respect to the rule base, learning can include the creation of new rules or the removal or modification of some of the existing rules. In summary, the outcome of applying learning techniques on a fuzzy system fundamentally consists of modifications on the parameters defining the membership functions for its linguistic variables and, incidentally, of changes in the structure of its rule base.

To define the validity criteria for the operation of a fuzzy system it is necessary to have information about the intended behavior of the system. The quantity and quality of this information determines the kind of learning that can be employed. This way, if several complete examples of correct behavior are available, it is possible to determine the "correctness" of the system by comparing these examples with the actual system behavior. Sometimes the only available information is not so concrete and is reduced to detecting an incorrect behavior. In the first case we are dealing with study-based learning (from the analysis of reliable data), while in the second case learning will be based on experience about system usage.

The following sections describe different learning algorithms based on the analysis of examples of correct behavior. This kind of learning, known as supervised learning, consists in a tuning process based on the knowledge of behavior patterns, that is, on specific examples defining the intended input/output performance for the system. A CAD tool that applies supervised learning is also described in this chapter.

The tuning process is also known as training, since it fits the system to a predefined behavior. The fundamental idea of this process is the comparison of the output generated by the system with the output defined by the training patterns. This allows defining the deviation of system behavior from the intended results. Therefore, the aim of supervised learning is to minimize this deviation. From this point of view, the problem of supervised learning is reduced to the mathematical problem of the minimization of a certain function that expresses the deviation of system behavior. To perform this minimization, different algorithms have been proposed. These algorithms can be grouped into three families: *gradient descent* algorithms, based on the calculation of the function derivatives with respect to system parameters; *simulated annealing* algorithms, based on statistical variations of system parameters, and *genetic algorithms*, based on processes of mutation and crossover among different system realizations.

5.1.1 THE ERROR FUNCTION

The first step in elaborating a supervised learning process supposes describing system deviation by means of a function, known as *error function*. There are different proposals for the definition of this function. A very extended option is the mean square error (MSE):

$$E = \frac{1}{N} \cdot \sum_{i} (y_i - \tilde{y}_i)^2 , \ i = 1, \dots N \tag{5.1}$$

where N is the number of data patterns, y_i is the output generated by the system for the i-th pattern, and $\tilde{y}_i$ is the correct output expressed by the training pattern. If the system has more than one output, the error could be defined as the arithmetic mean of the errors obtained for each output:

$$E = \frac{1}{N} \cdot \frac{1}{M} \cdot \sum_{i,j} (y_{ij} - \tilde{y}_{ij})^2 \quad , \ i = 1, \dots N; j = 1, \dots M \tag{5.2}$$

where M is the number of output variables in the system. This expression has the drawback of giving a higher weight to those variables defined on a wider numerical range, so the optimizing of some variables would influence the error function more than others. When it is desirable that all output variables have the same weight on the error function, it is necessary to normalize their deviations:

$$E = \frac{1}{N} \cdot \frac{1}{M} \cdot \sum_{i,j} \left(\frac{y_{ij} - \tilde{y}_{ij}}{range_j} \right)^2 \tag{5.3}$$

It can be useful to have the possibility of defining a different weight for each output variable on the error function, so the designer can select the relative influence of every output variable on the global deviation from its intended behavior. In this case:

$$E = \frac{1}{N} \cdot \sum_{i,j} w_j \cdot \left(\frac{y_{ij} - \tilde{y}_{ij}}{range_j}\right)^2 \tag{5.4}$$

where w_j is the weight of the j-th output variable on the global system error. These weights should be normalized so their sum is 1.

Another possible choice is the root mean square error (RMSE), defined as the square root of the mean square error given by any of the four previous expressions:

$$E = \sqrt{\frac{1}{N} \cdot \sum_{i,j} w_j \cdot \left(\frac{y_{ij} - \tilde{y}_{ij}}{range_j}\right)^2} \tag{5.5}$$

It is also possible to use the absolute value instead of the quadratic error, with the corresponding options of variable normalization and weight accounting:

$$E = \frac{1}{N} \cdot \sum_{i,j} w_j \cdot \left|\frac{y_{ij} - \tilde{y}_{ij}}{range_j}\right| \tag{5.6}$$

Another alternative for measuring system correctness is the maximum absolute error (XAE), although it has the drawback of depending only on one of the training patterns and, therefore, being difficult to tune:

$$E = max\left(\left|\frac{y_{ij} - \tilde{y}_{ij}}{range_j}\right|\right) \tag{5.7}$$

The above expressions assume a numerical output from the fuzzy system. However, it is possible to define fuzzy systems whose outputs are linguistic labels, as is the case of classifiers. In these systems, the output value is the linguistic label presenting the highest activation degree as a result of the inference process. A common definition for the deviation in the behavior of this kind of system is the number of classification errors:

$$E = \sum_i \delta_i \tag{5.8}$$

where δ_i is 1 when the classification of the pattern has been incorrect and 0 otherwise. This type of function considers equally all classification failures, without taking into account the distance from a correct classification. To consider this information it is necessary to add a new term like the following:

$$E = \sum_i \delta_i + \frac{1}{N} \cdot \sum_i (max\alpha_i - \tilde{\alpha}_i) \tag{5.9}$$

where N is the number of training patterns, $\tilde{\alpha}_i$ is the activation degree of the correct label, and $\max_i \alpha_i$ is the activation degree of the label selected by the system.

The choice of an adequate error function for the learning process depends both on the type of fuzzy system to be tuned and on the algorithm selected for performing the process. This way, one of the two latter options must be used for fuzzy classifiers, while the other former options are applicable in the case of fuzzy systems with numerical outputs, such as fuzzy controllers. If the system has a single output, the first option should be adequate, but for systems with several outputs a normalized error function should be applied. Finally, if the selected algorithm belongs to the family of gradient descent algorithms, the error function must be derivable. This last condition discards the use of the maximum deviation and of the number of classification errors.

5.1.2 GRADIENT DESCENT ALGORITHMS

Gradient descent algorithms are intended to lead the system to a local minimum of the error function, that is, a point where the derivatives of the error with respect to the parameters of the system (the gradient of the error function) become zero. To accomplish this, each step of the algorithm calculates the mentioned derivatives and the value of every parameter is calculated according to them. A graphical representation of the system evolution should show an error surface (a hypersurface to be more precise) where the system (a point on the surface, with a certain gradient) evolves descending through a valley (local minimum). This image gives its name to this kind of algorithm.

Different rules for updating system parameters in base to the values of the derivatives have been proposed, yielding the different gradient descent algorithms, described as follows.

- **Backpropagation (BP)**

This is the oldest of these algorithms and the best studied [Rumelhart,1986]. It defines an increment for each parameter (Δp) that is proportional to the value of the derivative, and given by:

$$\Delta p = -\eta \cdot \frac{\partial E}{\partial p} \tag{5.10}$$

where η, called the *learning rate*, is a configurable constant.

The main drawback of this algorithm is its slow convergence through the minimum of the error function. For not very sharp minima, or simply in the neighborhood of a minimum, the values of derivatives are small and, therefore, parameter variations are very slow. A high value for η can alleviate this effect but implies wider oscillations around the minimum, again causing a slow convergence. These defects lead to the definition of new algorithms, with the aim of improving convergence speed.

- **Backpropagation with Momentum (BPM)**

One of the reasons for the slow convergence of the backpropagation algorithm is that the value of the first derivative is insufficient for estimating the distance to the minimum. The values of higher order derivatives are needed to obtain this type of information, but the calculation of these derivatives is very costly. Instead of these expensive calculations, we can use information about the first order derivatives in other points the system has already traversed.

With this idea in mind, a simple modification of the BP algorithm was proposed, consisting of the introduction of a new term proportional to the previous increment for the parameter, somehow retaining historical information about the form of the error function. This new algorithm is given by the expression:

$$\Delta p(t) = -\eta \cdot \frac{\partial E}{\partial p} + \alpha \cdot \Delta p(t-1) \tag{5.11}$$

The second term in (5.11) is called *momentum term*. The constant α (the *momentum factor*) must be between 0 and 1. By expanding the above expression, the effect of derivatives in previously traversed points on the current increment can be evaluated:

$$\Delta p(t) = -\eta \cdot \sum_{n=0}^{N} \alpha^n \cdot \frac{\partial}{\partial p} E(t-n) \tag{5.12}$$

where $t-N$ is considered as the starting point at which $\Delta p(t-N)$ is zero.

A detailed explanation on how this modification speeds up the convergence process can be found in [Jacobs, 1988].

- **Quick Backpropagation (QuickProp)**

In the same line of estimating the form of the error function as a way of making the algorithm converge faster, Fahlman proposed the following formulation for the parameter increment [Fahlman, 1988]:

$$\Delta p(t) = \frac{S(t)}{S(t-1) - S(t)} \cdot \Delta p(t-1); \quad S(t) = \frac{\partial}{\partial p} E(t) \tag{5.13}$$

This expression corresponds to the step necessary to reach the minimum of a parabola, given the values of its derivative for two points. In other words, the error function is locally approximated by a parabola, obtaining a much faster convergence than in the case of the classical backpropagation algorithm.

However, the above expression has some disadvantages. The most important is that, when $S(t)$ and $S(t-1)$ take very close values, the increment can become too big and the algorithm degenerates into a chaotic state. Another problem is the need to define a certain initial step, and that, if the increment becomes zero, the algorithm stops. The first problem can be avoided by the introduction of a maximum growth factor μ, so non-divergence is guaranteed. To avoid the second problem, a gradient term is introduced, and the above expression is transformed into:

$$\Delta p(t) = g \cdot \Delta p(t-1) - \eta \cdot S(t); \quad g = min\left(\mu, \frac{S(t)}{S(t-1) - S(t)}\right) \tag{5.14}$$

- **Manhattan Algorithm**

A different line in looking for a faster convergence is to avoid the parameter increment depending on the absolute value of the derivative. The backpropagation algorithm evolves very slowly near a local minimum because it assigns a value for this increment proportional to the absolute value of the derivative, which is very small in this zone. The Manhattan algorithm [Anderson, 1986] avoids this dependence by using only the sign of the derivative of the error function with respect to the corresponding parameter:

$$\Delta p = \begin{cases} \Delta_0 & if \ \frac{\partial E}{\partial p} < 0 \\ -\Delta_0 & if \ \frac{\partial E}{\partial p} > 0 \\ 0 & ioc \end{cases} \tag{5.15}$$

This method uses a fixed step size (Δ_0) to increment the parameters, providing poor results for moderately complex systems, since a high value in the step size makes the algorithm converge promptly but produces a final result far from the minimum, while a lower value leads to a value closer to the minimum but requires a greater number of iterations. The algorithms that will be shown below try to avoid these problems by considering different mechanisms for adapting the step size.

- **Adaptive Step Size**

Following the above described way of avoiding the dependency on the absolute values of derivatives, Jang, designer of the ANFIS architecture [Jang, 1992], defined an algorithm that only uses the direction of the gradient vector, but not its modulus. The change in the parameter values is given by a vector of modulus *S* (the step size) in the gradient direction:

$$\Delta p = S \cdot \frac{\frac{\partial E}{\partial p}}{\sqrt{\sum_j \left(\frac{\partial E}{\partial p_j}\right)^2}} \tag{5.16}$$

In the implementation of this algorithm, Jang proposed a heuristic method for adapting the value of S, dividing by two its value in case the error increased and multiplying it by four when the error decreased for three successive iterations. The version implemented by the learning tool that will be described in this chapter uses an adaptive strategy similar to the one used by RPROP (described below): multiplying by a factor less than one (a decrement factor) when the error increases, or greater than one (an increment factor) if the error decreases.

• **Resilient Propagation (RPROP)**

Another defect of the backpropagation algorithm is the use of the same learning rate for all parameters. This may produce a slow convergence for some parameters, while the value of η is too high for others, causing oscillations around the optimal value (or even parameter divergence). Some authors have proposed assigning different learning rates to certain parameters, making use of heuristic rules, but these rules are only applicable for specific architectures.

Riedmiller and Braun proposed an algorithm called Resilient Propagation, joining the different lines for improving convergence speed into a simple formulation [Riedmiller, 1993]. The idea is based on the Manhattan algorithm (not using the absolute value of the derivative, since it delays convergence) but with a variable increment. These variations are assigned by studying the previous derivative values (collecting information about the function topology, as in the case of QuickProp). Furthermore, each parameter adapts its own value for Δ, so the problem of updating all parameters using the same factor is avoided. RPROP responds to the following expressions:

$$\Delta p = \begin{cases} \Delta_p(t) & if\ \dfrac{\partial E}{\partial p} < 0 \\ -\Delta_p(t) & if\ \dfrac{\partial E}{\partial p} > 0 \\ 0 & ioc \end{cases} \tag{5.17}$$

$$\Delta_p(t) = \begin{cases} \Delta_p(t-1)\cdot\eta^{+} & if\ \left[\dfrac{\partial}{\partial p}E(t)\right]\cdot\left[\dfrac{\partial}{\partial p}E(t-1)\right] > 0 \\ \Delta_p(t-1)\cdot\eta^{-} & if\ \left[\dfrac{\partial}{\partial p}E(t)\right]\cdot\left[\dfrac{\partial}{\partial p}E(t-1)\right] < 0 \\ \Delta_p(t-1) & ioc \end{cases} \tag{5.18}$$

where η^+ and η^- are the increment and decrement factors for Δ $(0 < \eta^- < 1 < \eta^+)$.

Thus far, RPROP is one of the fastest and most efficient gradient descent algorithms known and has been widely employed since its introduction.

•Approximate gradient algorithms

The strategy of the gradient descent algorithms is based on minimizing the value of the error function with respect to each system parameter. It is not always possible to perform these calculations efficiently, especially when the learning module is intended to be embedded in a hardware-based fuzzy system. To solve this problem, Jabri and Flower proposed in 1992 the use of first order finite-differences to estimate the value of derivatives instead of using exact calculations for them [Jabri, 1992]. These estimated values are then introduced in the different algorithms described above. The result is known as *weight perturbation*:

$$\frac{\partial E}{\partial p} \approx \frac{E(p+\delta)-E(p)}{\delta} \tag{5.19}$$

where the perturbation, δ, must take a small value for a correct estimation of the derivative. This scheme has the drawback of requiring a calculation of the error function for every parameter in the system, which means a considerable slowdown in each iteration of the algorithm. In 1997, Cauwenberghs proposed a new scheme for derivative estimates that can be performed using only two calculations of the error function, regardless of the number of system parameters [Cauwenberghs, 1997]. This scheme is known as *stochastic weight perturbation* and the derivative estimates are given by:

$$\frac{\partial E}{\partial p_j} \approx \pi_j \cdot \varepsilon; \qquad \varepsilon = \frac{1}{2\sigma^2} \cdot (E(\vec{p}+\vec{\pi})-E(\vec{p}-\vec{\pi})) \tag{5.20}$$

where $\pi_j = \pm\sigma$, with equal probability, p_j are the parameters under tuning, and $\vec{\pi}$ and $\vec{p}$ are the vectors with components π_j and p_j, respectively. This procedure notably decreases the time per iteration although it provides a coarser estimation, so convergence requires a higher number of iterations.

A question seldom discussed in texts about gradient descent algorithms regards the constraints that must be satisfied by system parameters. The results of a learning algorithm are the values of the changes in the system parameters after each iteration. However, the direct application of these changes may cause an incorrect specification of the system (for example, moving the apex of a triangular function beyond one of the points that define its basis). This kind of problem is solved by means of heuristic techniques that modify the value of changes in order to preserve the fulfillment of constraints on system parameters.

In summary, the algorithms described in this section are efficient for adjusting systems by quickly minimizing the error function, and there are schemes that allow their use in hardware applications and techniques for preserving the constraints that characterize system parameters. Conversely, the main criticisms made about these algorithms are that they are only applicable for differentiable error functions, and that the solution they provide is just a local minimum, that is, they do not guarantee that their outcome is (even close to) the global minimum of the error function.

5.1.3 OTHER SUPERVISED LEARNING ALGORITHMS

As stated in the previous section, the minimization of the error function by means of gradient descent algorithms has the problem of converging to local minima of this function. This problem is due to the way the function evolves, using the direction of the gradient: the process only explores system configurations inside a certain valley, obtaining as a result the local minimum for that valley. Therefore, wide areas of the parameter space are left unexplored, where some other valleys including lower minimum values for the error function can be located. The configuration of the system prior to the start of the learning process is fundamental in these algorithms, since it determines the trajectory the system will describe through the design space.

To solve this problem is necessary to establish procedures for exploring wider areas in the parameter space. These procedures must be based on statistical techniques able to generate system configurations covering different regions in this space. The use of random techniques has the additional advantage of exclusively grounding the evaluation of the system on the error function itself, and not on its derivatives. This allows us to expand the range of applicable error functions to non-differentiable ones. Furthermore, these algorithms do not require heuristic techniques for preserving the constraints on parameters, since those system configurations not satisfying the constraints can be simply rejected.

• **Simulated Annealing**

At the beginning of the 1980s the research on statistical techniques led to the definition of simulated annealing algorithms [Kirkpatrick, 1983]. These algorithms are based on an analogy between the learning process, which is intended to minimize the error function, and the evolution of a physical system, which tends to lower its energy as its temperature decreases. In a physical system in equilibrium, composed of a great number of particles, the probability of finding the system in a given state is given by its partition function. In general, this function has the following form:

$$z(r,p) = \frac{1}{Z} \cdot e^{\frac{-E(r,p)}{kT}} \tag{5.21}$$

where Z is a normalization constant, k is Boltzman's constant, T is the system temperature and E is an energy function depending on the type of processes (fluctuations) the system can experience. Following this model, Metropolis proposed in 1953 the use of an algorithm where the probability of acceptance for a new configuration is given by:

$$P = \begin{cases} 1 & if \Delta E < 0 \\ e^{\frac{-\Delta E}{kT}} & if \Delta E > 0 \end{cases} \tag{5.22}$$

where ΔE is the difference between the energy of the current configuration and the previous one. The properties of equilibrium of a physical system can be calculated by studying the series of configurations obtained this way. These procedures are known as *Montecarlo techniques* [Metropolis,1953].

Simulated annealing algorithms use the error function (or cost function, in the commonly used terminology in this field) as an energy measure and slowly anneal the system, thus leading the system to a global minimum for the error function. The different simulated annealing algorithms are characterized by their annealing schemes. These annealing schemes describe both the evolution of system temperature and the possible perturbations applied on the successive configurations [Geman, 1984; Szu, 1987; Mendoça, 1997].

These algorithms avoid the convergence of the system through local minima by using statistical techniques that widen the explored area inside the parameter space and that consider non-zero probabilities for accepting configurations raising the error function value. The probability of reaching the global minimum rises as the number of iterations of the algorithm grows. They have the additional advantage of being applicable on systems with non-differentiable error functions. Their main drawback is the requirement of a great number of iterations for ensuring an acceptable final result, making them rather slower than the gradient descent algorithms.

- **Genetic Algorithms**

The algorithms described up to now only store a system configuration in each of their iterations, limiting the area explored by the tuning process. As we have seen, gradient descent algorithms only explore a valley of the error function, determined by the initial state of the system. In the case of simulated annealing algorithms, initial perturbations are high and cover an important area of the parameter space, but as the temperature decreases the range of perturbations also decreases and the tuning process concentrates on a limited area of this space. The effect of the Metropolis algorithm is that the probability for finishing inside a valley is higher as the valley is deeper.

A simple way for broadening the explored area in the parameter space is to keep several system configurations, allowing both light perturbations on a given configuration (for exploring a certain valley) and heavy perturbations on some other configurations (for exploring areas not considered thus far). This idea constitutes the foundation for genetic algorithms.

Genetic algorithms [Goldberg, 1989; Davis, 1991; Buckles, 1992] originate in an analogy between the evolution of species, consisting of natural selection, and the evolution of a population of system configurations, consisting of the improvement of their error function. Every system configuration is described by means of a numerical list, called chromosome by analogy with natural selection. The evolution of the population is performed by mutations on these chromosomes (perturbations on a given configuration) and crossovers among different chromosomes (able to produce new configurations very far from the original ones, expanding the exploration area in the parameter space). The probability for a certain configuration contributing to generating new configurations by means of these mechanisms is lower as its error function is greater, so those configurations with higher errors are doomed to extinction.

The different genetic algorithms are distinguished by their way of representing the system configuration (the codification of this configuration in their chromosomal format), implementing the crossover and mutation operators, and by the selection strategies applied on the configurations participating as parents of the new generation.

In summary, genetic algorithms considerably enlarge the area explored inside the configuration space, thus allowing a faster way for finding the global minimum. Therefore, they are faster than simulated annealing algorithms, while keeping their advantages of convergence towards a global minimum and independence from the derivativeness of the error function. They have the drawback of requiring many more resources, since they need to keep multiple system configurations simultaneously.

- **Genetic Programming**

The different families of algorithms described until now are oriented to the modification of the parameters defining the system, but do not consider the modification of the structure of the system itself. In the case of fuzzy systems, these algorithms modify the membership functions defined for the linguistic variables intervening in the system, but the relations among these variables, represented by the rule base, are left unaltered by the optimization process.

A deeper learning on the data patterns used for training the system should allow the modification of these rules, generating new relations among variables or removing those relations that do not provide information, or even degrading system behavior [Herrera, 1995]. This would permit an important enlargement of the configuration search space for the system.

The genetic algorithms described in the above section can be modified to extend the tuning process by incorporating changes to the structure of the system. With this aim, new operators are introduced to modify the system structure. These operators complement mutation and crossover which, as we have seen, are oriented toward modifying the values of the system parameters. The new operators are not only able to change the contents of chromosomes, but can also modify their length, producing populations with chromosomes of different dimensions and thus leading to a redefinition of the crossover and mutation operators. In general, the optimization techniques based on the use of variable length chromosomes are known as *genetic programming* [Koza, 1992].

Genetic programming is a conceptually simple technique that allows a dramatic enlargement of the configuration space explored for system optimization, since it is directed to alter not only the values for the parameters of the system, but also its structure. It is also a very flexible technique, as a great number of potential operators can be employed. It has the drawback of being much slower than other procedures, due to the magnitude of its search space. It is necessary to provide mechanisms for reducing the number of structures to be considered and their complexity, since an adequate exploration of each structure itself requires an important number of computing resources.

5.2 LEARNING ON XFL-BASED SYSTEMS

The importance of learning techniques for fuzzy system identification and tuning has motivated a great number of works in the area of learning algorithms for this type of systems. Most of these algorithms were originally designed for neural networks, and their application to fuzzy inference systems has, generally speaking, suffered limitations both in the system to be tuned and in the learning process itself. These limitations include a small number of system variables (usually one output and not more than four inputs), a reduced number of membership functions per variable (up to seven in the best case), the applicability for very few shapes of membership functions and fuzzy operators, and lack of flexibility in the selection of the system parameters to be tuned.

The learning CAD tool we will describe below tries to resolve some of these shortcomings. This tool is *xfbpa*, integrated within *Xfuzzy* and designed for preserving the flexibility of system definition provided by the XFL specification language. The tool does not impose any limitations on the number of system inputs or outputs, or on the complexity of its rule base. XFL specifications tuned by *xfbpa* can include compound rule bases, with rules containing terms arbitrarily combined by means of the conjunction, disjunction, or negation connectives. Moreover, *xfbpa* supports all the membership function classes defined in Table 4.3 of Chapter 4.

The implementation of the different fuzzy operations can be arbitrarily selected by the user among a set of t-norms, s-norms, implication functions, and defuzzification methods. The learning algorithms used by *xfbpa* require the use of differentiable fuzzy operators. The set of fuzzy operators supported by *xfbpa* cannot be extended (as is the case with the other XFL-based tools) by the user. Although this is a limitation on the features of the tool, an important effort has been made to include in *xfbpa* the operators most widely used in fuzzy system design (Table 5.1), so this is more a theoretical than a practical limitation.

Table 5.1. Fuzzy operators supported by *xfbpa*.

T-norms	S-norms	Implications	Defuzzification Methods
min prod bounded_prod drastic_prod	max sum bounded_sum drastic_sum	min prod Diene Dubois_Prade Mizumoto Goguen Gödel Lukasiewicz Zadeh	CenterOfArea FuzzyMean WeightedFuzzyMean Quality E-Quality LevelGrading CenterOfSums Yager

xfbpa incorporates a set of supervised learning algorithms based on back-propagation. As discussed in the previous sections, this requires the availability of a training file and the selection of an error function quantifying the system deviation from its intended behavior. This behavior is expressed in terms of correct input/output tuples. Each line in the training file contains one of these tuples, where individual data are separated by any number of white space characters. The order in which these data are assigned to system variables corresponds to the order these variables are defined in the module *system* of the XFL specification.

The error function used by *xfbpa* corresponds to the definition in (5.4), included here again to clarify the discussion:

$$E = \frac{1}{N} \cdot \sum_{i,} w_j \cdot \left(\frac{y_{ij} - \tilde{y}_{ij}}{range_j}\right)^2, \; i = 1, \dots N, \; j = 1, \dots M; \qquad (5.23)$$

where N is the number of training patterns, M is the number of system outputs, w_j is the weight of the j-th output on the error, y_{ij} is the value obtained from the system for the j-th output with the i-th input pattern, $\tilde{y}_{ij}$ is the intended j-th output for the i-th input set, and $range_j$ is the range of definition of the j-th output variable. This function represents the weighted mean square error of the system outputs, normalized to their ranges. Its value is bound between 0 and 1. From equation (5.23), we can see that the tool provides a means for adjusting the error function, giving more significance to those variables that the designer considers more important in system deviation, by configuring the w_j parameters in the error function.

The current version of *xfbpa* incorporates five different supervised learning methods. For each one of them, the tool allows the definition of the parameters controlling its application. The selection of a certain algorithm or the values for its parameters applied on one execution of the tool do not impose any constraints on subsequent executions, either with the same learning data or with a different training file. This way, it is possible to use an algorithm parameterized for a fast convergence when looking for a reasonably good approximation, and then employ successive approximations with more selective configurations to perform a finer tuning of the system. Furthermore, since *xfbpa* uses XFL specifications as input and output, it is possible to use different instances of the tool in parallel for selecting the more adequate result. The learning algorithms provided by *xfbpa*, together with the configurable parameters for each one of them, are shown in Table 5.2.

The learning process can be executed in two modes: on-line or off-line. In the first case, the system parameters are modified each time a line in the training file is processed, while in the second mode the parameter updates are performed when the whole training data set has been exploited. The execution of the tool requires an end condition for the learning process to be specified. The different alternatives accepted by *xfbpa* are summarized in Table 5.3.

One of the most important characteristics of this learning tool is that it provides mechanisms for what we call *specific learning*, that is, for selecting which membership function parameters must be tuned and, to some extent, for establishing the relations among these parameters that must be preserved during the learning process. By default, *xfbpa* acts on all the parameters of all the membership functions in all the XFL types defined for the system. However, by means of a specific learning file (whose format will be described below, in the section for the graphical interface of the tool) it is possible to select the parameters to be tuned. Furthermore, as will be discussed for hardware synthesis tools, certain architectures require the verification of some properties in the distribution and overlapping of membership functions over the universe of discourse to be applicable. If a system has been designed with this aim, the application of the learning process can (almost certainly) break these con-

Table 5.2. Learning algorithms included within *xfbpa*.

Algorithm	**Parameters**
Backpropagation with Momentum (5.11)	p1: learning rate p2: momentum factor
QuickProp (5.14)	p1: learning rate p2: maximum growth factor
Adaptive Step Size (5.16)	p1: initial value p2: increment factor p3: decrement factor
Manhattan (5.15)	p1: initial update value
RProp (5.17), (5.18)	p1: initial value p2: increment factor p3: decrement factor

Table 5.3. End conditions for the learning process in *xfbpa*.

End Condition	
(upper) limit for iterations	i
(lower) limit for root mean square error	RMSE
(lower) limit for the maximum absolute error	XAE
(lower) limit for the relative error variation	$e = \frac{E(t) - E(t-1)}{E(t-1)}$

straints. To avoid this, the designer can direct *xfbpa* to execute the learning process in a way that preserves these constraints when tuning system parameters.

Finally, *xfbpa* offers facilities for simplifying the XFL specification obtained from the learning process. One of these facilities is a rule-pruning mechanism that permits the removal of those rules with an activation level under a specific threshold. If a rule is pruned because of this, a new tuning loop is automatically performed, since the modification of the rule base may affect the outcome of the learning process. The tool also includes a clustering algorithm, for minimizing the number of membership functions in rule consequents. It provides a second alternative for reducing the total number of rules in the system. As will be illustrated in Chapter 12, these facilities permit the definition of a strategy for fuzzy system identification with *xfbpa*.

5.2.1 COMMAND LINE INTERFACE

As with the other tools in the *Xfuzzy* environment, *xfbpa* can be executed both from the command line and by means of a graphical user interface. The syntax applicable for invoking the learning tool from the command line is summarized in Figure 5.1. The command requires as inputs the name of the file containing the XFL specification to be tuned and the name of the file holding the training data patterns. It is also possible to indicate the name of a file containing specific learning directives. The output is a file with the XFL specification obtained from the learning process and, optionally, a file containing information about its evolution. The remaining parameters of the command are intended for the selection of the learning algorithm and the definition of its parameters, the choice of the learning mode, the specification of the end conditions, and the assignment of weights in the error function for the output variables of the system.

5.2.2 GRAPHICAL USER INTERFACE

The graphical user interface for *xfbpa* is shown in Figure 5.2. The window is divided into four areas: a configuration area in the upper left part, a numeri-

```
Xfbpa: Supervised learning for XFL-based fuzzy systems

Synopsis: xfbpa XFLFile TrainingFile <-s SpecificTrainingFile>
              <-o OutputXFLFile> <-d LearningEvolutionFile>
              <-t[1|2|3|4|5]> <-p1 Value> <-p2 Value> <-p3 Value>
              <-i Value> <-e Value> <-m Value> <-M Value> <-r Value>
              <-n> <-E> <-W> <-w Value1 ... ValueN>


     XFLFile                    : XFL input file
     TrainingFile               : Training patterns file
     -s SpecificTrainingFile    : Specific learning directives file
     -o OutputXFLFile           : XFL output file
     -d LearningEvolutionFile   : Output file with data about learning evolution
     -t<1|2|3|4|5>              : Learning algorithm selection:
                                     1 - Backpropagation with Momentum
                                     2 - Adaptive Step Size
                                     3 - Manhattan
                                     4 - QuickProp
                                     5 - RProp
     -p1 ... -p3                : Parameters for the learning algorithm (see Table 5.2)
     -i, -e, -m, -M             : End conditions (see Table 5.3)
     -r Value                   : Threshold for rule pruning
     -n                         : On-line learning flag
     -E                         : Do not show error messages
     -W                         : Do not show warning messages
     <-w Value1 ... ValueN>     : Weights for the outputs in the error function (5.23)
```

Figure 5.1. Command line interface for *xfbpa*.

cal information area in the upper right part, a graphical information area occupying the center of the window, and the usual control buttons area in the lower part.

The configuration area provides a file selection dialog for setting the training file (*Training File*) and buttons for defining the algorithm to be applied (*Algorithm*) and configuring its parameters (*Control Variables*). The text areas labelled *Iteration Limit*, *Err. Rel. Var. Limit*, *Mean Dev. Limit* and *Max. Dev. Limit* are intended for establishing the end conditions, while the one labelled *Rules Threshold* is for the rule-pruning threshold. The button *Learning Type* is for setting the learning mode.

In the bottom line of the configuration area, the button *Output Weights* opens a window for defining the weights of the output variables in the error function, while the button *Evolution to File* opens a file selection dialog for setting where to store the information produced during the learning process. Clicking on the button labelled *Specific Learning* gives access to the window

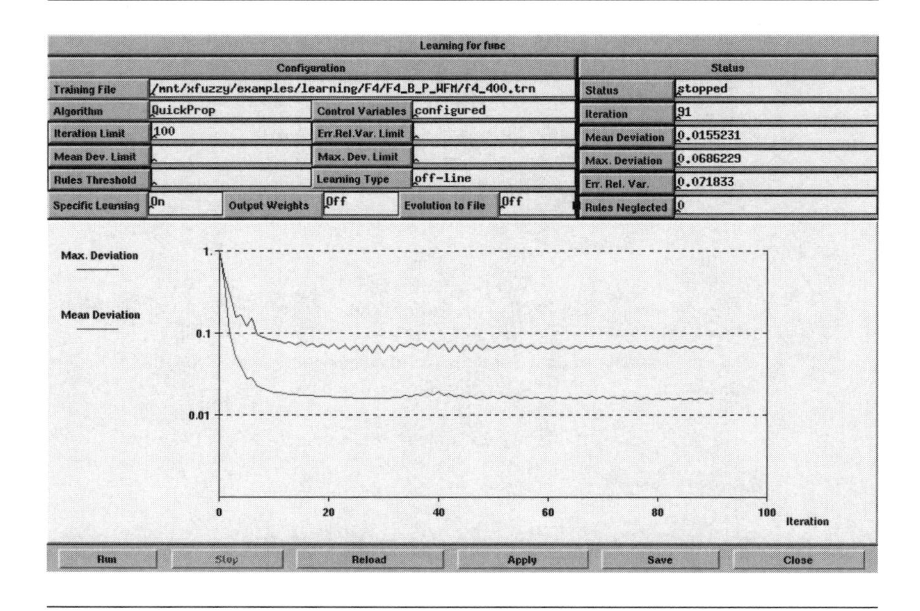

Figure 5.2. Graphical user interface for *xfbpa*.

shown in Figure 5.3, which permits the handling of the specific learning mechanisms. This window contains three selectable lists with the types defined in the XFL specification, the different shapes of membership functions, and the tunable parameters for each class of function. The designer can select which types, classes, or parameters are going to be tuned (*Enable*) and which will stay unaltered (*Disable*). Furthermore, with the option *Linked* it is possible to set the tool for preserving, in the selected type(s), the synthesis requirements imposed by certain digital implementations (discussed in Chapter 9). The remaining buttons in the window permit loading or saving specific learning files, and going back to the main window.

The numerical information area offers information about the status of the learning process showing the values of the different variables controlling it (error measurements and number of iterations) during its execution.

When the learning process is active, the graphical information area permits an intuitive visualization of the system evolution. The area automatically rescales as the number of iterations (X-axis) increases, while the Y-axis uses a logarithmic scale so the information can be easily read as the system evolves to lower error values.

The buttons in the control area permit starting (*Run*) and stopping (*Stop*) the execution of the learning process, reloading again the original specification

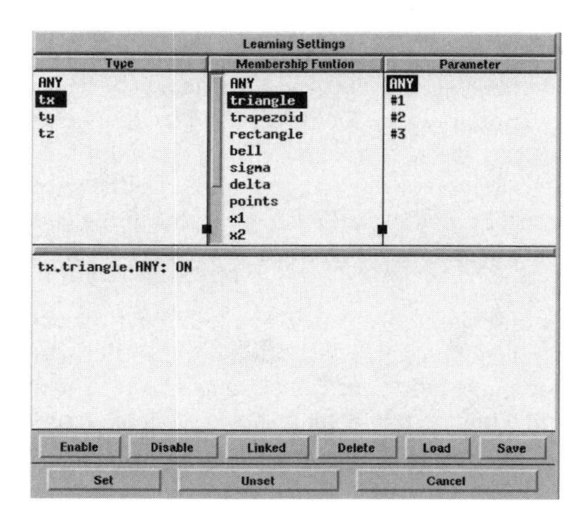

Figure 5.3. Specific learning window.

discarding the currently applied changes (*Reload*), updating the in-memory representation of the specification with the learned parameters (*Apply*), saving the tuned specification in an XFL file (*Save*), and closing the window (*Close*).

5.3 SIMULATION OF FUZZY SYSTEMS

In general terms, simulation techniques are applied whenever it is necessary to evaluate the capacity of a system or its components and predict its behavior within its operational environment, and actual experimentation with the system is impossible, very costly, or implies distortions on it. This is especially true when the system is under development and, therefore, there is not yet an available implementation. In these cases a model of the system can be employed, a representation of the system that must be detailed enough to permit extracting valid conclusions from its use.

When simulating the behavior of a fuzzy system, it is necessary to use two models: one for the fuzzy system itself and another for its operational environment. Obtaining the first model is immediate using a tool for producing a software implementation of the system. The model corresponding to the system operational environment is, however, completely dependent on the type of application under development. In problems such as decision-making or pattern recognition, the fuzzy system environment usually consists of an interface easing the provision of input data and the access to output data. A software implementation of this interface is the most frequent choice.

On the other hand, in control applications the fuzzy system will interact with one or more external devices (known as "the plant"). The model for this plant can be produced, in this case, either by using an algorithm coded in a high level programming language, or by taking advantage of the approximation capabilities of fuzzy logic for reproducing the input-output behavior of the plant from its linguistic description. A third possibility consists of including the plant itself in the simulation loop of the fuzzy system. For this purpose, the simulation environment must provide mechanisms for easing the interaction between the software model of the fuzzy system and the electrical signals used and accepted by the plant.

Providing a simulation tool is a basic requirement for any CAD environment, since it permits a thorough verification of the system prior to the start of the implementation phase, when the detection and correction of possible errors is much more demanding both in resources and time. Moreover, in some cases (such as model verification or rapid prototyping) the results of the simulation tool can fulfill the purposes of system development.

5.4 SIMULATION OF XFL-BASED SYSTEMS

The *Xfuzzy* design environment incorporates an XFL-based simulation tool, *xfsim*, able to combine into a single executable program any number of XFL specifications and other modules for simulating the behavior of the fuzzy system under development inside its operational environment. The output of *xfsim* permits the reconstruction of this executable program as many times as necessary, simplifying the iteration of the simulation and correction processes as the system evolves to fit its intended behavior. The simulation program, once generated, can be executed with different initial conditions and offers different output formats, and it can even be connected to other software elements to provide more sophisticated methods for exploiting and presenting simulation data, like real-time animations.

The input to *xfsim* is a file containing the necessary information for building the simulation program. This information is structured in a series of sections that begin with a section identifier (*%variables*, *%behavior*, *%output*, *%options*) and end with the symbol %%. Figure 5.4 shows a sample simulation file that corresponds to the truck parking control problem commented on in Chapter 4. The semantics of the different sections is described below.

%variables: This section defines the set of state variables. Every variable is defined by means of an identifier and a range of acceptable values. This range has two main purposes: first, establishing the default value for the variable and, second, the simulation is interrupted if any variable takes a value out of its range. Apart from those variables that are explicitly defined, *xfsim* defines two internal variables: *_t,* representing simulation time, and *_n*, holding the number of iterations of the simulation loop that has been performed.

```
%variables
ang -90 < 270 deg
x 0 < 100 m
y -100 < 0 m
wheel -30 < 30 deg
%%
%behavior
c:truck.mod.c:model (x, y, ang, wheel)
fuzzy:truck.xfl:truck (x, ang, wheel)
%%
%output
gnuplot (x, y)
stdout (_t, x, y, ang)
file (_t, ang)
%%
%options
$simwhile _n < 200
$simtic 10.0
$timeunit ms
%%
```

Figure 5.4. A sample input file for *xfsim*.

%behavior: Defines the behavior of the simulated system (the system under development plus its environment), expressed in terms of a combination of different modules. Each of these modules may correspond to: (1) a fuzzy system defined using XFL (keyword: *fuzzy*); (2) a function written in C (keyword: *c*); and (3) data stored by a file (keyword: *raw*). The different modules are called into the simulation loop in the same order that they appear in this section. The interaction among modules is accomplished by their acting upon the state variables.

%output: Establishes the set of results to be obtained from the simulation. As many output elements as necessary can be defined, including connections of the simulation program with external programs through its standard output (*stdout*). Apart from this, there are two output formats that *xfsim* can include into the simulation program: a numerical data output, compatible with the *raw* modules described above, and a format accepted by the free software graphics tool *gnuplot*.

%options: Sets a number of options controlling the execution of the simulation program, such as the program end condition and the duration (in simulation time) of each simulation step.

The output of *xfsim* consists of two files: a C source file containing the main function of the simulation process and a *Makefile* that permits the reconstruction of the simulation executable whenever any of the modules defining its behavior changes, without using *xfsim* again, just by calling the Unix standard tool *make*. This last file is built from the definition of the modules that specify the behavior of the whole system, and from a series of options that indicate, for example, how the XFL to C compiler must be called or which C compiler must be used.

A series of numerical values can be passed as parameters to the simulation program when it is started, which will be assigned as initial values for the state variables. The assignment is made in a positional way: the first value is assigned to the first defined variable and so on. If one variable must not be assigned and must use its default initial value (the center of its definition range) the character '-' should be used instead of a value in the corresponding position. The execution of the simulation program is performed following the steps below:

- The state variables are initialized, either with the values provided when the program was called or with the default values derived from the simulation definition. The output mechanisms are also initialized.
- As long as the end condition is not reached, the simulation loop is iterated. The body of the simulation loop consists of successive calls to the procedures defining the simulation behavior. State variables are passed by reference, so any of the values defining simulation state can be read or modified by these procedures. At the end of the body, the values of the appropriate state variables are sent to the output functions.
- When the simulation loop ends, the program closes the output mechanisms and starts the applicable programs to visualize the simulation results.

5.4.1 COMMAND LINE INTERFACE

The syntax applicable for invoking *xfsim* from the command line is summarized in Figure 5.5. Apart from specifying the name of an input file, the *xfsim* command accepts three optional parameters. The first one is for the selection of an alternative file for fuzzy operator definitions. The other two control which output files will be generated by the tool.

To show the use of the tool, let us consider that the definitions shown in Figure 5.4 are stored in a file called '*truck*'. Executing the command:

```
# xfsim truck
```

will yield as output two files, one called '*truck.sim.c*', and another called '*truck.make*'. By means of this last file, it is possible to build the simulation program '*truck.sim*', calling the standard program *make* in the following way:

Xfsim: Build simulations for XFL-based fuzzy systems

Synopsis: xfsim <-O OperationsFile> <-m> <-p> DefinitionFile

DefinitionFile	: File containing the simulation program definition
-O OperationsFile	: Fuzzy operator definitions file
-m	: Only generate the Makefile for building the simulation program
-p	: Only generate the main module source file of the simulation program

Figure 5.5. Command line interface for *xfsim*.

```
# make -f truck.make
```

Once the program has been built, the simulation can be executed as many times as necessary, just by invoking '*truck.sim*'. For example, the following two commands:

```
# truck.sim
# truck.sim 180 80 -75 -
```

correspond to two different executions of the simulation. The first one uses the default values for the state variables, while in the second one the initial angle of the truck axis with the horizontal is fixed to 180, the initial x-position is set to 80, the initial y-position is fixed to –75, and only the last state variable (*wheels*) takes its initial default value (0). Figure 5.6 shows the results of the visualization programs once the simulation has finished for the second of the commands shown above. In this case, the goal of parking the truck at the desired loading dock, with position (50, 0), and arriving at a right angle is accomplished.

5.4.2 GRAPHICAL USER INTERFACE

The graphical user interface for *xfsim* is a window (Figure 5.7) that permits handling the contents of a simulation definition. The text areas labelled *Variables*, *Behavior*, and *Output* correspond to the sections for variables, behavior and outputs in the definition. There is a selectable list with two buttons (*Add* and *Delete*) for the section dealing with non-XFL source files and the rest of buttons and text areas correspond to the different options for building and executing the simulation.

As it is common for all *Xfuzzy* windows, the bottom area of the window contains a set of buttons providing mechanisms for handling the window and the simulation definition contained within it. *Load* and *Save* open file selection dialogs for reading and writing, respectively, a simulation definition from/to a

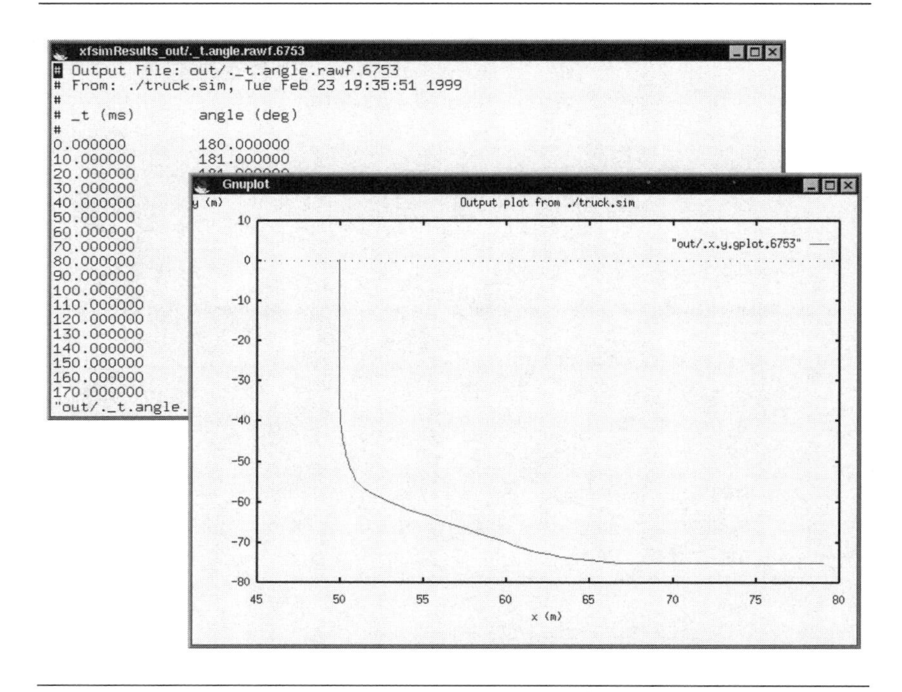

Figure 5.6. Visualization programs showing the results of a simulation generated by *xfsim*.

file, while *New* removes the current definition and restores a set of default initial values for the elements of the window. *Make* generates the simulation program *Makefile* and runs it through *make*, leaving a new simulation program ready for use. *Run* launches the execution of the simulation program. Before the program is executed, a dialog window for assigning initial values to state variables is shown. During the execution of the simulation program a control window offers information about the status of the simulation and the option of interrupting the execution at any moment. The last button, *Close*, stops the interface, closing its window.

5.5 ON-LINE VERIFICATION OF XFL-BASED SYSTEMS

Another important aspect of the analysis capabilities of a CAD tool is the possibility of emulating the system under development within its real operational environment. Although the cautions discussed at the beginning of this chapter with respect to the limitations imposed by the use of approximations are still important, the use of these mechanisms can be very valuable, especially for tuning the interfaces between the system and its environment. In the spe-

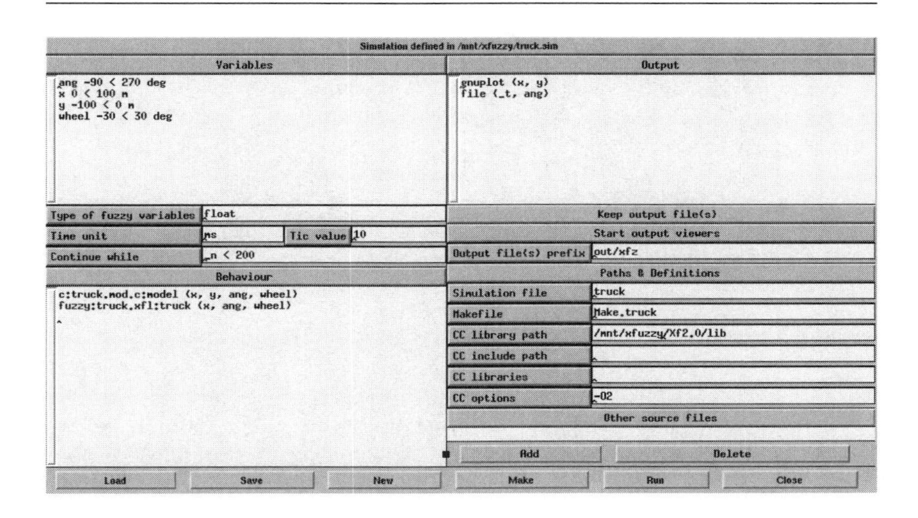

Figure 5.7. Graphical user interface for *xfsim*.

cific case of control applications, fitting the controller to sensors and actuators is a critical task in the development of the system, which can be significantly enhanced if these emulation capabilities are available.

To cover the on-line verification of fuzzy systems, the *Xfuzzy* design environment provides a tool for connecting a software implementation of an XFL-defined system to its real operational environment, including its interface circuits, through a data acquisition board connected on a standard PC bus. The current version of this tool, called *xflab*, is specifically designed for the National Instrument boards Lab-PC-1200 and Lab-PC-1200/AI, although it can be easily adapted for other similar boards by means of its specific interface library. Figure 5.8 graphically illustrates the plant-controller interaction provided by *xflab*.

The *xflab* tool basically consists of two components. The first one is a library of drivers containing functions, written in C, for managing the different input and output ports of the data acquisition board. The second component is a graphical user interface offering a series of windows for configuring the verification process and controlling its execution. More precisely, this graphical interface is oriented to the following tasks:

- Configuring the data acquisition board. The board is initialized by defining the operational mode of the input/output ports, the configuration of analog and digital signals, etc.

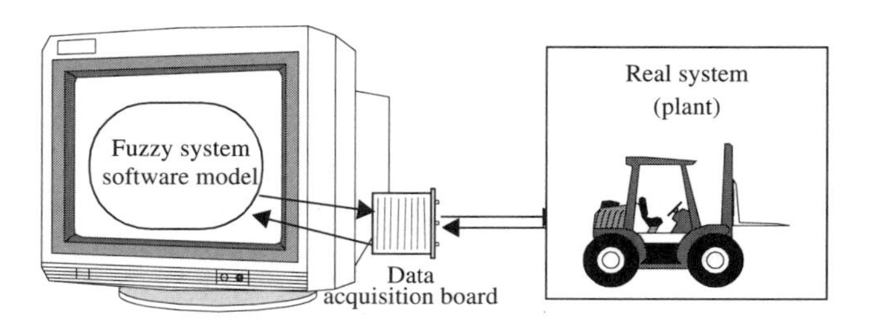

Figure 5.8. Use of *xflab* for the on-line verification of fuzzy systems.

- Connecting the plant and the simulation environment through the data acquisition board. At this stage, the channels connecting the plant with the software model of the inference system are assigned, and the signals to be monitored during the simulation execution are specified. The use of a software-hardware mixed model permits the designer to include pre- and post-processing routines for signal adaptation.
- Executing the simulation. Under the control of the graphical interface, the verification process is started. First, the plant is monitored by reading the input ports of the data acquisition board and applying the data preprocessing routines (if required). Afterwards, the data obtained from the plant are stored into intermediate variables and provided as inputs to the software module modeling the fuzzy system. Finally, the outputs of the inference engine are postprocessed (by the mechanisms specified by the designer) and fed to the plant through the output ports of the data acquisition board. This process continues until the end condition specified by the user is reached. For each simulation cycle the tool registers and stores the values of the selected signals. These values can be graphically represented or stored into a file to process them later.

The use of this on-line verification scheme has a series of limitations constraining its use. The main limitation is imposed by the system response times, since it must be ensured that the inference speed of the tool permits the plant operation in real time. Other constraints are related to the limits introduced by the interface board itself (limited number and type of ports, ranges of allowed values at the ports that often require specific signal conditioning circuits, etc.). Nevertheless, this kind of system offers a great flexibility, since any modification can be simply accomplished by recompiling a program without redesigning the operational platform.

As stated above, the tool offers a series of windows for performing the data acquisition board configuration: input/output modes, channel assignment, configuration of analog and digital signals, etc. One of them is shown in Figure 5.9. Once the board has been configured, two execution modes can be chosen. In the *Monitor mode*, the tool registers and stores the values of the input signals. The *Control mode* uses an implementation of the fuzzy system obtained from the XFL to C compiler (*xfc*) connected to the inputs and outputs defined for the data acquisition board.

The features of *xflab* can be used for other tasks apart from the functional verification of the system in its real environment. Among them we can consider those tasks previous to the design phase, such as data acquisition for the learning or simulation procedures.

Finally, although in this text the on-line verification has been considered as a previous stage to the hardware implementation of a fuzzy system, the system emulation provided by *xflab* may constitute the final system implementation for certain practical applications. These are the cases where the complexity of the inference system makes its implementation on an integrated circuit difficult, or the cases where the projected number of devices does not justify its production in silicon. Conversely, when the previous considerations are not applicable and when the requirements in terms of power consumption and area demand the use of dedicated hardware, the hardware implementation of fuzzy inference systems by means of some of the techniques discussed in the following chapters is a sensible alternative.

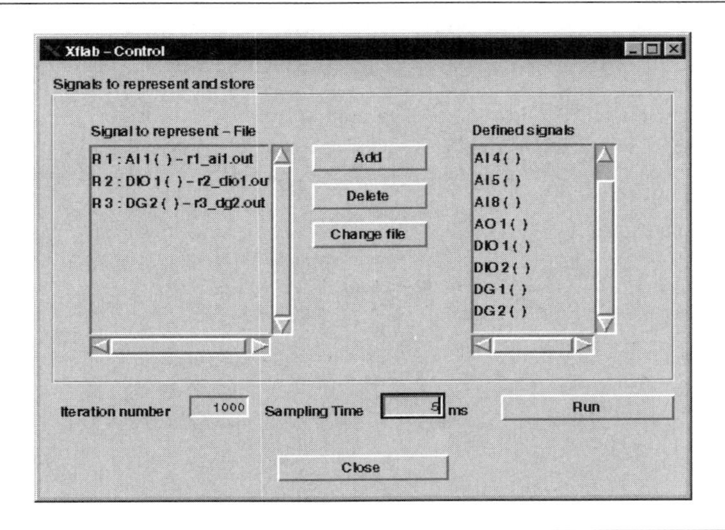

Figure 5.9. Graphical user interface for *xflab*.

REFERENCES

1. **Anderson, C.**, Learning and problem solving with multilayer connectionist systems, Technical report, University of Massachusetts, 1986.
2. **Buckles, B. P., Petry, P. E.**, Eds., *Genetic Algorithms*, IEEE Computer Society Press, 1992.
3. **Cauwenberghs, G.**, Analog VLSI stochastic perturbative learning architecture, *Analog Integrated Circuits and Signal Processing*, N. 13, pp. 195-209, 1997.
4. **Davis, L.**, *Handbook of Genetic Algorithms*, Van Nostrand Reinhold, 1991.
5. **Fahlman, S., E.**, Faster-learning variations on back-propagation: an empirical study, in *Proc. of the 1988 Connectionist Models Summer Schools,* pp. 38-51, Morgan Kauffmann, 1988.
6. **Geman, S., Geman, D.**, Stochastic relaxation, Gibbs distributions and the Bayesian restoration of images, *IEEE Transactions on Pattern Analysis and Machine Intelligence*, Vol. 6, pp. 721-741, 1984.
7. **Goldberg, D. E.**, *Genetic Algorithms in Search, Optimization and Machine Learning*, Addison Wesley, 1989.
8. **Herrera, F., Lozano, M., Verdegay, J. L.**, Generating rules from examples using genetic algorithms, in *Fuzzy Logic and Soft Computing*, Bouchon, B., Yager, R., Zadeh, L., Eds., pp. 11-20, World Scientific, 1995.
9. **Jabri, M., Flower, B.**, Weight perturbation: an optimal architecture and learning technique for analog VLSI feedforward and recurrent multilayer networks, *Neural Computation,* Vol. 3, pp. 546-565, 1992.
10. **Jacobs, R.**, Increased rates of convergence through learning rate adaptation, *Neural Networks*, Vol. 1, N. 4, pp. 295-308, 1988.
11. **Koza, J. R.**, *Genetic Programming*, The MIT Press, 1992.
12. **Jang, J. S.**, ANFIS: Adaptative-network-based fuzzy inference systems, *IEEE Transactions on Systems, Man, and Cybernetics*, Vol. 23, pp. 665-685, 1993.
13. **Kirkpatrick, S., Gelatt, C. D., Vecchi, M. P.**, Optimization by simulated annealing, *Science*, Vol. 220, pp. 671-680, 1983.
14. **Mendoça, P. R. S., Caloba, L. P.**, New simulated annealing algorithms, in *Proc. IEEE Int. Symposium on Circuits and Systems*, pp. 1668-1671, Hong-Kong, 1997.
15. **Metropolis, N., Rosenbluth, A. W., Rosenbluth, M. N., Teller, A. H.**, Equation of state calculation by fast computing machines, *The Journal of Chemical Physics*, Vol. 21, N. 6, pp. 1087-1092, 1953.
16. **Riedmiller, M., Braun, H.**, RPROP: A fast and robust backpropagation learning strategy, in *Proc. 4th Australian Conference on Neural Networks,* pp. 169-72, Sydney, 1993.
17. **Rumelhart, D. E., Hinton, G. E., Williams, R. J.**, Learning representations by back-propagation, *Nature*, Vol. 323, pp. 533-536, 1986.
18. **Szu, H., Hartley, R.**, Fast simulated annealing, *Physics Letters A*, Vol. 122, pp. 157-162, 1987.

Chapter 6

HARDWARE REALIZATION OF FUZZY SYSTEMS

As shown in the previous chapters, most development environments for designing fuzzy systems implement them with a high-level programming language. Software implementations of fuzzy systems are easily obtained and provide a high flexibility since they usually support fuzzy systems with an arbitrary number of rules, without any limitations concerning the number and type of membership functions, and with a wide range of inference mechanisms. On the other hand, software approaches are not adequate for applications demanding fuzzy systems with a small size, low power consumption, and high inference speed. To verify the requirements imposed by these kinds of applications, hardware approaches have to be adopted. This issue is addressed in the first section of this chapter by briefly describing the different ways of implementing fuzzy systems depending on the application. The rest of the chapter reviews the different hardware approaches, also known as fuzzy hardware. There are two basic procedures for developing fuzzy hardware. One approach is to employ general-purpose microprocessors, occasionally expanded with new instructions or new circuitry. The other approach is to design dedicated hardware, that is, application-specific integrated circuits (ASICs), optimized for fuzzy logic–based inference systems. The last sections of the chapter focus on the second procedure. Several strategies to implement fuzzy systems with dedicated hardware are analyzed and the fundamental concepts concerning the design and fabrication of integrated circuits (ICs) are discussed. This material serves as an introduction to the following chapters where microelectronic realizations of fuzzy systems with different design techniques are described in greater detail.

6.1 FUZZY SYSTEM IMPLEMENTATIONS DEPENDING ON THE APPLICATION

Design and implementation of a fuzzy system obviously depends on the requirements of the application towards which it is addressed. Three requirements are particularly relevant to decide how to implement the fuzzy system: the time available for rule processing, the size it can occupy, and the power it can consume. As will be shown in this chapter, the development of dedicated hardware becomes more important as these requirements grow more restrictive (less time, size, and power available).

Let us first consider applications like data analysis or decision making. The requirements of these applications concerning time, size, and power are not usually restrictive. What is required is a high flexibility and that the fuzzy system features the capability of interacting with different software tools like databases. As a consequence, the most adequate implementation of these applications is usually a software implementation on a personal computer (PC) or a workstation.

Other important application areas of fuzzy systems are process control and industrial automatization. Both software and hardware approaches have been employed traditionally in these fields. Among the hardware approaches, we can cite the use of programmable logic controllers (PLCs) or programmable automata as well as the use of electronic proportional-integral-derivative (PID) controllers aimed at controlling particular processes like temperature control, liquid level regulation, etc. PLCs offer the advantage of directly interacting with the process to be controlled because they contain analog/digital input/output interfaces and their programmability is very simple. These automata have been employed in many fuzzy logic–based applications by programming them to perform fuzzy inferences. Response time in the range of 10 to 500 ms has been achieved [von Alstrock, 1998]. Klockner-Moeller was the first company that included a dedicated processor optimized to carry out fuzzy logic inferences within a PLC, thus obtaining what is known as a fuzzy PLC. This solution performs by hardware the mathematical operations related to fuzzy logic that cannot be performed by a conventional PLC. As a consequence, the inference speed can be increased significantly and a response time somewhat superior to 1 ms has been reported [von Alstrock, 1998]. Another company like Omron has included a fuzzy module, the FZ001, in their PLCs. This module contains a digital fuzzy processor (the FP3000) which admits up to 128 rules with 8 inputs and 2 outputs and that achieves a minimum inference time of 125 μs per rule. Concerning electronic PID controllers, several companies like Omron offer products whose control parameters are internally adjusted by fuzzy logic.

Another significant area in the application of fuzzy logic–based systems is embedded control, which requires more restrictive area occupation and power consumption. Conventional hardware approaches employed in this field are microcontrollers because they integrate random access and read-only memories (RAMs and ROMs) with communication circuitry on the same chip. The microcontrollers usually employed in industrial applications are not as suitable for mathematical calculation as the microprocessors included as the central processing unit (CPU) of current computers. Hence, they should be adequately programmed to implement efficient fuzzy systems. This was the objective pursued by Motorola and Aptronix when they developed the FIDE system (*Fuzzy Inference DEvelopment language*) [Yen, 1995]. This language allows the user to define a fuzzy system with a particular number of inputs, outputs, membership functions and rules, and to translate this definition in real time into the specific code of Motorola microcontrollers (like 6805, 68HC05, and 68HC11). Intel and Inform Software also worked together to develop FuzzyTECH development software, which generates assembly code to program Intel microcontrollers (like 8051, 8096, and 8C196) and microcontrollers of other firms. In a similar way, Togai InfraLogic developed MicroFPL for microcontrollers of Motorola, Intel, Oki, etc. Since the code generated by these software tools is optimized, response times below the millisecond can be achieved for fuzzy systems of medium complexity [von Alstrock, 1998].

In order to improve the inference speed, several microcontrollers have been expanded with a set of dedicated fuzzy logic instructions. Another solution, which allows better improvements, is to employ external fuzzy processors that cooperate with the main microcontroller and off-load from it all the computation required by the fuzzy inferences. These solutions are adequate for applications demanding response times of hundreds or tens of microseconds. In this case, inference speed is usually limited by the bottleneck produced by the data buses which connect the main CPU with the external processor. Higher speed can be obtained if data communication is performed on chip by adding dedicated fuzzy logic circuitry to the microcontroller circuitry. Several implementations of these solutions will be described in detail in Section 2.

A different way for taking advantage of existing hardware to implement fuzzy systems is to expand either with software or hardware digital signal processors (DSPs) that offer more computing power than general-purpose microcontrollers [Rovatti, 1998; 1999]. As an example of this approach, it has been reported that a DSP of 32 bits (the TMS320C6201 of Texas Instruments) operating at 200 MHz can perform a fuzzy inference with 6 inputs in about 1.5 μs [Rovatti, 1999].

Finally, several applications related to robotics, aeronautics, voice and image processing, automotive electronics, etc. demand response times of a few microseconds or below the microsecond with even more restrictive area-occupation and power-consumption requirements than in the former applications. The adequate solution for these cases is to implement the whole fuzzy system as a stand-alone system in an application specific integrated circuit (ASIC). A fuzzy ASIC features the best area occupation and power consumption as well as the best response time, as shown in Figure 6.1. Its cost grows inversely proportional to the production volume so that it is the cheapest solution for applications that require a massive production of units. As a drawback, fuzzy ASICs currently offer the least flexibility. For instance, they support the least complexity regarding the number of rules (as illustrated in Figure 6.1) and communication interfaces. Market guidelines in this field, as in many others, tend towards an increasing demand of systems on a chip, so that microelectronic research is focused on improving the flexibility of fuzzy ASICs. One of the main objectives of the following chapters is to show the latest advances in the design and development of whole fuzzy systems on a chip.

As an introduction to the following sections that focus on hardware approaches for implementing fuzzy systems, Figure 6.2 summarizes current application fields where fuzzy hardware has been employed. The response time ranges and the number of rules required nowadays are depicted.

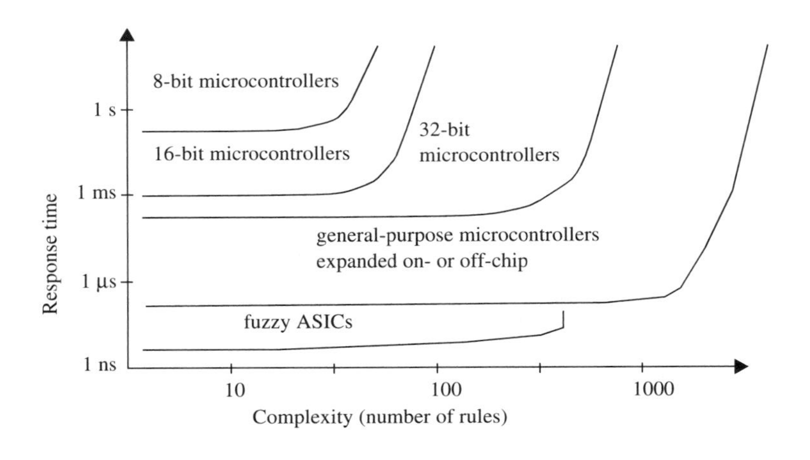

Figure 6.1. Comparison between fuzzy ASICs and general-purpose microcontrollers.

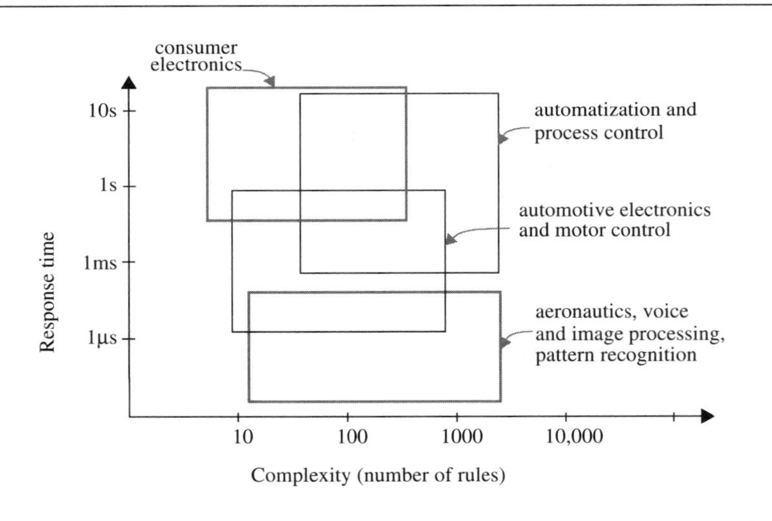

Figure 6.2. Application fields of fuzzy systems.

6.2 FUZZY SYSTEM REALIZATIONS WITH GENERAL-PURPOSE PROCESSORS

As concluded in the section above, the need for fuzzy hardware increases as the specifications in terms of response time, area occupation, and power consumption of the fuzzy system grow more restrictive. The main reason for that is that fuzzy systems have particular features to which general-purpose hardware is not suited. Among these features, we can cite the inherent parallelism of fuzzy systems and the employment of several non-standard operations (like maximum, minimum, defuzzification, etc.), that is, operations which are seldom used by other systems.

6.2.1 PARALLELISM OF FUZZY SYSTEMS

The three basic processing stages of a fuzzy system, namely fuzzification, inference, and defuzzification, have to be performed serially. The membership degrees of the input signals to the fuzzy sets that represent the antecedents of the rules are computed during fuzzification stage. Once they are available, it is possible to proceed with the inference stage in which the partial conclusion of each rule is calculated. Finally, once the partial conclusions are available, it is possible to proceed with the defuzzification stage to obtain the system output. However, the operations required within each stage feature a high level of parallelism. Parallelism appears in the fuzzification stage because several membership degrees have to be computed for each input signal and because there are usually multiple input signals. Parallelism in the inference and defuzzification stages is primarily caused by two facts. On the one hand, several rules have to be processed since more than one rule can be activated simultaneously, as opposed to a conventional expert system. On the other hand, many elements of the output universe of discourse have to be swept to calculate the partial conclusions of the rules, the global conclusion of the rule base, and the output defuzzified value.

General-purpose processors are exclusively sequential so that all these operations have to be performed serially, thus resulting in a high response time. Two alternatives are mainly used to reduce the response time. One of them is to implement simplified inference methods that avoid sweeping all the elements in the output universe of discourse. The other solution is to consider that not all the rules but only a portion of them are activated for each set of input signals. Since only the active rules have to be processed instead of the whole rule base, the response time can be reduced.

6.2.2 NON-STANDARD OPERATIONS OF FUZZY SYSTEMS

General-purpose microelectronic systems are not efficient for performing certain basic operations of fuzzy inference methods, as has been shown by several authors [Watanabe, 1993; Ungering, 1994a,b; Surmann, 1995]. To illustrate this issue, let us comment on the work in [Surmann, 1995], which analyzes different realizations of a fuzzy system that implements the Min-Max inference method with the Center-of-Gravity defuzzification method. This work concludes that if an 8-bit resolution fuzzy system is realized with an 8-bit microprocessor, the defuzzification stage consumes the highest portion of computing time, 77%. The reason is that Center-of-Gravity defuzzification requires carrying out accumulation and multiplication operations with integers of 16 or 32 bits. Since only 8-bit resolution operators are available, these operations have to be performed in several cycles, with the intermediate results being stored into registers. On the other hand, if the same fuzzy system is realized with a 64-bit microprocessor, the most time-consuming stage is the inference stage, which takes 83% of the total computing time. In this case, rule evaluation, in particular the calculus of maxima and minima, becomes the bottleneck.

There are two basic solutions to reduce the response time due to the emulation of operators like defuzzification, maximum, or minimum. The first of them, which can be named as "*software expansion*", consists in adding new instructions, implemented as microprograms, to the microprocessor [Ungering, 1994b; Costa, 1995; von Alstrock, 1998]. An example of this solution is the 68HC12 Motorola microcontroller, which achieves an inference speed ten times higher than the 68HC11 thanks to its four fuzzy logic dedicated instructions: one instruction to fuzzify the input signals through trapezoidal membership functions, two instructions to evaluate the rules by employing a Min-Max inference method and by applying or not a weight to the rules, and another instruction to calculate sums of products and sums of weights in order to speed up the defuzzification stage. Another example is the Fuzzy-166 microcontroller developed by Inform from a Siemens CISC processor. This microcontroller achieves a response time of hundreds of microseconds [Costa, 1995]. The second solution to reduce the response time of general-purpose microprocessors will be referred to as "*hardware expansion*". It consists of introducing several changes into the microprocessor circuitry so as to include new components. This solution accepts different possibilities that are discussed in detail in the following section.

6.2.3 HARDWARE EXPANSION OF GENERAL-PURPOSE PROCESSORS

The functionality of a general-purpose microprocessor can be expanded by including the required fuzzy logic components on chip, with the advantage that data communication is performed on chip. Several proposals following this idea have been reported in the literature:

(a) One of these proposals is related to implementing an 8-bit fuzzy microprocessor by expanding the architecture of a 6502 microprocessor [Ungering, 1994a]. In particular, the arithmetic logic unit (ALU) is expanded so as to carry out operations with 20-bit resolution. New instructions implemented as microprograms are also included to control the new features of this ALU. These modifications produce a hardware increase of approximately 20% with the advantage of achieving 6 times higher inference speed. The inference time is estimated as 437 clock cycles (21.85 μs, if working at 20 MHz).

(b) Another solution is to add a fuzzy ALU to the conventional functional blocks (input/output unit, control program, control unit, memory unit, and ALU) of a standard microprocessor. The resulting processor is named a fuzzy/scalar RISC processor because it can implement both fuzzy and scalar operations. The proposal reported in [Patyra, 1996] includes a fuzzy ALU in a 32-bit microprocessor and codes the fuzzy vectors with 5×64 = 320 bits. The main problem of this proposal is the high area occupation of the resulting chip. It is estimated to be above 1 cm^2 considering a 1.2-μm CMOS technology.

(c) Nakamura et al. have proposed another approach that also adds dedicated fuzzy functional blocks to a conventional microprocessor [Nakamura, 1993]. These additional components are an antecedent and a consequent block. The antecedent block calculates the membership degrees of the input signals to the antecedent fuzzy sets. The membership functions employed are trapezoids defined by 6 parameters which are stored in a 16-bit register. The consequent block has two sections. One of them is the min-max section, which computes the shape of the consequent membership function. The other is the accumulation section, which calculates the numerator and denominator of the Center-of-Area defuzzification expression by a structure consisting of two adders and two accumulators. This processor has been developed with 12- and 16-bit resolutions for the inputs and the outputs, respectively. Its inference speed is 200 kilo fuzzy logic inferences per second (KFLIPS) considering 20 rules with 2 inputs and 1 output and operating at 25 MHz.

The functionality of a general-purpose microprocessor can be also expanded if it is connected to other chips dedicated to carrying out some of the fuzzy logic operations. Contrary to the previous proposals, data communication is now performed off chip. Some examples of this idea are the following:

(a) Kim and Lee-Kwang connect several processing elements to a bus that is controlled by a conventional microprocessor [Kim, 1997]. The system they proposed, named KAFA (KAist Fuzzy Accelerator), consists of two parts: a system control unit and a set of fuzzy processing units (each of them performing the fuzzy logic operations over one element of the universe of discourse). In particular, they describe a system with 128 fuzzy processing units, implemented with field programmable gate arrays (FPGAs), and connected to a PC486 by an ISA bus. The inference time obtained for a rule base with 100 rules of 2 inputs and 1 output is 59.8 μs.

(b) Lee and Bien propose another architecture whose main advantages are modularity and expansion capability [Lee, 1994]. This is a consequence of being constituted by two types of blocks: IF and THEN blocks. The IF blocks are dedicated to computing the activation degrees of the rules while the THEN blocks evaluate the consequents and the result of the inference. IF and THEN blocks are interconnected to form the rule base and connected to a VME bus. Architecture expansion is possible by adding IF and THEN blocks. The measured inference speed for a system implementing 256 rules with IF blocks that admit 8 input signals, THEN blocks that provide 4 output signals, and operating at 16 MHz is 20 KFLIPS.

Finally, a third approach for hardware expansion is to connect the general-purpose microprocessor to a single chip that carries out all the operations required by the fuzzy logic inferences. These chips are usually known as fuzzy coprocessors since their task is similar to that performed by the mathematical coprocessors of the first-generation personal computers. Several research proposals and commercial products (of Fujitsu, Siemens, Omron, SGS-Thomson, Togai Infra Logic, or Toshiba) follow this idea:

(a) As an example, the first work realized by Siemens in the field of fuzzy coprocessors resulted in an 8-bit coprocessor named SAE 81C99 [Eichfeld, 1993]. It occupies a silicon area of 6.4 mm^2 (excluding the pad ring) in a 1.0-μm CMOS technology. Its objective was to be integrated as a macrocell within conventional microcontrollers so as to obtain a fuzzy microcontroller. It manages up to 256 inputs, 64 outputs, and 16.384 rules. Further research on this coprocessor gave as a result the SAE 81C99A, which incorporates four ways of operation depending on the defuzzification method employed [Eichfeld, 1995]. The latter work has focused on developing a 12-bit fuzzy coprocessor named SAE 81C991 with the capability of being included into the Siemens microcontroller family C16x [Menke, 1995]. The structure of this coprocessor shows two groups of functional blocks interconnected by an internal bus of 16 bits. One group contains the blocks that perform the fuzzy logic algorithm (fuzzifier, inference block, and defuzzifier) while the other block contains the interface, control, and timing blocks. The coprocessor has two buses for external communication with the microcontroller and with the memory that stores the knowledge base. This memory can be also implemented as an internal ROM, thus reducing the circuitry required to build the system. The membership degrees are computed with an arithmetic method [Eichfeld, 1996] while the defuzzification methods employed are simplified. Its silicon area is 17.9 mm^2 (excluding the ring pad) in a 0.8-μm CMOS technology and its inference time depends on the number of both rules and input signals. Considering, for example, 80 rules with 4 inputs, 1 output, and operating at 25 MHz, the inference time is 16 μs.

(b) Among the Omron coprocessors we can cite the FP1000, FP3000, and FP5000. The latter, FP5000, is outstanding for admitting a high number of inputs (up to 128), outputs (up to 128), and rules (up to 32,639) [Shimizu, 1992]. Its structure consists of three main blocks dedicated to input/output, control, and processing tasks. The input/output block has 2 RAMs for both the input and the output, so that if one of them is being addressed by the CPU, the other can be addressed by the coprocessor itself. The control block is in charge of selecting the rules and starting the inference. The processing block has a pipeline architecture that performs the writing of output data, maximum and minimum operations, rules evaluation, etc. Operating at 10 MHz, this coprocessor achieves an inference time of 0.1 μs for every antecedent plus a fixed time of 1.3 μs.

(c) Another example of fuzzy coprocessors is the WARP coprocessor developed by SGS-Thomson. Its main feature is the use of memories that work in parallel in order to obtain a high inference speed. Among the 6 internal memories, 4 of them are employed to store the antecedent membership functions, another stores the consequent membership functions, and the last one stores the microcode. The coprocessor admits up to 16 inputs covered by a maximum of 16 membership functions, up to 16 outputs covered by a maximum of 128 membership functions, and up to 256 rules with 4 inputs and 1 output. It has two operational modes named off-line and on-line. During the off-line mode, the memories are charged while during on-line mode, the circuit executes the fuzzy inference. The defuzzification method employed is the Center of Area, which is implemented with a technique that precomputes the consequents so as to reduce the required operations. Regarding global performance, this coprocessor achieves a response time of less than 1.85 μs for an inference of 32 rules with 5 inputs and 2 outputs [Pagni, 1998].

6.3 FUZZY SYSTEM REALIZATIONS WITH DEDICATED HARDWARE

For certain applications, fuzzy systems realizations with general-purpose microprocessors are not efficient in terms of cost, size, and inference speed, even considering the modifications described in the above section. In these cases, requirements are only met by using dedicated hardware whose architecture and constituent elements are customized to the particular application.

Implementations of fuzzy systems with dedicated hardware are highly influenced by several facts. To begin with, the choice of a realization strategy conditions the storage and computing resources that will be required. In addition, the implementation technique and the design methodology employed to realize the integrated circuits are two highly relevant facts that determine the development time and the final cost of the system. Finally, the realization of the basic building blocks of the fuzzy system depends greatly on the analog or digital design style employed. All these influential facts are analyzed in detail in the following sections.

6.3.1 HARDWARE REALIZATION STRATEGIES

Three different strategies can be followed when considering the microelectronic implementation of a fuzzy system. These strategies differ in the type of hardware resources employed (more use of memory against computing elements or vice versa) and in the system performance achieved (more flexibility against simplicity or vice versa).

- **Off-line strategy**

A first alternative to implement a fuzzy system consists of precomputing all the output values that should be provided by the system for every possible combination of the input signals. This results in the generation of a look-up table that can be implemented by an integrated circuit. An advantage of this solution is that all the problems related to the parallelism or to the non-standard operations of fuzzy systems are eliminated. Another advantage is that this approach does not impose any limitation on the system complexity, that is, it allows full freedom when choosing the number of rules, the number and type of membership functions, the type of operators, etc. As a drawback, these look-up table implementations are practical only when the number of inputs and outputs and their quantization degrees are low, otherwise the total number of bits to store would be very large. If the fuzzy system has u inputs and v outputs, coded by a and b bits, respectively, the $v \cdot b$ bits corresponding to the $2^{a \cdot u}$ possible combinations of the inputs have to be precomputed. The microelectronic circuits suitable to this strategy are standard digital memories and programmable logic devices (PLDs) [Leong, 1993]. The response time in this approach is very short because only one data has to be retrieved from a memory and this can be performed in a few nanoseconds.

- **Semi-off-line strategy**

When the input/output relation of the system verifies certain conditions, it is not necessary to precompute the output values for every possible combination of input values. Let us suppose, for instance, a fuzzy system whose u input variables are covered by L triangular and normalized membership functions with an overlapping degree of two. This means that L–1 intervals can be distinguished for each input and that the sum of the membership degrees of the input to the two activated fuzzy sets within each interval is always '1', as shown in Figure 6.3. Let us also suppose that the "and" connective among the antecedents is the product, that the defuzzification method is the Fuzzy Mean and that the complete rule base is implemented (L^u rules). These fuzzy systems are piecewise multilinear interpolators [Zeng, 1996; Rovatti, 1998; Baturone, 1999], as we will see in Chapter 12. This means, for instance, that if two inputs (x_1, x_2) are considered, the system output is: $ax_1x_2 + bx_2 + cx_1 + d$, where the values of the constants a, b, c and d depend on which of the $(L–1)^u$ grid cells is activated by the input signals. The total number of grid cells to which an input combination may belong is $(L–1)^u$ because each of the u inputs may belong to L–1 different intervals within its universe of discourse. In this situation, a semi-off-line strategy consists in implementing the $(L–1)^u$ multilinear functions that

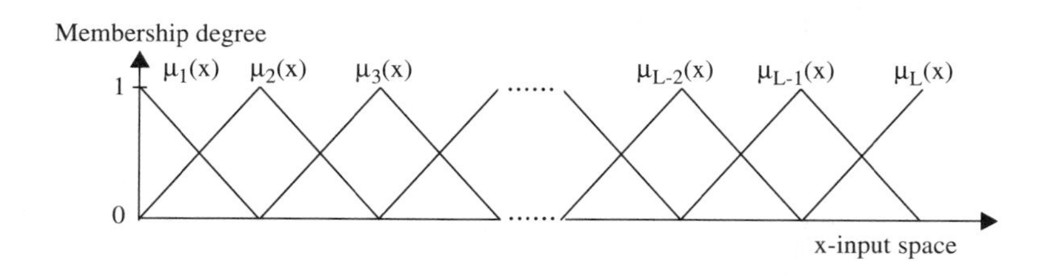

Figure 6.3. Triangular and normalized membership functions with an overlapping degree of 2.

the system has to provide. In order to meet this objective, a solution is to use a programmable circuit which implements the required multilinear function ($ax_1x_2 + bx_2 + cx_1 + d$) and which is adequately programmed with the parameters *a*, *b*, *c*, and *d* corresponding to the grid cell activated by the inputs. Considering the above situation of two inputs, the operations to be realized are a two-input multiplication (x_1x_2), three scalings (by *a, b, c*, and *d*) and a sum, and the memory size is of $4(L-1)^u$ words (to store the parameters *a*, *b*, *c*, and *d* of each grid cell). This strategy has been employed in digital realizations with DSPs, like those commented on in Section 1 [Rovatti, 1998; 1999], and also in analog/digital ASIC realizations, like those presented in [Matas, 1997; Baturone, 1999]. These realizations implement piecewise multilinear and multiquadratic fuzzy systems.

The use of this strategy also eliminates the problems of parallelism and non-standard operations of the fuzzy systems. Besides, it reduces the memory size (the number of data to store) if compared with the previous strategy. The cost of this reduction is an increase in hardware complexity because additional blocks have to be included not only to perform the operations required (multiplication, scaling, sum, etc.) but also to identify the grid cell activated by the inputs.

• **On-line strategy**

The main drawback of the off-line strategy is the exponential growth of the memory size as the number of inputs or the number of elements in the universes of discourse increase. On the other hand, the semi-off-line strategy provides low flexibility to implement a fuzzy system since it is constrained to particular realizations such as piecewise multilinear or multiquadratic systems. For these reasons, the strategy usually employed to implement generic and complex fuzzy systems is an on-line strategy, in which the inference process is concurrently performed as the input changes. All the fuzzy system realizations briefly described in Section 1, except for those of DSPs, follow this on-line strategy. The dedicated hardware solutions that are described in Chapters 7, 8, and 9 also focus mainly on this strategy.

The advantages of a reduced memory size and high flexibility are coupled with the drawback of an increase in the number of operations to carry out. Consequently, the inference speed achieved is lower, and the problems of parallelism and non-standard operations of fuzzy systems cannot be avoided. The operations to be performed can be grouped into three stages: input stage (fuzzification), rule processing stage, and output stage (rule aggregation and defuzzification, if needed). The input stage should contain operators to fuzzify the inputs, that is, operators that define the fuzzy sets of the rule antecedents. These operators are usually known as membership function circuits (MFCs). The rule processing stage should contain operators like a minimum or product that implement the antecedent connectives (t-norms/s-norms) so as to obtain the activation degrees of the rules. Implication operators (again, like the minimum or product) are also required in the rule processing stage to connect the activation degree of each rule to its consequent. Finally, the membership function or the singleton value of the rule consequents (depending on the inference method) have to be generated. The operators which perform this function are usually known as membership function generators (MFGs). The output stage should contain operators that implement the rule aggregation (like maximum or sum operators) and the defuzzification operator (if required by the inference method). All these operators are detailed in Figure 6.4.

6.3.2 IMPLEMENTATION TECHNIQUES OF INTEGRATED CIRCUITS

The design process of an integrated circuit basically consists in translating the high-level system description into a set of geometrical patterns, called a *layout*, which meets the specifications once it is implemented on silicon. Throughout this process, the system is described by different representations corresponding to different abstraction levels. The particular design steps that draw from the specifications to the layout depend primarily on the circuit and implementation techniques employed.

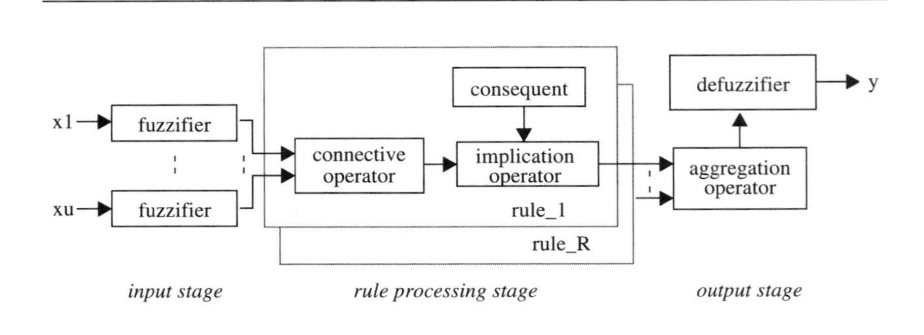

Figure 6.4. Operators required if the fuzzy system is realized with an on-line strategy.

Having selecting the starting architecture of the system, the first design stage consists of partitioning the system into a set of functional blocks that implement all the operations required, as shown in Figure 6.5. The next step in the process design is aimed at detailing the structure of the functional blocks. If analog circuitry is used, this step means the identification of the type of operational amplifiers (OPAMPs), transconductor amplifiers (OTAs), current mirrors, etc. which are to be used. However, if digital circuitry is employed, this step will provide a logic level description, which describes the system by means of logic gates, flip-flops, and bit lines. It is also usual in the digital case to pass through intermediate representations that describe the system with a register transfer language (RT level). The next step of both analog and digital design processes is to obtain an electrical level description, which represents the system as the interconnection of the basic elements provided by the fabrication technology selected. The last step is to detail the layout, that is, the geometrical masks to physically realize all the constituent layers of the integrated circuit.

The geometrical information resulting at the end of the design process is the starting point for the fabrication stage of the integrated circuit. At the present time, while digital integrated circuits are mostly designed and fabricated in metal-oxide-semiconductor (MOS) technologies, analog integrated circuits employ not only MOS but also bipolar technologies, as well as technologies that combine the former ones in one process (BICMOS). However, the necessity of combining complex digital functions with analog func-

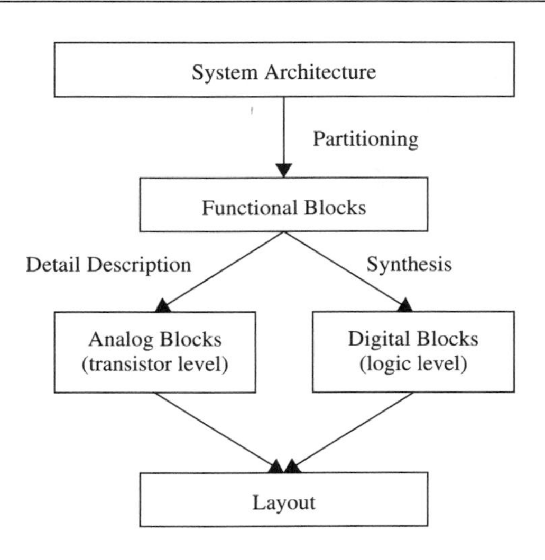

Figure 6.5. Design process of fuzzy systems with dedicated hardware.

tions on the same integrated circuit has resulted in an increased use of MOS technologies, in particular of CMOS technologies, which provide two basic types of devices, an enhancement-mode n-channel transistor and an enhancement-mode p-channel transistor, called NMOS and PMOS transistors, respectively. This is why this book mainly describes CMOS realizations of fuzzy systems.

The fabrication process of integrated circuits consists of a sequence of photomasking, diffusion, ion implantation, oxidation, and epitaxial growth steps applied to a slice of silicon wafer. The first process steps are related to the diffusion layer, thus defining the channel of the MOS transistors. The following steps are associated with the polysilicon layer, which defines the gate of the transistors. In the final steps, metallization (one or several layers) is deposited to implement the connectivity of the circuit, in particular to implement connections between diffusion and polysilicon layers [Gray, 1993].

The two main factors that determine the final cost of an integrated circuit are the design cost and the fabrication cost. Considering application specific integrated circuits (ASICs) for which the demand is not massive, the design cost is a relevant factor. As a consequence, the development of structured design techniques attracts a large interest because they allow minimizing the time-to-market and they make it easier to handle the inherent complexity of very large scale integrated (VLSI) systems. The different design techniques establish a trade-off between the optimization degree of a product and the cost invested in its development. The choice of a particular technique depends, on the one hand, on the number of units to be fabricated and, on the other hand, on the circuit techniques employed.

In a *full-custom* design technique, the designer realizes personally all the circuit layout tasks, that is, the selection of all device geometries, their placement and routing. This results in a very high performance: integrated circuits with the least silicon area and with high quality behavior (high speed, low power consumption, etc.), with the cost of a very long design time. In the digital domain, this type of design is only employed for systems of a very high performance, like high-density memories or certain blocks of microprocessors. Yet, this is the design style usually employed in the analog domain.

Whenever the development cost of an integrated circuit is not justified by the production of a large number of units, or the design time should be shortened, *semi-custom* design technique can be helpful. These techniques can be grouped in two main categories, *standard-cells* and *gate-arrays*.

Semi-custom design with standard cells is based on the use of basic functional elements completely characterized and organized in a cell library, whose layouts are provided by the vendor of that technology. As a consequence, the generation of the layout is reduced to select, place, and route the required cells. Considering an analog circuit, these cells can be current mirrors, biasing circuits, amplifiers, A/D-D/A converters, or filters. If the cell library is digital, it usually contains logic gates, flip-flops, registers, counters, combinational subsystems (like coders, decoders, adders, etc.), or memories (RAMs, ROMs, etc.). In both cases, customizing the design affects all the masks required to fabricate the integrated circuit. While there are CAD tools in the digital design that ease the automatic generation of the layout (they usually perform the placement and routing steps), automatization of the analog circuit design is currently an open research field.

The use of gate arrays is an alternative to reduce the fabrication cost of digital systems. A gate array is an integrated circuit made of an array of predefined logic gates which has passed through all the fabrication steps except for the final steps related to metallization deposition. Therefore, the generation of the layout is reduced to detailing the metallization layers, that is, the connectivity. The interconnection between the devices of each cell defines the cell functionality while the interconnection between different cells permits us to define the system. Silicon wafers can be massively fabricated, so that only the metallization masks are needed to complete a particular integrated circuit. Hence, the fabrication time and cost of the wafer and, in turn, of the integrated circuit are reduced.

An implementation technique that greatly reduces the realization cost of a digital system is the use of preconnected devices like *Field Programmable Gate Arrays* (FPGAs). FPGAs are integrated circuits already fabricated and afterwards adequately programmed to obtain a required functionality. They consist of a set of input/output blocks and configurable logic blocks whose interconnection is programmable. Although depending on the vendor and on the particular device family, the logic blocks of an FPGA usually contain memory cells (flip-flops) together with small combinational circuits capable of performing Boolean functions of a few variables. The objective of the design process in this case is not to generate the layout but to obtain the needed information to program the blocks and their interconnection. The design based on FPGAs is usually aided by CAD tools provided by the vendors themselves, which follow conventional digital design techniques. The features of these devices make them ideal candidates to realize small system series and prototypes that will later be fabricated as ASICs. FPGA and ASIC implementations of fuzzy systems are described in the next chapters of this book.

6.3.3 DESIGN METHODOLOGIES OF INTEGRATED CIRCUITS

Whenever the complexity of a system is high, there is a need for the use of a systematic design procedure (a *design methodology*) that permits us to obtain a reliable solution as soon as possible. This necessity appears independent of the implementation technique employed, although it is greatly influenced by it. A design methodology is an ordered way of passing through all the different levels to represent a system. The particular order employed defines the type of methodology. In the analog domain, for example, the starting point is usually the electrical level design of the basic building blocks of the system (current mirrors, operational amplifiers, etc.), which are then interconnected so as to form the functional blocks of the system architecture, thus following a so-called *bottom-up methodology*. This methodology for designing analog fuzzy ASICs is described in Chapters 7 and 8. However, *top-down methodologies* are used in the digital domain. They start from an architectural description of the system either by a schematic representation or by a hardware description language (HDL) and pass down the subsequent design levels in a hierarchical way. This methodology for designing digital fuzzy ASICs and FPGAs is described in Chapters 9 and 10.

The design tasks within each level consist of a sequence of synthesis, analysis, and verification steps. The synthesis steps are devoted to finding solutions to the problems that appear at each level. Analysis tasks measure the system characteristics according to the system metrics at each design level. Finally, verification contains the set of testing methods that evaluate if the solutions proposed at each level meet the specifications and whether or not they alter the global system behavior.

Before finishing this brief review about the design process of integrated circuits it is important to point out the key role of the computer-aided design tools. CAD tools can be classified according to different concepts. On the one hand, they are usually associated with a particular abstraction level (functional, logical, electrical, or geometrical). On the other hand, most of these tools are clearly identified with a particular design step (synthesis, analysis, or verification). The synthesis tools ease the process of generating new designs while the analysis and verification tools permit their validation. The progress of CAD tools has run parallel with the evolution of techniques for integrated circuits. The increase of system complexity and the highly competitive market have motivated the development of a large number of CAD tools, usually embedded into so-called *design environments*. These tools often combine a considerable computing power with user-friendly interfaces. CAD tools for suitable design of fuzzy integrated circuits are described in Chapter 10.

6.3.4 ANALOG AND DIGITAL DESIGN STYLES

The differences among the possible styles of integrated circuit design lie in the nature of the signals selected to represent information. In the continuous-time analog design styles, the circuits work with voltages or currents that are continuous in the time domain and whose amplitudes can take a continuous value within a defined range. This idea is shown in Figure 6.6a. In the discrete-time analog design styles, the signals are not continuous in time but discrete and their amplitudes maintain a continuous range of values (Figure 6.6b-d). In the digital design styles, the signals are discrete both in time and amplitude (because they can only take two values, typically represented by a Boolean '0' or '1'). Clock signals, which do not appear in continuous-time techniques, are fundamental in discrete-time techniques. They control when to sample signals and when these samples should be transferred from one block to another. If the signal amplitude remains constant during a time interval (T/2, in Figure 6.6c

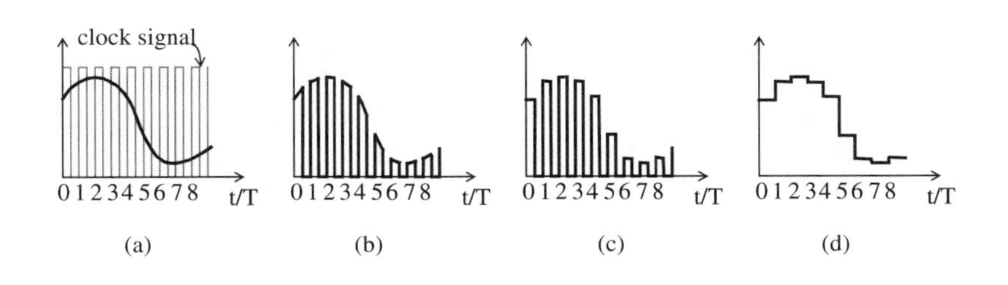

Figure 6.6. (a) Continuous-time signal of continuous amplitude. Discrete-time signals of continuous amplitude: (b) sampled; (c) of constant amplitude; and (d) sampled and held.

and T, in Figure 6.6d), the transference may take place at any instance of that interval. Once an instance is chosen, each processing block introduces a constant delay of T.

Digital circuitry currently supports most microelectronic systems. This is so because it provides greater precision than analog circuitry due to its higher robustness against noise, distortion, and fabrication process variations. Another reason is that the relations between specifications and design parameters are simple, so that the design process of digital circuitry is systematic and structured, making it possible to have many CAD tools, which greatly reduce development costs by quickly and automatically passing through all the design steps. In addition, digital design techniques offer efficient ways to store information, to program the resulting system, and to connect it to conventional digital processing environments.

The cost of developing analog circuitry is very much higher. On the one hand, hierarchical approaches are difficult, and the relations between specifications and design parameters are more complex, not only because the number of specifications is bigger but also because of the large number of design styles capable of performing the same functionality. Most analog realizations of fuzzy systems reported in the literature have been designed manually. Only a few works like [Lemaitre, 1994; Manaresi, 1996a; Franchi, 1998] describe an automatic design. As in many other analog design fields, automatic synthesis methods, which are already available in the digital domain, still have to be developed. A fact that illustrates this situation is that the analog fuzzy integrated circuits reported in the literature implement very much fewer rules (4 [Miki, 1993], 9 [Baturone, 1997], 15 [Manaresi, 1996b], 49 [Oehm, 1996]) than their digital counterparts (102 [Watanabe, 1990], 128 [Katashiro, 1993], 960 [Sasaki, 1993], 512 [Jiménez, 1995]).

Analog approaches are also less efficient concerning rearrangement and programmability. As an example, the implementation described in [Peters, 1995], admits programmability of the shapes and positions of the antecedent membership functions and the singleton values of the consequents by externally connecting analog voltage supplies. However, standard application environments usually do not contain many analog voltage supplies.

Another drawback of analog implementations is their relatively low precision. However, this does not seem a handicap since typical fuzzy logic applications [King, 1977; Li, 1989; Mizumoto, 1991; Boverie, 1991] employ universes of discourse with only 32 points, which is equivalent to working with 5 bits, and the resolution of the internal signals is 3 or 4 bits (a resolution of 10% is said to be enough in [Yamakawa, 1993]). On the other hand, it should be considered that the sensors used in many applications provide low resolution (5 or 6 bits).

As advantages of analog realizations, we can cite that they are preferred for their high speed and low area and power consumption when designing fully parallel and multivalued systems like fuzzy systems [Vittoz, 1994]. Analog fuzzy integrated circuits achieve response times below the microsecond (0.60 μs, in [Manaresi, 1994]; 0.10 μs, in [Lemaitre, 1994]; 0.16 μs, in [Peters, 1995]; 0.47 μs, in [Vidal-Verdú, 1996]; or 0.50 μs, in [Huertas, 1996]). An additional advantage is that these realizations process the rules in parallel so that these response times do not depend on the number of rules. The capability of analog hardware to obtain low area and power consumption fuzzy chips becomes apparent when it is noticed that the building blocks can be designed in general with very few transistors. An illustrative example is the block which connects the rule antecedents to a maximum or minimum operation. While the digital approach reported in [Watanabe, 1990] requires 98 transistors to connect 2 digital inputs of 4 bits, the analog approach reported in [Baturone, 1994] requires only 7 transistors to connect 2 inputs and may achieve equivalent resolutions of up to 8 bits. Examples of very low area occupation analog fuzzy chips are those reported in [Lemaitre, 1994] (0.4 mm^2 in a 1.2-μm CMOS technology) and in [Manaresi, 1994] (0.33 mm^2 in a 1.5-μm CMOS technology).

Finally, it is interesting to point out that the sensors and actuators employed in many applications are continuous, so that analog fuzzy chips are preferred because digital fuzzy chips would require A/D and D/A converters.

The solution adopted by several authors is to employ mixed-signal circuitry to combine the best of analog and digital domains [Huertas, 1992; 1993; 1994; Miki, 1993; Fattaruso, 1994; Baturone,1997; 1998]: analog circuitry to implement the fuzzy algorithm with parallelism, low area and power consumption, and digital circuitry to allow easy programmability, and an efficient mechanism to store the programmable values into on-chip memories, thus allowing the chip to stand alone and even to be tuned to dynamically changing environments.

- **Analog Design Styles**

As previously mentioned, the analog domain is characterized by the high number of possible design styles for the same functionality. Continuous or discrete-time techniques can be employed depending on the nature of the signal that holds information. Having chosen one of these techniques, current- or voltage-mode design styles can be used if the signals are currents or voltages, respectively. An intermediate design style is the transconductance-mode, which works with currents as inputs and voltages as outputs.

Continuous time techniques usually feature the best area occupation and inference speed, avoiding the need for control signals. Conventional continuous time schemes can be classified according to their operation mode: voltage, current, or transconductance mode. There are two voltage mode schemes. One scheme, which is called RC-Active, is based on resistors (R), capacitors (C) and single-ended amplifiers. The other scheme, widely known as MOSFET-C, employs MOS transistors, capacitors (C), and fully differential amplifiers as basic elements. A well-known transconductance-mode design style is the Gm-C technique, which relies on capacitors (C) and transconductance amplifiers (OTAs). The basic elements in the current-mode design are the MOS transistors mainly grouped into current mirrors.

Current-mode circuits are especially suitable for realizing fuzzy systems because the basic fuzzy logic operators can be implemented with a very few transistors. A clear example is the realization of addition and substraction which can be reduced to simple wire connections. Implementations of multi-input minimum/maximum and bounded-difference operations are also very simple with current-mode circuits. Besides, current mode designs offer several advantages such as being independent in a first approximation from temperature, robust to scaling of technologies, capable of operating at low voltage supplies, and adequate to interact with certain sensors (of temperature, light, etc.). This technique is currently in progress so that research is being done to alleviate its main drawbacks, in particular its low accuracy.

Another worthwhile feature of a fuzzy integrated circuit is its capability of interfacing with current devices. In this sense, although some sensors and actuators admit current-mode signals, most of the devices employed in the current applications use voltages as input/output signals. Hence, although the internal processing of the fuzzy chip is best performed in the current domain, the input/output blocks should, respectively, admit and provide voltage signals.

The two discrete-time techniques currently employed are the switched capacitors (SC) and the switched currents (SI) techniques. SC techniques correspond to voltage-mode designs. They were proposed in the 1960s to solve the problems that developed when designing filters with RC-Active continuous-time techniques [Allen, 1984]. SC techniques replace resistors by switches controlled by a clock signal so that the behavior of the resulting circuit is determined by the period of the clock signal and the capacitance ratios. These ratios can be easily programmed and allow a precision superior to that obtained with other analog techniques [Nakayama, 1987]. As an example, the works in [McCreary, 1981; Shyu, 1984] about device mismatching (which in the end is the factor that limits the accuracy of analog integrated circuits) conclude that capacitance ratios of small area occupation allow resolutions of 9 bits, somewhat bigger than the 8-bit resolution that can be achieved by the geometric ratios between transistors (upon which the programmability of many current-mode circuits is based) [Pelgrom, 1989]. In addition, there are many error cancellation techniques (amplifier and comparator offsets, influence of parasitic capacitances, switches feedthrough, etc.) that make it possible to have SC circuits with resolutions ranging from 10 to 16 bits [McCreary, 1975; Allen, 1984; Li, 1984; Gregorian, 1986; Piedade, 1990]. Since the accuracy demanded by many fuzzy logic–based applications is not usually so high, no complex design techniques are required in the realizations of SC fuzzy integrated circuits. Anyway, they are far from the simplicity of current-mode fuzzy chips that exploit the potentiality of MOS transistors to realize the required operations. SC implementations employ more complex blocks like operational amplifiers or comparators to realize operations, thus resulting in a higher area and power consumption. Another drawback of SC circuits is that they need linear capacitors, which are fabricated with two polysilicon layers so that special fabrication technologies need to be used. Research on this topic is aimed at developing structures based only on MOS transistors. Unlike digital circuits, SC circuits offer the advantage of being able to implement multiplication, addition, and delay operations in a single clock cycle, making it possible to have a high signal processing speed with power consumption and area occupation inferior to that of digital circuits.

SC techniques are being successfully applied to several engineering applications [Huertas, 1985; Rodríguez-Vázquez, 1987; Maundy, 1991]. This is justified not only by the development of efficient analysis and synthesis techniques but also by the advances in software tools, such as high-level simulators and layout tools. An example is Switcap2 [Suyama, 1992], which permits the simulation of complete systems in a few seconds. As a result, the design time cost of an SC fuzzy integrated circuit can be very low.

Switched current (SI) techniques began to be developed at the end of the 1980s to reduce the above-mentioned limitations of SC circuits. They are current-mode techniques, which are based on storing the electrical charge in the gates of MOS transistors (this is the basis of a memory cell), so that they are compatible with standard digital MOS technologies. Since they are recent techniques, they are not yet well established and several simulators and testing tools are being developed. Their main drawback is the low precision obtained in comparison with SC circuits. The objective of current research is to obtain memory cells with greater accuracy and better signal to noise ratio. For this to happen, the main causes of error, which are typical of both current-mode designs (transistor mismatching and low output resistances in the current mirrors) and discrete-time designs (switch feedthrough), have to be reduced [Toumazou, 1993].

REFERENCES

1. **Allen, P. E., Sánchez Sinencio, E.**, *Switched Capacitor Circuits*, Van Nostrand Reinhold, 1984.

2. **Baturone, I., Huertas, J. L., Barriga, A., Sánchez-Solano, S.**, Current-mode multiple-input maximum circuit, *Electronics Letters*, Vol. 30, N. 9, pp. 678-680, 1994.

3. **Baturone, I., Sánchez-Solano, S., Barriga, A., Huertas, J. L.**, Implementation of CMOS fuzzy controllers as mixed-signal IC's, *IEEE Transactions on Fuzzy Systems,* Vol. 5, N. 1, pp. 1-19, 1997.

4. **Baturone, I., Barriga, A., Sánchez-Solano, S., Huertas, J. L.**, Mixed-signal design of a fully parallel fuzzy processor, *Electronics Letters*, Vol. 34, N. 5, pp. 437-438, 1998.

5. **Baturone, I., Sánchez-Solano, S., Barriga, A., Huertas, J. L.**, Design issues for the VLSI implementation of universal approximator fuzzy systems, in *Proc. World MultiConference on Circuits, Systems, Communications and Computers*, pp. 6471-6476, Athens, 1999.

6. **Boverie, S., Demaya, B., Titli, A.**, Fuzzy logic control compared with other automatic control approaches, in *Proc. 30th Conference on Decision and Control*, pp. 1212-1216, Brighton, 1991.

7. **Costa, A., De Gloria, A., Faraboschi, P., Pagni, A., Rizzotto, G.**, Hardware solutions for fuzzy control, *Proceedings of the IEEE*, Vol. 83, N. 3, pp. 422-434, 1995.

8. **Eichfeld, H., Künemund, T., Klimke, M.**, An 8b fuzzy coprocessor for fuzzy control, in *Proc. IEEE Int. Solid-State Circuits Conference*, pp. 180-181, San Francisco, 1993.

9. **Eichfeld, H., Klimke, M., Menke, M., Nolles, J., Künemund, T.**, A general-purpose fuzzy inference processor, *IEEE Micro*, Vol. 15, pp. 12-17, 1995.

10. **Eichfeld, H., Künemund, T., Menke, M.**, A 12b general-purpose fuzzy logic controller chip, *IEEE Transactions on Fuzzy Systems*, Vol. 4, N. 4, pp. 460-475, 1996.

11. **Fattaruso, J. W., Mahant-Shetti, S. S., Barton, J. B.**, A fuzzy logic inference processor, *IEEE Int. Journal of Solid-State Circuits*, Vol. 29, N. 4, pp. 397-402, 1994.

12. **Franchi, E., Manaresi, N., Rovatti, R., Bellini, A., Baccarani, G.**, Analog synthesis of nonlinear functions based on fuzzy logic, *IEEE Int. Journal of Solid-State Circuits*, Vol. 33, N. 6, pp. 885-895, 1998.

13. **Gray, P. R., Meyer, R. G.**, *Analysis and Design of Analog Integrated Circuits*, John Wiley & Sons, 1993.

14. **Gregorian, R., Temes, G. C.**, *Analog MOS Integrated Circuits for Signal Processing*, Wiley-Interscience Pub., 1986.

15. **Huertas, J. L., Rodríguez-Vázquez, A., Rueda A., Chua, L. O.**, Nonlinear switched-capacitors networks: basic principles and piecewise-linear design, *IEEE Transactions on Circuits and Systems*, Vol. 32, pp 305-319, 1985.

16. **Huertas, J. L., Sánchez-Solano, S., Barriga, A., Baturone, I.**, Serial architecture for fuzzy controllers: hardware implementation using analog/digital VLSI techniques, in *Proc. 2nd. Int. Conference on Fuzzy Logic and Neural Networks*, pp. 535-538, Iizuka, 1992.

17. **Huertas, J. L., Sánchez-Solano, S., Barriga, A., Baturone, I.**, A fuzzy controller using switched-capacitor techniques, in *Proc. 2nd. IEEE Int. Conference on Fuzzy Systems*, pp. 516-520, San Francisco, 1993.

18. **Huertas, J. L., Sánchez-Solano, S., Barriga, A., Baturone, I.**, A hardware implementation of fuzzy controllers using analog/digital techniques, *Int. Journal of Computers and Electrical Engineering*, Vol. 20, N. 5, pp. 409-419, 1994.

19. **Huertas, J. L., Sánchez-Solano, S., Baturone, I., Barriga, A.**, Integrated circuit implementation of fuzzy controllers, *IEEE Int. Journal of Solid-State Circuits*, Vol. 31, pp. 1051-1058, 1996.

20. **Jiménez, C. J., Sánchez-Solano, S., Barriga, A.**, Hardware implementation of a general purpose fuzzy controller, in *Proc. 6th IFSA World Congress*, pp. 185-188, Sao Paulo, 1995.

21. **Katashiro, T.**, A fuzzy microprocessor for real-time control applications, in *Proc. 5th IFSA World Congress*, pp. 1394-1397, Seoul, 1993.

22. **Kim Y. D., Lee-Kwang, H.**, High speed flexible fuzzy hardware for fuzzy information processing, *IEEE Transactions on Systems, Man, and Cybernetics*, Vol. 27. N, 1, pp. 45-55, 1997.

23. **King, P. J., Mamdani, E. H.**, The application of fuzzy control systems to industrial processes, *Automatica*, Vol. 13, pp. 235-242, 1977.

24. **Lee, S. H. Lee, Y., Bien, Z.**, FLEXi: Fuzzy logic controller with expandable architecture, in *Proc. 3rd Int. Conference on Fuzzy Logic, Neural Nets and Soft Computing*, pp. 549-550, Iizuka, 1994.

25. **Lemaitre, L., Patyra, M. J., Mlynek, D.**, Analysis and design of CMOS fuzzy logic controller in current mode, *IEEE Int. Journal of Solid-State Circuits*, Vol. 29, N. 3, pp. 317-322, 1994.

26. **Leong, J. Y., Lim, M. H., Lau, K. T.**, A general approach to encoding heuristics on programmable logic devices, in *Proc. 5th IFSA World Congress*, pp. 917-920, Seoul, 1993.

27. **Li, P. W., Chin, M. J., Gray, P. R., Castello, R.**, A ratio-independent algorithmic analog-to-digital conversion technique, *IEEE Int. Journal of Solid-State Circuits*, Vol. 19, pp. 828-836, 1984.

28. **Li, Y. F., Lau, C. C.**, Development of fuzzy algorithms for servo systems, *IEEE Control Systems Magazine*, Vol. 9, N. 2, pp. 65-72, 1989.

29. **Manaresi, N., Franchi, E., Guerrieri, R., Baccarani, G.**, A modular analog architecture for fuzzy controllers, in *Proc. ESSCIRC'94*, pp. 288-291, Ulm, 1994.

30. **Manaresi, N., Rovatti, R., Franchi, E., Guerrieri, R., Baccarani, G.**, Automatic synthesis of analog fuzzy controllers: a hardware and software approach, *IEEE Transactions on Industrial Electronics*, Vol. 43, N. 1, pp. 217-225, 1996a.

31. **Manaresi, N., Franchi, E., Guerrieri, R., Baccarani, G.**, A field programmable analog fuzzy processor with enhanced temperature performance, in *Proc. ESSCIRC'96*, pp. 152-155, Neuchatel, 1996b.

32. **Matas, J., García de Vicuña, L., Castilla, M.**, A synthesis of fuzzy control surfaces in CMOS technologies, in *Proc. IEEE Int. Conference on Fuzzy Systems*, Vol. 2, pp. 641-647, Barcelona, 1997.

33. **Maundy, B. J., El-Masry, E. I.**, Feedforward associative memory switched-capacitor artificial neural networks, *Analog Integrated Circuits and Signal Processing*, Vol. 1, pp. 321-338, 1991.

34. **McCreary, J. L., Gray, P. R.**, All-MOS charge redistribution analog-to-digital conversion techniques-Part I, *IEEE Int. Journal of Solid-State Circuits*, Vol. 10, N. 6, pp. 371-379, 1975.

35. **McCreary, J. L.**, Matching properties and voltage and temperature dependence of MOS capacitors, *IEEE Int. Journal of Solid-State Circuits*, Vol. 16, N. 6, pp. 608-616, 1981.

36. **Menke, M., Brunner, R., Eichfeld, H., Künemund, T.**, A general purpose 12b fuzzy coprocessor, in *Proc. ESSCIRC'95*, pp. 118-121, Lille, 1995.

37. **Miki, T., Matsumoto, H., Ohto, K., Yamakawa, T.**, Silicon implementation for a novel high-speed fuzzy inference engine: mega-flips analog fuzzy processor, *Journal of Intelligent and Fuzzy Systems*, Vol. 1, N. 1, pp. 27-42, 1993.

38. **Mizumoto, M.**, Min-max-gravity method versus product-sum-gravity method for fuzzy controls, in *Proc. 4th IFSA World Congress*, pp. 127-130, Brussels, 1991.
39. **Nakamura, K., Sakashita, N., Nitta, Y., Shimomura K., Tokuda, T.**, Fuzzy inference and fuzzy inference processor, *IEEE Micro*, Vol. 13, N. 5, pp. 37-48, 1993.
40. **Nakayama, K., Kuraishi, Y.**, Present and future applications of switched-capacitor circuits, *IEEE Circuit & Devices Magazine*, Vol. 3, pp. 10-21, 1987.
41. **Oehm, J., Grafe, M., Kettner, T., Schumacher, K.**, Universal low cost controller for electric motors with programmable characteristics curves, *IEEE Int. Journal of Solid-State Circuits*, Vol. 31, N. 7, pp. 1041-1045, 1996.
42. **Pagni, A.**, Digital approaches, in *Handbook of Fuzzy Computation*, Institute of Physics Publishing, 1998.
43. **Patyra, M. J., Braun, E.**, Fuzzy/Scalar RISC processor: architectural level design and modeling, in *Proc. IEEE Int. Conference on Fuzzy Systems*, pp. 1937-1943, New Orleans, 1996
44. **Pelgrom, M. J., Duinmaijer, A. C. J., Welbers, A. P. G.**, Matching properties of MOS transistors, *IEEE Int. Journal of Solid-State Circuits*, Vol. 24, N. 5, pp. 1433-1440, 1989.
45. **Peters , L., Guo, S.**, A high-speed, reconfigurable fuzzy logic controller, *IEEE Micro*, Vol. 15, pp. 1-11, 1995.
46. **Piedade, M. S., Pinto, A.**, A new multiplier-divider circuit based on switched capacitor data converters, in *Proc. IEEE Int. Symposium on Circuits and Systems*, pp. 2224-2227, New Orleans, 1990.
47. **Rodríguez-Vázquez, A., Huertas, J. L., Rueda, A. Pérez-Verdú, B., Chua, L. O.**, Chaos in switched-capacitors: discrete maps, *Proceedings IEEE,* Vol. 75, pp 1090-1107, 1987.
48. **Rovatti, R., Vittuari, M.**, Linear and fuzzy piecewise-linear signal processing with an extended DSP architecture, in *Proc. IEEE Int. Conference on Fuzzy Systems*, pp. 1082-1087, Anchorage, 1998.
49. **Rovatti, R.**, High speed implementation of piecewise-quadratic Takagi–Sugeno systems with memory saving, in *Proc. World MultiConference on Circuits, Systems, Communications and Computers*, pp. 6451-6454, Athens, 1999.
50. **Sasaki, M., Ueno, F., Inoue, T.**, An 8-bit resolution 140 KFLIPS fuzzy microprocessor, in *Proc. 5th IFSA World Congress*, pp. 921-924, Seoul 1993.
51. **Shimizu, K., Osumi, M., Imae, F.**, Digital fuzzy processor FP-5000, in *Proc. 2nd Int. Conference on Fuzzy Logic & Neural Networks*, pp. 539-542, Iizuka, 1992.
52. **Shyu, J. B., Temes, G. C., Krummenacher, F.**, Random error effects in matched MOS capacitors and current sources, *IEEE Int. Journal of Solid-State Circuits*, Vol. 19, N. 6, pp. 948-955, 1984.
53. **Surmann, H., Ungering, A. P.**, Fuzzy rule-based systems on general-purpose processors, *IEEE Micro*, Vol. 15, N. 4, pp. 40-48, 1995.
54. **Suyama, K., Fang, S.C.**, Users' manual for Switcap2. Version 1.1, University of Columbia, 1992.
55. **Toumazou, C., Hughes, J. B., Battersby, N. V.**, Eds, *Switched-Currents: An Analogue Technique for Digital Technology*, IEE Circuits and Systems Series 5, Peter Peregrinus Ltd. 1993.
56. **Ungering A. P, Bauer, H., Goser, K.**, Architecture of a fuzzy-processor based on an 8-bit microprocessor, in *Proc. IEEE Int. Conference on Fuzzy Systems*, pp. 297-301, Orlando, 1994a.
57. **Ungering A. P, Goser, K.**, Architecture of a 64-bit fuzzy inference processor, in *Proc. IEEE Int. Conference on Fuzzy Systems*, pp. 1776-1780, Orlando, 1994b.

58. **Vidal-Verdú, F.**, Una aproximación al diseño de controladores neuro-difusos usando circuitos integrados analógicos, Ph.D. dissertation, Universidad de Málaga, 1996.

59. **Vittoz, E.A.**, Analog VLSI signal processing: why, where, and how? in *Analog ICs and Signal Processing,* Vol. 6, pp. 27-44, Kluwer Academic Pub., 1994.

60. **von Altrock, C.**, Adapting existing hardware for fuzzy computation, in *Handbook of Fuzzy Computation*, Institute of Physics Publishing, 1998.

61. **Watanabe, H., Dettloff, W., Yount, K. E.**, A VLSI fuzzy logic controller with reconfigurable, cascadable architecture, *IEEE Int. Journal of Solid-State Circuits*, Vol. 25, N. 2, pp. 376-382, 1990.

62. **Watanabe, H.**, Hardware approach to fuzzy inference - ASIC and RISC, in *Proc. 5th IFSA World Congress*, pp. 975-976, Seoul, 1993.

63. **Yamakawa, T.**, A Fuzzy inference engine in nonlinear analog mode and its application to a fuzzy logic control, *IEEE Transactions on Neural Networks*, Vol. 4, N. 3, pp. 496-522, 1993.

64. **Yen, J., Langari, R., Zadeh, L. A.**, Eds., *Industrial Applications of Fuzzy Logic and Intelligent Systems*, IEEE Press, 1995.

65. **Zeng, X. J., Singh, M. G.**, Approximation accuracy analysis of fuzzy systems as function approximators, *IEEE Transactions on Fuzzy Systems*, Vol. 4, N. 1, pp. 44-63, 1996.

Chapter 7

CONTINUOUS-TIME ANALOG TECHNIQUES FOR DESIGNING FUZZY INTEGRATED CIRCUITS

Continuous-time analog techniques offer a very good relation between area occupation and inference speed. This is achieved, on the one hand, by eliminating the need for control signals and, on the other hand, by fully exploiting the potentiality of the basic element of a MOS integrated circuit which is the MOS transistor. Continuous-time analog techniques were employed to implement the first analog fuzzy integrated circuits. This is described in Section 7.1 The chapter continues in Section 7.2 with the description of three basic architectures of fuzzy integrated circuits that process the rules in parallel. These architectures can be implemented with analog or digital techniques. The advantage of using continuous-time analog techniques is that the operations required within each rule can also be realized in parallel with circuits that admit multiple inputs, consume low power, and occupy small silicon area. After the different architectures have been explained, the next three sections are dedicated to describing the design of the basic building blocks in these architectures. Section 7.3 shows the circuits (MFCs) that can be employed in the fuzzification stage to generate the membership functions of the fuzzy sets covering the input universes of discourse. Section 7.4 describes the circuits that can be employed in the rule processing stage to implement the basic t-norm and s-norm operators of a fuzzy system, such as minimum, maximum, product (scaling), and addition. Finally, the circuits required in the output stage of a fuzzy system are analyzed in Section 7.5. As a summary of the chapter, Section 7.6 describes the design of a CMOS programmable prototype that implements a simplified fuzzy inference system with 9 rules and 2 inputs. Experimental results of this prototype illustrate the operations performed in each stage as well as the input/output surfaces obtained for different programmable codes. This chapter focuses on MOS technologies, which are cheaper than bipolar ones and fully compatible with digital designs. Also, some bipolar circuits are described to provide the reader with an unified view of this design style.

7.1 THE FIRST ANALOG FUZZY INTEGRATED CIRCUITS

The first microelectronic realizations of fuzzy logic–based systems with analog techniques were carried out by Yamakawa [Yamakawa, 1987; 1988]. These ICs implement Min-Max Mamdani's method with the Center-of-Area defuzzification method. Hence, they manage fuzzy sets to represent both the antecedents and the consequents of the rules.

The membership functions that define the antecedent fuzzy sets are trapezoidal. They are generated by the input/output transfer function of blocks called MFCs (Membership Function Circuits). This type of MFCs will be referred to as arithmetic-type MFCs because they implement several arithmetic operations to obtain the membership degree. Their constituent elements are bipolar transistors, resistors, diodes, and programmable switches that allow changing the location of the membership function in the universe of discourse, its support (that is, the input values whose membership degree is not null), and its slopes (see Figure 7.1a).

The membership functions that define the consequent fuzzy sets are also trapezoidal. They are provided by blocks called MFGs (Membership Function Generators). These MFGs are not arithmetic but will be referred to as memory-based MFGs because they are memories that store the membership degrees corresponding to all the discrete points into which the output universe of discourse is divided (see Figure 7.1b). These data are provided in parallel by a 25-wire bus.

One of the main features of the first analog fuzzy ICs (a one-rule chip with three antecedents and one consequent, and a defuzzifier chip) is their fully parallel architecture, as shown in Figure 7.2. Both the rules and the operations within each rule are processed in parallel. One of the MIN blocks calculates in parallel the minimum of the MFC outputs. This value, which represents the activation degree of the rule, is employed by the other MIN block to cut the consequent membership function provided by the MFG via the 25-wire bus. The inference results of each rule are combined in parallel by an array of MAX operators. The outputs of this array enter into adder and adder/scaler circuits, which are connected in turn to an analog divider. The response time of the rule chip is 1μs while that of the defuzzifier chip is 5μs. The silicon area occupied by both circuits is rather large (64 mm^2 in a 5-μm BiCMOS technology) due to the fully parallel architecture, in particular due to the MAX operator array and to the 25 lines that handle the discrete output universe of discourse. This is why

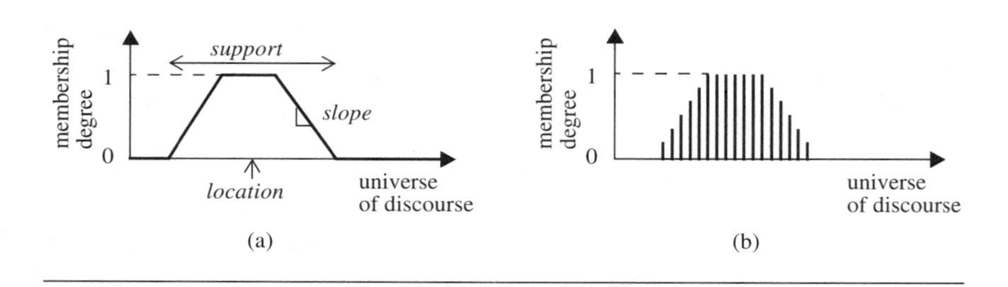

Figure 7.1. (a) Parameters that define the trapezoidal membership functions generated by arithmetic MFCs. (b) Trapezoidal membership functions generated by memory-based MFGs.

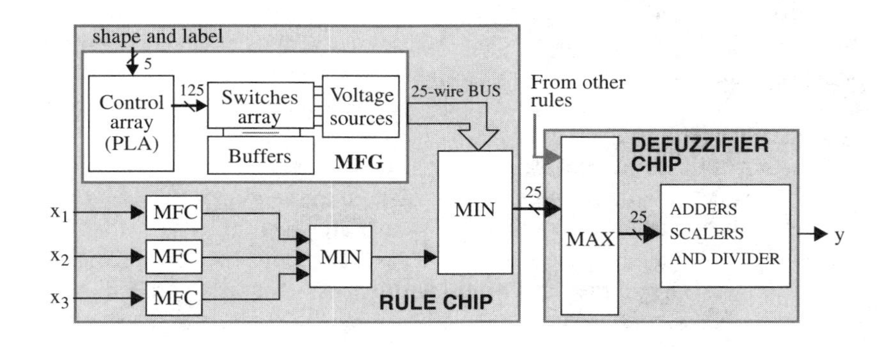

Figure 7.2. Parallel architecture proposed in [Yamakawa, 1987].

the solutions to solve control problems reported in [Yamakawa, 1993] rely on the connection of several rule chips.

The following section describes in greater detail this and other types of fully parallel architectures.

7.2 FULLY PARALLEL ARCHITECTURES

From an architectural point of view, a relevant feature of fuzzy systems is their inherent parallelism, as mentioned in the previous chapter. On the one hand, several rules are activated for each input combination. On the other hand, several data can be processed in parallel within each rule, like the antecedents and the discrete points of the output universe of discourse. To avoid the need of sweeping the points of the output universe of discourse, most analog fuzzy ICs implement a zero-order Takagi–Sugeno or singleton fuzzy system. If we recall Section 3.6 in Chapter 3, this type of system employs constant or singleton values, c_i, to define their rule consequents and provides the following output:

$$y_{out} = \frac{\sum_{i=1}^{R} c_i \cdot h_i}{\sum_{i=1}^{R} h_i} \tag{7.1}$$

where h_i represents the activation degree of the i-th rule.

Besides parallelism, another important issue at the system level is the chip programmability. The parameters that can be programmed in a fuzzy IC are the following: (a) the number of fuzzy sets to cover the rule antecedent; (b) the parameters that define their membership functions; (c) the number of fuzzy sets employed for the rule consequents; (d) the parameters that define their membership functions; and (e) the parameters that outline the rule base. If the fuzzy IC is not programmable, all these parameters are fixed during the fabrication process, otherwise it requires programmability circuitry whose complexity increases as the number of programmable parameters is bigger.

Three types of architectures are described in the following. The first type contains a data path for each rule (rule by rule architecture). The second one shares circuitry among different rules, and the third architecture only processes the active rules. One of the figures that will be considered when evaluating the efficiency of these architectures is the programming capability of the resulting fuzzy IC.

7.2.1 RULE BY RULE ARCHITECTURE

This architecture fixes the maximum number of rules, R, and inputs, u, that the IC will be able to manage and associates a data path for each rule. The latter means that each rule is implemented with its own MFCs, connective circuits, and MFGs, that is, $u \cdot R$ MFCs, R connective circuits, and R MFGs are required, as shown in Figure 7.3. The first analog fuzzy ICs employ this architecture, as was described in the section above, and also later realizations, such as those reported in [Sasaki, 1992a; Landolt, 1993; Miki, 1993; Manaresi, 1994; Franchi, 1998].

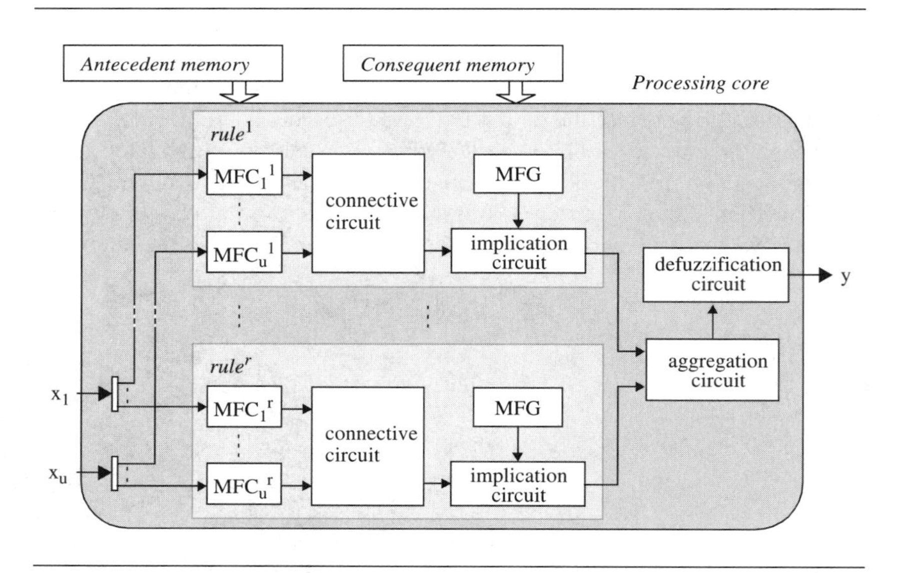

Figure 7.3. Rule by rule architecture.

The ICs described in [Miki, 1993] and [Franchi, 1998] are programmable realizations that contain memory circuitry to store the programmable parameters of the MFCs and MFGs. The memory circuitry employed in [Miki, 1993] is based on shift registers with a master-slave configuration, so that the chip can operate with a program while another is being loaded. Shift registers are much more area and power consuming than RAMs, although this is not a problem if the chip contains a low number of rules. The work in [Franchi, 1998] shows a chip that employs RAMs to program the antecedents and consequents. The programmable values are generated with a resistor ladder. These values are 75 voltage levels for the singleton values of the consequents (because the chip implements a zero-order Takagi–Sugeno or singleton fuzzy system) and 100 voltage levels to program the antecedent membership functions. Each of these values is connected to either all the MFCs (if corresponding to the antecedents) or to all the MFGs (if corresponding to the consequents), as shown in Figure 7.4. The RAM stores all the bits that control these switches. Considering a general case in which the a parameters that define the antecedents are programmed by p bits, thus giving 2^p possible values, and that the singletons are programmed by q bits, which means 2^q possible values, the number of bits and hence the number of control bits is $u \cdot R \cdot a \cdot 2^p + R \cdot 2^q$.

This architecture offers the highest flexibility to define the rule base. In addition, it is very modular and, consequently, suitable for admitting an automatic IC design process, like that described in [Franchi, 1998]. On the other hand, one of the main drawbacks is that several rules usually employ the same antecedent fuzzy sets or consequent values, so that some MFCs and MFGs would be repeated, thus managing a redundant information. Another important drawback is that the programming circuitry (the shift registers and in particular the array of switches) occupies a very large area (for instance, the array of switches of the chip described in [Franchi, 1998] occupies 93% of the total chip area).

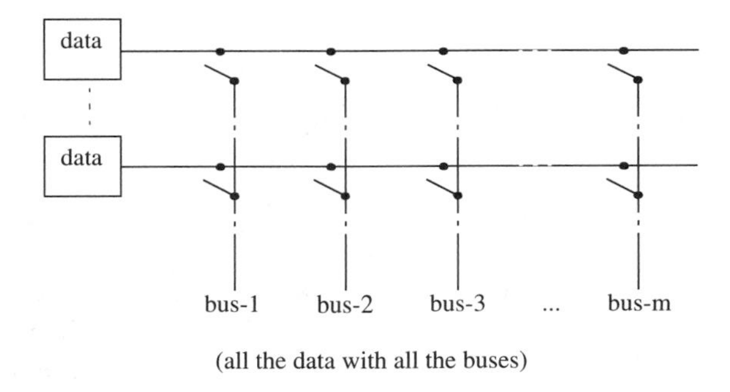

Figure 7.4. Array of programmable switches.

7.2.2 ARCHITECTURE THAT SHARES CIRCUITRY

Sharing the MFCs and MFGs among the rules is the most efficient solution regarding area and power consumption for non-programmable ICs. Let us consider that L is the number of fuzzy sets per input (the same number is taken for all the inputs for the sake of simplicity), u is the number of inputs, and M is the number of fuzzy sets (or singleton values) for the output. Compared with a rule by rule architecture, an architecture that shares circuitry reduces the number of MFCs from $u \cdot R$ to $u \cdot L$, and the number of MFGs from R to M (with L and M being smaller than R). If this architecture that shares circuitry is used to implement programmable fuzzy ICs, complexity increases because memory circuits are required to define not only the antecedent and consequent membership functions (like in a rule by rule architecture) but also the rule base, that is, the MFCs and MFGs associated with each rule. Besides, if the same flexibility of a rule by rule architecture is pursued, additional circuitry should be added to program the number of rules of the knowledge base (R or inferior) and the number of antecedents of each rule (u or inferior).

An architecture of this type has been proposed in [Baturone, 1997a] to implement current-mode fuzzy ICs that work with the Fuzzy Mean as defuzzification method and the minimum operator as antecedent connective. The block diagram of this architecture is shown in Figure 7.5. The $\overline{\text{MFCs}}$ provide the complement of the membership degrees. The combination MAX-COMPL is used to implement a minimum operation, because a maximum (MAX) circuit is simpler to realize with current-mode techniques than a minimum (MIN) circuit, as will be shown later in this chapter. The block called COMPL, which implements a complement operation, is placed after the MAX block. Thus, the output signal of the k-th COMPL block is given by:

$$Ih_k = \overline{max[\overline{I\mu_k}(x_i), \ldots, \overline{I\mu_k}(x_j)]} = I_{ref} - max[\overline{I\mu_k}(x_i), \ldots, \overline{I\mu_k}(x_j)] \quad (7.2)$$

where $\overline{I\mu_k}(x_i)$ is the output signal provided by the $\overline{\text{MFC}}$ block associated with the i-th antecedent of the k-th rule and where I_{ref} is the value of the current representing the maximum membership degree, '1'.

By applying De Morgan's Laws, it follows that:

$$Ih_k = \overline{max[\overline{I\mu_k}(x_i), \ldots, \overline{I\mu_k}(x_j)]} = min[I\mu_k(x_i), \ldots, I\mu_k(x_j)] \quad (7.3)$$

so that Ih_k represents the activation degree of the k-th rule.

In order to fix a number of R rules with u inputs, R connective circuits (MAX-COMPL) with u inputs should be employed. The array of the A-switches, which appears at the top of Figure 7.5, is included because the u antecedents of each rule can be represented by any one of the L fuzzy sets. There are $u \cdot L$ switches per rule which are controlled by a digital word with no active bit or only one active bit. The position of the active bit indicates the fuzzy set that represents that antecedent.

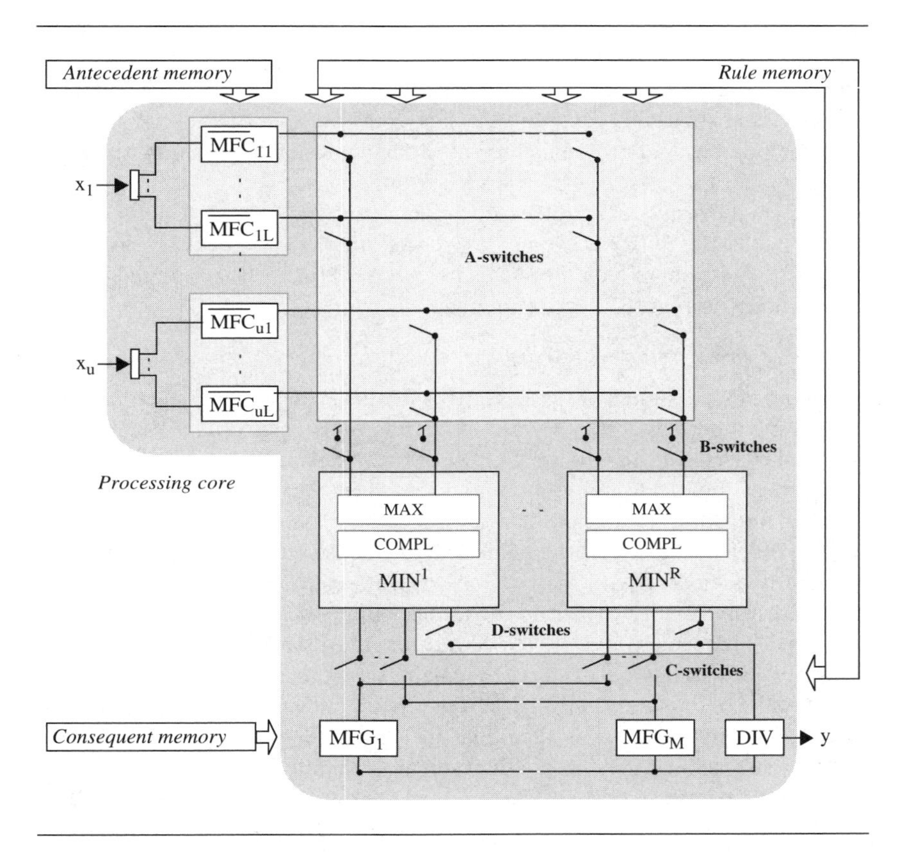

Figure 7.5. Architecture that shares circuitry to implement a programmable fuzzy IC. (©1997 IEEE. Reprinted with permission from [Baturone, 1997a]).

The purpose of the B-switches shown in the middle of Figure 7.5 is to program the number of antecedents in the rules. For instance, if switch B_i^n is closed, the antecedent i is not considered in the rule n, because it makes the input current to the MAX block be null (that is, $\overline{I\mu} = 0$, and consequently, the membership degree to that antecedent is maximum). When a B-switch is closed, all the A-switches connected to the same node have to be opened (i.e., the previously mentioned word of $u \cdot L$ bits must not have any active bit).

The C-switches shown at the bottom of Figure 7.5 are in charge of selecting the consequent singleton value of each rule. A special case is that all C-switches are open, which forces the selected singleton value, c_k, to be zero.

Finally, the D-switches are included to eliminate a rule from the knowledge base. A rule is eliminated when its corresponding C- and D-switches are open (so that $c_k = 0$ and $Ih_k = 0$).

This architecture is programmed by three buses, as shown in Figure 7.5. Considering the same notation as in the rule by rule architecture, the antecedent bus that programs the antecedents has a width of $u \cdot L \cdot a \cdot p$ bits, the width of the consequent bus that programs the singleton values has a width of $M \cdot q$ bits, and the rule bus that programs the knowledge base (switches A, B, C, and D) has a width of $u \cdot L \cdot R + u \cdot R + M \cdot R + R$ bits.

Since the arrays of switches can again occupy a large area, a good trade-off between flexibility and complexity is to implement the whole rule base ($R = L^u$) and to share only the MFCs but not the MFGs because in this case all the switches in Figure 7.5 are eliminated.

7.2.3 ACTIVE RULE-DRIVEN ARCHITECTURE

One of the features of fuzzy systems that has been already commented on is that the input variables only activate a few rules of the knowledge base. In particular, if α is the maximum number of fuzzy sets per input that may overlap each other, the maximum number of active rules is α^u. An active rule-driven architecture only contains the circuitry to identify the active rules and the computing blocks needed for processing only the active rules. Hence, if this architecture is compared with the previous one, the number of MFCs required is reduced from $L \cdot u$ to $\alpha \cdot u$, and the number of connective circuits and MFGs required is reduced from L^u to α^u. As a result, the complexity of the processing core depends mainly on u and α but remains basically the same independent of the number of labels per input, L, and of the total rule number, L^u (as an example, the processing core is the same for ICs implementing 3×3=9 rules or 13×13=169 rules, if u and α are the same in both cases).

A fully parallel architecture to implement programmable fuzzy ICs following this idea of active rules has been proposed in [Baturone, 1998a]. Figure 7.6 illustrates the block diagram of this architecture for a singleton fuzzy system that has two inputs with a maximum overlapping of two and one output. The data that define the membership functions of the $2L$ labels are stored in the memories X_i-Mem, while those that define the singleton values of the consequents are stored in the memory Y-Mem. As will be shown in the following, these memories are adequately organized to provide all the required parameters in parallel without replicating information or employing costly arrays of switches.

The membership degrees of each input variable to the active fuzzy sets are obtained by the transfer functions of α MFCs. The global size of the memory X-Mem that stores the parameters of the L membership functions of each input is $a \cdot L$ words, where a is the number of parameters to program the membership functions. Each memory X_i-Mem is divided into α conventional RAMs that store the parameters of those membership functions that are never active simultaneously. As an example, Figure 7.7 shows the partition of the X_1-Mem for different coverings of the input universe of discourse, x_1. Figure 7.7a illus-

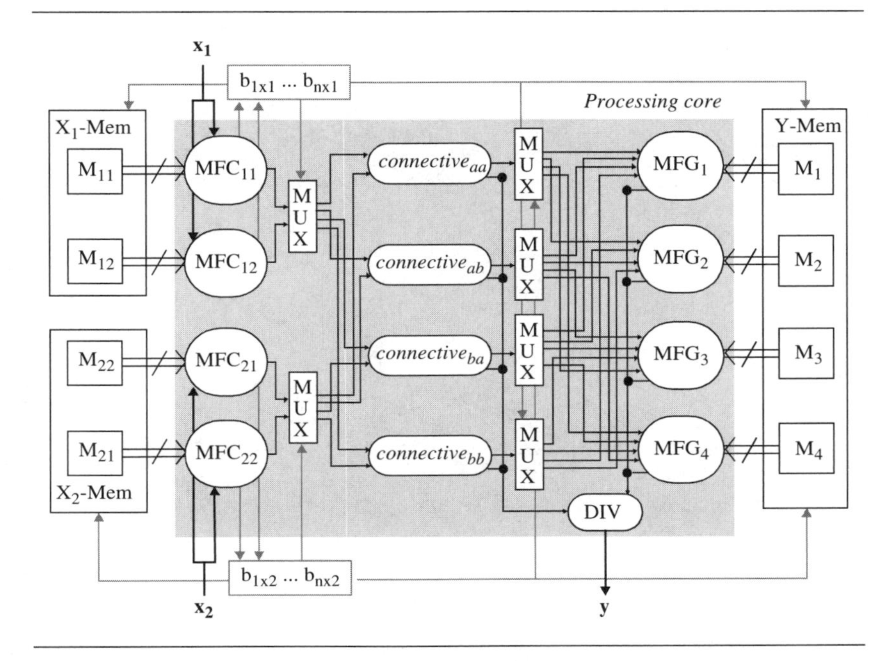

Figure 7.6. Fully-parallel active rule-driven architecture.
(©1998 IEE. Reprinted with permission from [Baturone, 1998a]).

trates the case of 5 fuzzy sets with a maximum overlapping degree of 2, L=5 and α=2. In this case, the X_1-Mem consists of two RAMs. One of them, called M_{11}, which stores the $3 \cdot a$ words associated with the membership functions L_1, L_3, and L_5, and the other, called M_{12}, which stores the other $2 \cdot a$ words, as shown at the upper part of Figure 7.7.

The possible combinations of active fuzzy sets are $L+1-\alpha$. For instance, if L=5 and α=2, there are 4 combinations: L_1 with L_2, L_2 with L_3, L_3 with L_4, and L_4 with L_5 (bottom part of Figure 7.7a); if L=9 and α=3, there are 7 combinations (bottom part of Figure 7.7b) or if L=9 and α=4, there are 6 combinations (bottom part of Figure 7.7c). Therefore, given an input x_i, the required words that define the α active fuzzy sets are retrieved from X_i-Mem by a code of n bits, $\{b_{1xi}, ..., b_{nxi}\}$ (shown in Figure 7.6), where n is the integer number greater than or equal to $\log_2(L+1-\alpha)$. In the example of Figure 7.7a, for instance, two bits, b_1b_2, address the global memory X-Mem. Thus, the membership degree calculation is performed in parallel because the set of parameters that define the α active fuzzy sets per input are obtained with only one access to the memories, X_i-Mems.

Each code $\{b_{1xi}, ..., b_{nxi}\}$ results from comparing the corresponding input x_i with the centers of the membership functions covering the i-th input space.

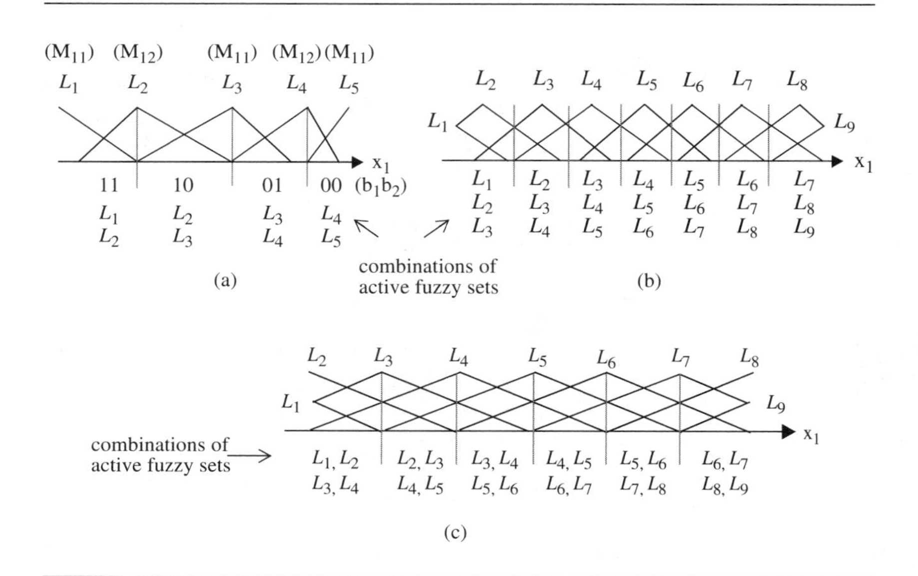

Figure 7.7. Examples of active fuzzy sets for different coverings of the input space (different number of labels, L, and overlapping degrees, α): (a) L=5, α=2; (b) L=9, α=3; (c) L=9, α=4.

Let us call these centers I_{aux}. In the examples in Figure 7.7, the input should be compared with $\{I_{aux2}, I_{aux3}, I_{aux4}\}$ (Figure 7.7a), with $\{(I_{aux2}+I_{aux3})/2, ..., (I_{aux7}+I_{aux8})/2\}$ (Figure 7.7b), or with $\{I_{aux3}, I_{aux4}, I_{aux5}, I_{aux6}, I_{aux7}\}$ (Figure 7.7c).

Let us consider that $\{L_{ax1}, L_{bx1}\}$ and $\{L_{ax2}, L_{bx2}\}$ are the two combinations of fuzzy sets that are activated by the inputs x_1 and x_2 (the I_{aux} of L_a is supposed to be smaller than that of L_b). The 4 connective circuits connect these labels orderly, that is, $connective_{aa}$ connects L_{ax1} with L_{ax2} and so on. The blocks called MUX, which follow the MFCs, identify which of these MFCs represents L_a and which represents L_b. The MUX are controlled by the $\log_2\alpha$ least significant bits of the corresponding code $\{b_{1xi}, ..., b_{nxi}\}$. In the example of Figure 7.7a, $\log_2\alpha$ is 1, so that the least significant bit, b_2, controls which MFC provides L_{ax1} or L_{bx1}. If b_2 is '1', MFC_{11} provides L_{ax1} while MFC_{12} provides L_{bx1} and vice versa if b_2 is '0'.

The L^u digital words that define all the singleton values are stored in the memory Y-Mem. This memory is divided into α^u conventional RAMs, each of them storing the singleton values that are never active simultaneously. Let us consider the rule base illustrated in Figure 7.8, which corresponds to a system with 2 inputs, 5 labels per input (L^u = 25 rules) and a maximum overlapping degree of 2 (as the one indicated in Figure 7.7a). Since α^u = 4, each of the 4

RAMs of Y-Mem stores 9, 6, 6, and 4 digital words, respectively (M_4, for instance, stores c_7, c_9, c_{17}, and c_{19}). Given a set of inputs $\{x_1, ..., x_u\}$, the α^u required singleton values are retrieved from the α^u RAMs by the complete code $\{b_{1x1}, ..., b_{nx1}, ..., b_{1xu}, ..., b_{nxu}\}$. Thus, the required α^u singleton values are retrieved in parallel, that is, with only one access to the global memory Y-Mem. As an example, the singleton values of the RAM M_4 would be selected, according to Figure 7.8a, by the following signals: c_7 by $b_{1x1} \cdot b_{1x2}$, c_9 by $\bar{b}_{1x1} \cdot b_{1x2}$, c_{17} by $b_{1x1} \cdot \bar{b}_{1x2}$, and c_{19} by $\bar{b}_{1x1} \cdot \bar{b}_{1x2}$.

The MUX blocks that follow the connective circuits identify which activation degree of the α^u active rules have to be multiplied (scaled) by which active singleton value. These blocks are controlled by the $u \cdot \log_2\alpha$ least significant bits of the complete code $\{b_{1x1}, ..., b_{nx1}, ..., b_{1xu}, ..., b_{nxu}\}$. Figure 7.8b illustrates the scheme of the MUX that follows the block *connective*$_{bb}$. For the example of active rules shown in Figure 7.8a, the code $\{b_{1x1}, b_{2x1}, b_{1x2}, b_{2x2}\}$ is {0101}. According to Figure 7.8a, the block *connective*$_{bb}$ should be connected

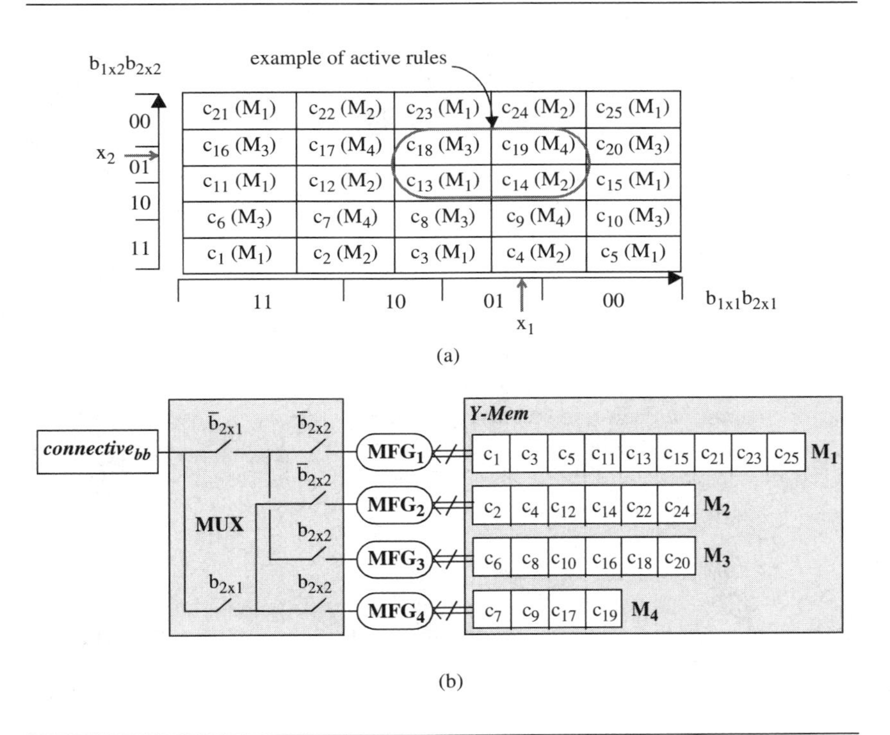

Figure 7.8. (a) Example of how the consequent memory is partitioned into 4 RAMs; (b) Scheme of the MUX that follows the block *connective*$_{bb}$.

Table 7.1. Resources required by singleton fuzzy systems.

u	α	L	rules	MFCs	MFGs	memory sizes (number of words)									
						each antecedent		all the singleton values							
						M1	M2	M1	M2	M3	M4	M5	M6	M7	M8
2	2	5	25	4	4	3	2	9	6	6	4				
2	2	8	64	4	4	4	4	16	16	16	16				
3	2	9	729	6	8	5	4	125	100	100	100	80	80	80	64
3	2	10	1000	6	8	5	5	125	125	125	125	125	125	125	125

to MFG_4, which in turn should be programmed by the singleton value c_{19}. Since $\{b_{2x1}, b_{2x2}\}$ is $\{11\}$, Figure 7.8b shows that the block $connective_{bb}$ is connected correctly to MFG_4 and since $\{b_{1x1}, b_{1x2}\}$ is $\{00\}$, the data which receives MFG_4 is c_{19}, according to what is stated in the paragraph above.

Table 7.1 shows the number of computing blocks and the size of the digital memories required to implement different singleton fuzzy systems with this active rule-driven architecture. The number of words stored in each antecedent memory is always L and that stored in the consequent memory is always L^u, so that no information is replicated to achieve parallel retrieval. In addition, the number of computing blocks is small. In order to implement a complex fuzzy system with an IC, this architecture is the most efficient of the three described in this section.

The three architectures described herein process the rules in parallel and their processing core can be realized with analog or digital techniques. The advantage of using continuous-time analog techniques is that the operations to be performed within each rule can be realized in parallel in a very efficient way thanks to the use of multiple input circuits that offer low area and power consumption. The design of these circuits is detailed in the following sections.

7.3 FUZZIFICATION STAGE

The analog MFCs generate a membership function with their input/output transfer characteristic, so that they are arithmetic MFCs. Two different MFCs are described in this section: transconductance-mode MFCs, which work with a voltage as the input signal and provide a current as the output signal, and current-mode MFCs, whose input and output are currents. Selection of the most suitable will depend on the nature of the input signals to the fuzzy IC. An input signal represented by a voltage can be drawn to several transconductance-mode MFCs with the same wire, without need of replication. On the contrary, an input signal represented by a current only can drive a single current-mode MFC, that is, the fan-out is 1. In order to connect a current-mode input signal to L MFCs the current should be replicated L times.

7.3.1 TRANSCONDUCTANCE-MODE MFCS

Generic bell-shaped membership functions, like those illustrated in Figure 7.9a and b, can be obtained by adding pieces usually known as S or Z pieces [Yamakawa, 1985]. The result can be a shifted membership function, from which some offset should be subtracted (Figure 7.9a), or the complement of a membership function, which should be complemented later (Figure 7.9b). S or Z pieces can be generated with two MOS transistors connected as a source-coupled current-biased differential pair (Figure 7.10a). Most of the transconductance-mode MFCs reported in the literature are based on this type of differential pair. The differences among them are the consequence of employing the two transistors operating in strong or weak inversion.

- **Transistors operating in weak inversion**

Pair differential-based circuits with MOS transistors operating in weak inversion have been employed to generate non-linear generic functions [Turchetti, 1993] and Gaussian-like functions. The latter have been employed in the field of both analog neural networks [Mead, 1989; Watkins, 1992] and analog fuzzy ICs [Landolt, 1996]. The equation that models the behavior of a MOS transistor working in weak inversion is the following [Mead, 1989]:

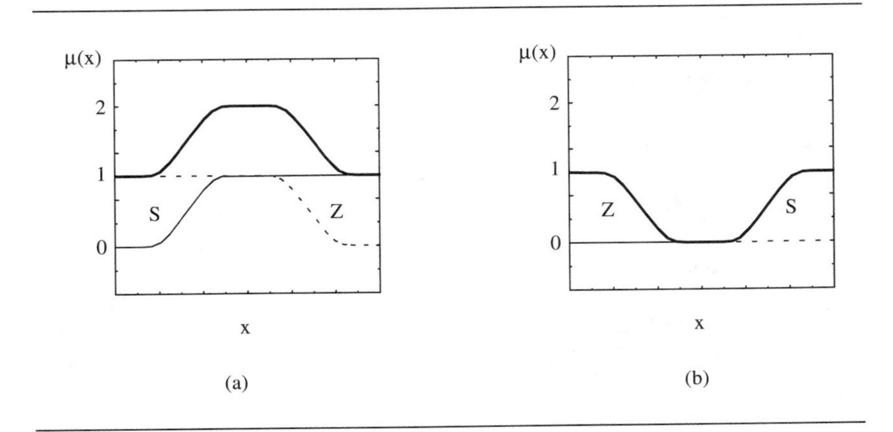

Figure 7.9. The bold lines show a bell-shaped function resulting from adding S and Z pieces: (a) a shifted function and (b) a complemented function.

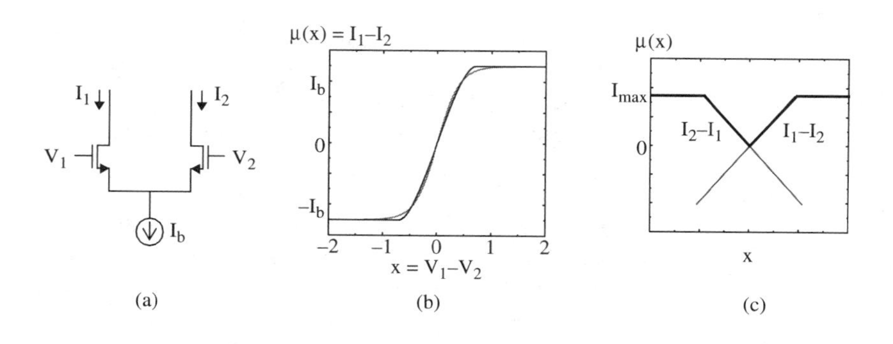

Figure 7.10. (a) Differential pair; (b) Transfer functions generated by a differential pair working in weak inversion (discontinuous line) and in strong inversion (continuous line); (c) Complemented triangular function resulting from calculating the maximum of I_1–I_2, I_2–I_1, and I_{max}.

$$I_{DS} = I_o \cdot e^{\frac{k \cdot V_{GS}}{U_T}} \cdot \left(1 - e^{\frac{-V_{DS}}{U_T}}\right) \cong I_o \cdot e^{\frac{k \cdot V_{GS}}{U_T}} \tag{7.4}$$

where k and I_o are constants that depend on the process while U_T is a voltage equal to $K_B \bullet T / q$, where K_B is Boltzmann's constant, T is the absolute temperature and q is the charge of an electron. V_{GS} represents the gate-to-source voltage of the transistor and I_{DS} is the current entering through the drain terminal into the transistor.

Applying the former equation and Kirchhoff's current law ($I_b = I_1 + I_2$) to the differential pair in Figure 7.10a, the following expressions are obtained:

$$I_{1,2} = \frac{I_b}{2} \pm \frac{I_b}{2} \cdot \tanh\left[\frac{k(V_1 - V_2)}{2U_T}\right] \tag{7.5}$$

$$I_1 - I_2 = I_b \cdot \tanh\left[\frac{k(V_1 - V_2)}{2U_T}\right] = f_1(I_b, V_1 - V_2) \tag{7.6}$$

The shape of function f_1 is illustrated in Figure 7.10b with a discontinuous line. It corresponds to an S piece. A Z piece is obtained from the difference I_2–I_1.

- **<u>Transistors working in strong inversion</u>**

Transistors working in strong inversion can be biased in either ohmic or saturation regions. The MFCs reported in the literature employ mostly the latter ones [Sasaki, 1992b; Chen, 1992; Tsukano, 1995; Vidal-Verdú, 1995; 1996], so we will focus on them in the following. The equation that models the behavior of an MOS transistor working in the saturation region is not exponential but quadratic:

$$I_{DS} = \frac{\beta}{2} \cdot (V_{GS} - V_T)^2 \tag{7.7}$$

This is Schichmann–Hodges' model in which V_T is the threshold voltage of the transistor and $\beta = C_{ox} \cdot \mu \cdot W/L = K_P \cdot W/L$ is its transconductance (with μ being the carrier mobility; C_{ox}, the gate capacitance per unit of area, and W and L, respectively, the width and length of the transistor channel).

Applying this equation to the transistors of the differential pair in Figure 7.10a, the following expressions are obtained:

$$I_{1,2} = \frac{I_b}{2} \pm I_b \sqrt{\frac{\beta}{4I_b}} (V_1 - V_2) \sqrt{1 - \frac{\beta}{4I_b}(V_1 - V_2)^2} \quad \text{if } |V_1 - V_2| \leq \sqrt{\frac{2I_b}{\beta}} \tag{7.8}$$

$$I_1 - I_2 = \sqrt{\beta I_b}(V_1 - V_2) \sqrt{1 - \frac{\beta}{4I_b}(V_1 - V_2)^2} = f_2(I_b, V_1 - V_2) \tag{7.9}$$

Over the upper bound $|V_1 - V_2| = (2I_b / \beta)^{1/2}$, one of the transistors is off and all the current I_b flows through the other. The shape of the function f_2 is shown with a continuous line in Figure 7.10b. It corresponds to an S piece.

The non-linearities generated by a differential pair with MOS transistors operating in weak inversion (function f_1 in Equation (7.6)) or MOS transistors operating in the saturation region of strong inversion (function f_2 in Equation (7.9)) carry out a qualitatively similar covering of the universe of discourse, as shown in Figure 7.10b. The difference lies in the range of values of I_b for which the circuit operates in weak inversion (range of nanoamperes or inferior) or in strong inversion (range of microamperes or superior). Consequently, working in strong inversion permits higher operation speed, a larger dynamic range, and greater robustness against noise, although the power consumption is superior. Due to these advantages, many analog fuzzy ICs are designed to operate in strong inversion.

In order to obtain triangular membership functions, some authors have proposed working with the central lineal portions of the S and Z pieces that are obtained from the current differences I_1–I_2 and I_2–I_1 of a saturated differential pair [Sasaki, 1992a], as shown in Figure 7.10c. Maximum and complement operators are employed in these realizations.

The location within the universe of discourse and the width of a membership function generated by a transconductance-mode MFC can be programmed by the voltage values at the gates of the transistors of the pair. Since the gate node is a high impedance node, there are no interface problems with the programming circuitry. The slopes of the functions can be controlled either by the geometries of the transistors of the pair or by analog signals if additional circuitry is added to the basic differential pair. The reader is referred to [Chen, 1992; Tsukano, 1995; Vidal-Verdú, 1995; 1996] for more details on these issues.

7.3.2 CURRENT-MODE MFCS

Other authors [Yamakawa, 1985; Huertas, 1992; Baturone, 1992; Kettner, 1993a; Lemaitre, 1994] have adopted current-mode designs in which the input variables to the MFCs as well as those that control the membership functions provided are represented by currents. As mentioned previously, the fan-out when working in current mode is 1. Consequently, if a block is to drive several other blocks, its output current should be replicated the same number of times. The current mirror is a basic block in a current-mode design since it permits the replication of currents, the inversion of their direction (from incoming to outgoing or vice versa), and their scaling. The simplest current mirror is shown in Figure 7.11. The input/output relation provided is:

$$I_o = \frac{\beta_2}{2} \cdot (V_{GS1} - V_T)^2 = \frac{\beta_2}{\beta_1} I_i = \frac{W_2 \cdot L_1}{L_2 \cdot W_1} I_i = w \cdot I_i \qquad (7.10)$$

Current mirrors that offer better performance than simple current mirrors are summarized in Figure 7.12 [Babanezhad, 1987; Sackinger, 1990; Crawley, 1992; Nairn, 1995].

Current-mode MFCs are able to generate easily any piecewise linear membership function. Considering the membership functions more widely employed such as trapezoidal and triangular functions (which can be seen as a special case of trapezoidal ones), the current-mode MFCs reported in the literature can be basically classified into three general types.

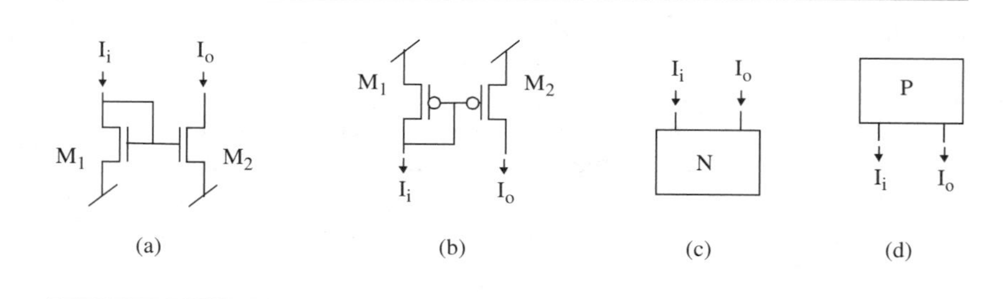

Figure 7.11. (a) n-type and (b) p-type simple current mirrors; (c) Blocks employed in this book to represent n-type and (d) p-type current mirrors.

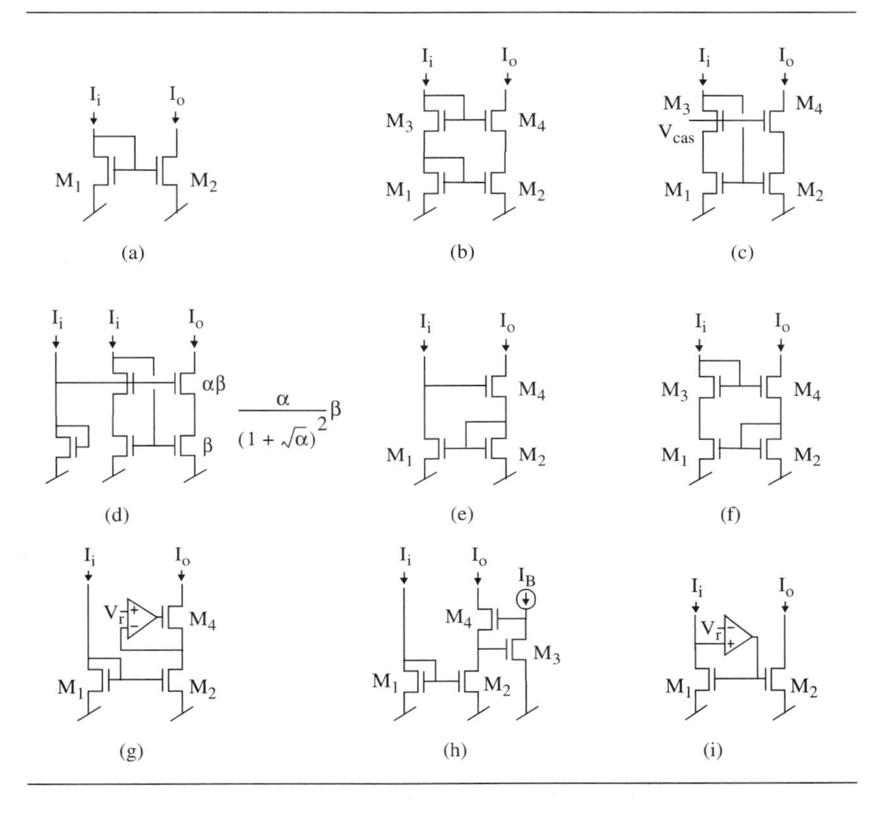

Figure 7.12. Summary of n-type current mirrors: (a) simple; (b) conventional cascode; (c) externally and (d) adaptively biased cascode; (e) Wilson; (f) improved Wilson; (g) and (h) regulated and (i) active.

The first type performs the fuzzification operation according to the following expression:

$$\mu\,(I_{in}) = I_{ref}\,\Theta\,[\,m_2(I_{in}\,\Theta\,I_2) + m_1(I_1\,\Theta\,I_{in})\,] \qquad (7.11)$$

where the bounded difference operator, Θ, is defined as:

$$x\,\Theta\,y = \begin{cases} x - y & \text{if } x > y \\ 0 & \textit{ioc} \end{cases}$$

Figure 7.13a shows a trapezoid which results from adding the functions m_2 $(I_{in}\,\Theta\,I_2)$ and m_1 $(I_1\,\Theta\,I_{in})$, and subsequently performing the bounded difference with I_{ref}. Figure 7.13b illustrates the schematics of the MFC proposed in [Yamakawa, 1985] whose transfer function corresponds to Equation (7.11).

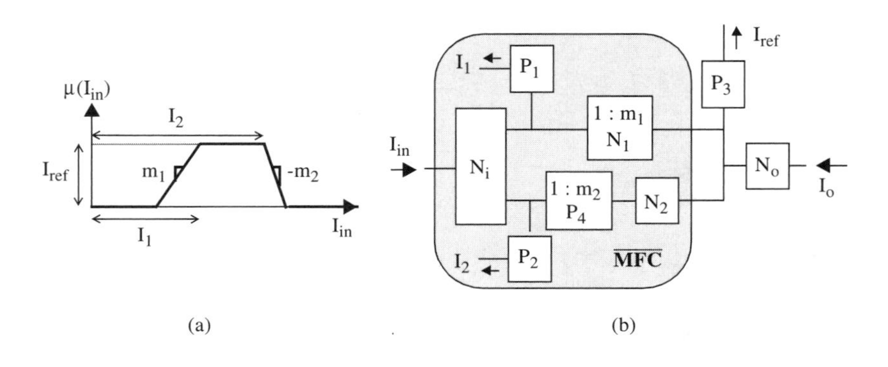

Figure 7.13. (a) Parameters that define a trapezoidal membership function; (b) MFC that implements Equation (7.11).

The blocks that contain an "N" represent n-type current mirrors, while those that contain a "P" refer to p-type mirrors. Current mirrors are very useful not only for replicating and scaling currents but also to implement bounded differences taking advantage of the fact that the input current to a mirror can only take one direction. For instance, the mirror N_1 in Figure 7.13b only accepts as input an entering current. Hence its input current is $(I_1 \ominus I_{in})$. On the other hand, the p-mirror P_4 only admits a leaving current as input so that its input current is $(I_{in} \ominus I_2)$. In addition, these mirrors N_1 and P_4 realize the scaling by m_1 and m_2, respectively. The output current of P_4, $m_2(I_{in} \ominus I_2)$, is inverted by the mirror N_2 to be added to $m_1(I_1 \ominus I_{in})$. Finally, the mirror N_o performs the bounded difference with I_{ref}.

The second type of MFCs employs the technique used in [Kettner, 1993a], which combines bounded-difference expressions like $m_x(I_{in} \ominus I_x)$, one expression per each breakpoint I_x. This is illustrated in Figure 7.14a for the case of trapezoidal functions. The fuzzification operation performed corresponds to the following expression:

$$\mu(I_{in}) = [\, m_1(I_{in} \ominus I_1) - m_1(I_{in} \ominus I_2)\,] \ominus m_2(I_{in} \ominus I_3) \tag{7.12}$$

Figure 7.14b shows how the equation above can be realized with current mirrors. The bounded differences $I_{in} \ominus I_2$ and $I_{in} \ominus I_3$ are implemented by the mirrors P_4 and P_5, respectively. The mirrors P_6 and N_1 are used to implement the scaling by m_1 and m_2, respectively. In addition, P_6 also performs at its input the operation $(I_{in} \ominus I_1) - (I_{in} \ominus I_2)$. Finally, the mirror N realizes the bounded difference between $m_1[(I_{in} \ominus I_1) - (I_{in} \ominus I_2)]$ and $m_2(I_{in} \ominus I_3)$.

The third type of MFCs employs the structure proposed in [Huertas, 1992; Baturone, 1992] to generate trapezoidal or triangular membership functions with the following expression:

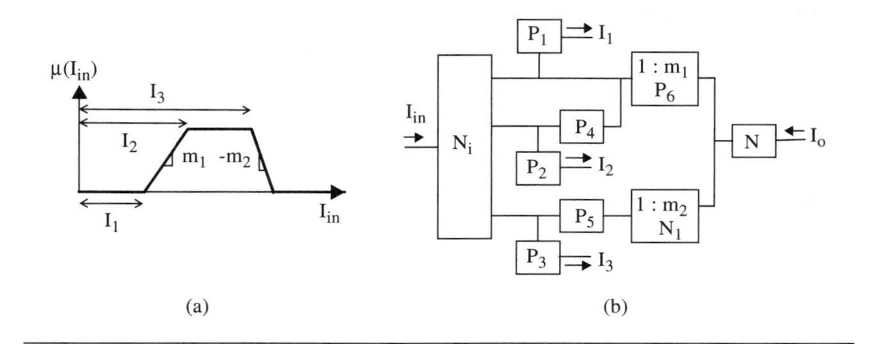

Figure 7.14. (a) Parameters that define a trapezoidal membership function; (b) Realization of Equation (7.12) with current mirrors.

$$\mu\,(I_{in}) = I_{ref}\ \Theta\ m_b[\ m_a(\ I_{in}\ \Theta\ I_{aux}\ ) + (\ I_{aux}\ \Theta\ I_{in}\ )\ \Theta\ I_{sat}\ ] \tag{7.13}$$

where the parameters I_{ref}, I_{aux}, I_{sat}, m_a, and m_b are depicted in Figure 7.15a.

Figure 7.15b shows the block diagram of this structure. The first operation that is carried out is the double bounded difference operation between I_{in} and I_{aux}, $I_{in}\ \Theta\ I_{aux}$, and $I_{aux}\ \Theta\ I_{in}$. This is realized by a current switch, which consists of the NMOS and the PMOS transistors shown in Figure 7.15b [Wang, 1990a]. If I_{in} is bigger than I_{aux}, the PMOS transistor is off and the current $I_{in}\ \Theta\ I_{aux}$ flows through the NMOS transistor. If I_{in} is smaller than I_{aux}, the NMOS transistor is off and the current $I_{aux}\ \Theta\ I_{in}$ flows through the PMOS transistor. Thus,

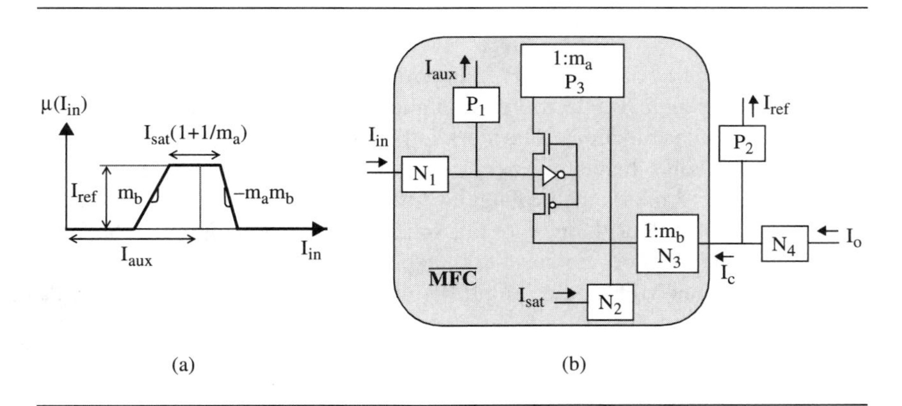

Figure 7.15. (a) Parameters that define a trapezoidal function; (b) MFC that implements Equation (7.13).

the current I_{aux} controls the location of the membership function. The mirror P_3 scales the signal $I_{in} \Theta I_{aux}$ by m_a (m_a is 1 for symmetric trapezoids) and draws its output to the input of the mirror N_3, where it is added with the current $I_{aux} \Theta I_{in}$. The mirror N_3 is employed to realize the bounded difference with I_{sat} and to scale its input signal by m_b. The mirror N_4 realizes the bounded difference with I_{ref}. As a consequence of these arithmetic operations, the current I_{sat} determines the upper basis of the trapezoid while the current I_{ref} determines its height (that is, the maximum membership degree) and, indirectly, the bottom basis.

These three types of MFCs have been compared in terms of accuracy, area occupation, power consumption, and speed in [Sánchez-Solano, 1992; Huertas, 1994]. The conclusion of this work is that the third MFC described is the most efficient one. The main reasons for this result are that the third MFC replicates a smaller number of currents, that linearity is required for a lower range of currents (below $I_{ref} + I_{sat}$), and that it is more robust against errors. The basic building blocks are current mirrors and a current switch. Precision is limited by the random or systematic mismatching of the transistors that form the current mirrors. Random mismatching is primarily due to variations in the threshold voltages of supposedly identical transistors [Nairn, 1990] and can be reduced by using large devices [Pelgrom, 1989]. Systematic mismatching is caused by the finite output resistance of the mirrors and by the difference between the voltages values at the input and output nodes of the mirror. This is the most influential error of the third MFC described. If the transistors that constitute the current switch are biased with the same gate voltage, as in [Wang, 1990a], the voltage swing at the input node is at least $V_{Tn}+|V_{Tp}|$, where V_T is the threshold voltage of the transistor. Voltage swing can be reduced if the transistors are biased with different voltages, V_{bn} and V_{bp}, provided that $V_{bn} - V_{bp} < V_{Tn}+|V_{Tp}|$. Another solution proposed in [Domínguez-Castro, 1992] is to include negative feedback by means of a simple inverter, as shown in Figure 7.15b.

The structure of this third MFC can be simplified if its flexibility in covering the input space is reduced. For example, the current I_{sat} is not needed to generate triangular functions. The work in [Manaresi, 1997] proposes a daisy-chain fuzzification scheme to generate a covering of the input space with regularly spaced and overlapped triangular functions, such as those indicated in Figure 7.16a. The block diagram of this scheme is shown in Figure 7.16b-c. Its advantage is that I_{in} does not have to be replicated to drive different MFCs.

Current-mode MFCs admit digital programmability by adding very little circuitry, because the outputs of the mirrors can be controlled by switches, that is, the mirrors that determine the location, saturation, or slopes of the membership functions can be directly transformed into simple D/A converters, as shown in Figure 7.17.

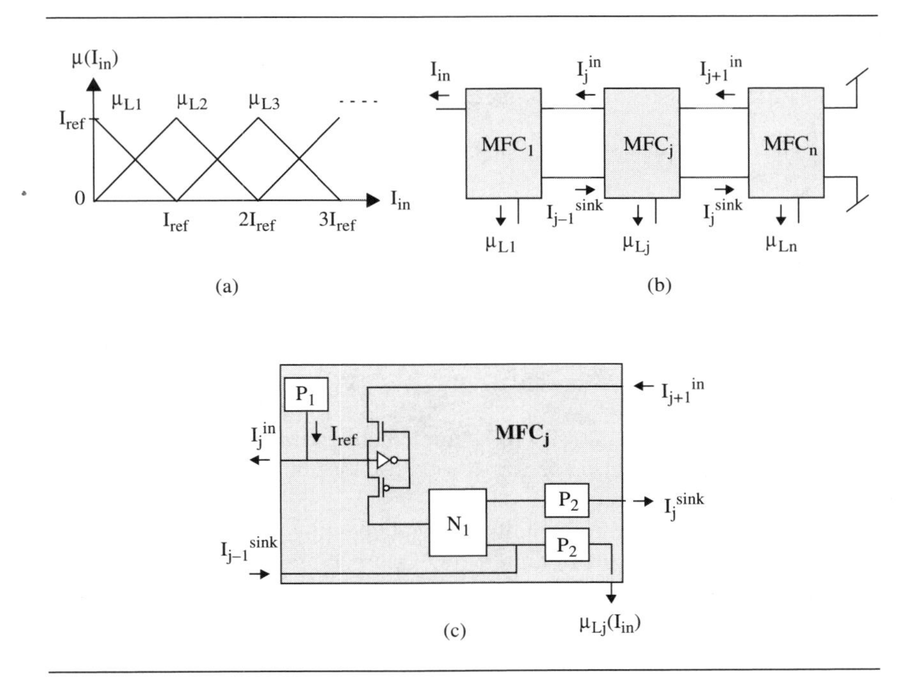

Figure 7.16. (a) Regular covering of the input space; (b) Daisychain fuzzification scheme proposed in [Manaresi, 1997]; (c) Block diagram of each chain link.

7.4 RULE PROCESSING STAGE

The operations that have to be implemented in the rule processing stage are the antecedent combination to obtain the activation degrees of the rules, the implication between antecedents and consequents, and the rule aggregation. Depending on the inference method, Min-Max or Product-Sum, the usual implication operators are the minimum and the product while the aggregation operators are the maximum or the sum. The connective operator among antecedents is a t-norm. Among the t-norms that have been implemented with analog hardware we can cite the widely used minimum, the product (which is more difficult to implement), and the normalized conjunction. The latter was proposed in [Sasaki, 1994] for the case of two inputs, and was extended to multiple inputs in [Rovatti, 1998]. Its definition for two inputs is as follows:

$$T(\mu_1, \mu_2) = \frac{1}{2}[min(\mu_1, \mu_2) + max(0, \mu_1 + \mu_2 - 1)] \tag{7.14}$$

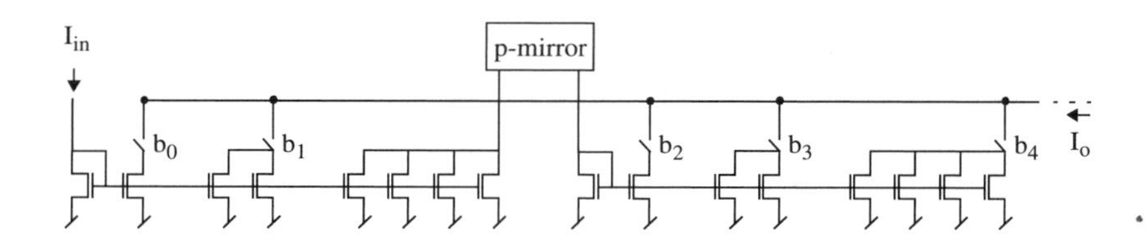

Figure 7.17. Digitally programmable current mirror(D/A): $I_o=(b_0+2b_1+4b_2+8b_3+16b_4+...)I_{in}$.

$$\text{with } (\mu_1, \mu_2) \in [0, 1]$$

As explained in [Manaresi, 1997], implementing this operator is not practical for more than two inputs because its expression and hence the circuitry required is very complex.

Continuous-time analog circuits for implementing these basic operations in the rule processing stage: minimum, maximum, product (scaling as a particular case), and addition are described in the following sections.

7.4.1 MINIMUM AND MAXIMUM OPERATORS

MIN/MAX circuits are discussed jointly in this section because their realizations are either complements or one of them can be obtained from the other by applying De Morgan's Laws:

$$Min(x_i) = \overline{Max(\bar{x}_i)} = x_{ref} - Max(x_{ref} - x_i) \tag{7.15}$$

where x_{ref} represents the maximum membership degree, '1'.

- **<u>Two-input circuits</u>**

The first fuzzy CMOS ICs employed two-input MIN/MAX circuits [Yamakawa, 1985; Zhijian, 1990]. These circuits are very easily designed with current-mode techniques because they require only addition/subtraction and bounded-difference operations:

$$I_{min}(I_1, I_2) = I_2 - (I_2 \Theta I_1) \tag{7.16}$$

$$I_{max}(I_1, I_2) = I_2 + (I_1 \Theta I_2) \tag{7.17}$$

As we have seen when current-mode MFCs were explained, addition/subtraction is easily implemented by connecting wires (the outputs of current mirrors, for instance) while bounded-difference operations can be implemented with current mirrors or current switches. An example that illustrates the use of

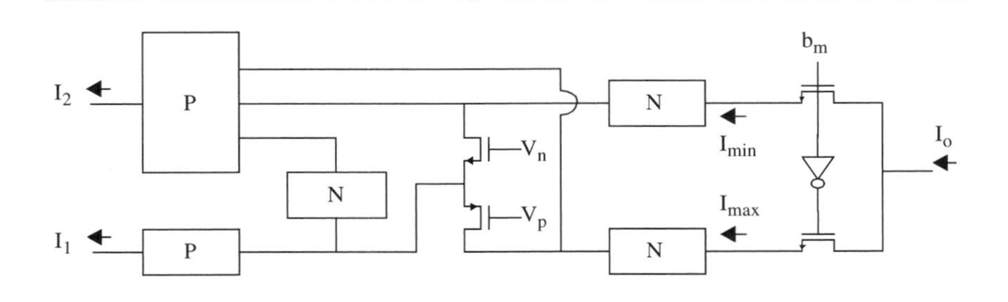

Figure 7.18. Programmable MIN/MAX circuit proposed in [Huertas,1994].

current mirrors and current switches for implementing minimum/maximum operations is the circuit shown in Figure 7.18. This circuit provides the minimum or the maximum of the two input currents depending on the value of the control bit b_m.

• **Multi-input circuits**

There are several cases where multi-input MIN/MAX circuits are required. Mamdani's systems, for instance, require multi-input MAX circuits to aggregate all the rules and any fuzzy system also needs multi-input MIN/MAX circuits whenever many antecedents need to be connected with "and"/"or" connectives within each rule. Multi-input circuits can be obtained from two-input circuits by placing them in series or forming a binary tree. The problem is the accumulation of errors and delays.

Two algorithms have been proposed in the literature to calculate the maximum value of several inputs that eliminate the above problem. They are the following:

$$y = \Sigma_i \, [x_i \ominus (y - \xi_i)] \text{ with } \xi_i = x_i \ominus \Sigma_{j \neq i} \, \xi_j \tag{7.18}$$

$$y = A \, \Sigma_i \, (x_i \ominus y) \tag{7.19}$$

The first algorithm was described in [Sasaki, 1990] and the current-mode realization shown in Figure 7.19 was proposed. In this structure, each current is replicated to inhibit all the other currents. The result of this competition is that all the diode-connected transistors are off except the one that receives the maximum input current. All the transistors connected to this one by its gate (excluding the output transistor) operate in the ohmic region and conduct the rest of the inputs that do not have another conducting path. The resulting structure is of $O(N^2)$ complexity (being N the number of inputs) so that it occupies a large area and it offers a low operation speed due to the numerous parasitic capacitances.

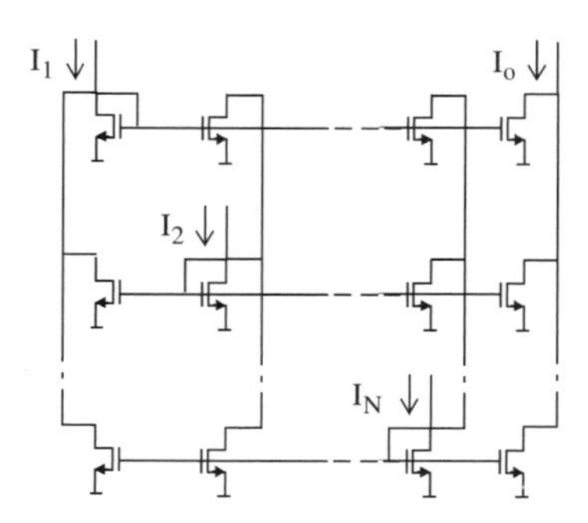

Figure 7.19. Multi-input MAX circuit proposed in [Sasaki, 1990].

The second algorithm is proposed in [Inoue, 1991]. The inhibition signals are more easily generated in this case because they do not involve internal variables such as ξ_i in Equation (7.18). Since this structure allows order-N complexity, the rest of this section focuses on realizations of this algorithm.

Conventional diode circuit

The conventional structure of a multi-input MAX circuit is the diode circuit shown in Figure 7.20a, which is used to implement OR operations for Boolean logic. The maximum input controls the output voltage because its corresponding diode is directly biased while the others are inversely biased. If the orientation of the diodes is inverted, the resulting structure is a MIN circuit. Bipolar transistors biased in the direct-active operation region behave as diodes between their base and emitter terminals. Hence, they can be employed to realize MIN/MAX circuits as shown in Figure 7.20b [Yamakawa, 1987; Gilbert, 1990]. The same structure, but replacing bipolar transistors with MOS transistors working in strong inversion, has also been used as a multi-input MAX circuit (Figure 7.20c) [Peters, 1995]. However, since the equations that govern the behavior of bipolar and MOS transistors in strong inversion are different, the performance of a bipolar circuit cannot be generally reproduced by using MOS instead of bipolar transistors. In this case of MIN/MAX circuits, the errors of the MOS circuit are even one order of magnitude higher than those of the bipolar circuit. The errors of the MOS circuit can only be reduced by reducing the ratio I_{bb1} / β, where β is the MOS transistor transconductance. The problem is that the speed of the circuit is also decreased considerably. As a consequence, other techniques are more adequate for improving the precision of a MAX circuit with MOS transistors working in strong inversion. The traditional solution for achieving greater resolution with the conventional diode circuit is to include the diodes within the feedback loop of an amplifier, as shown in Figure

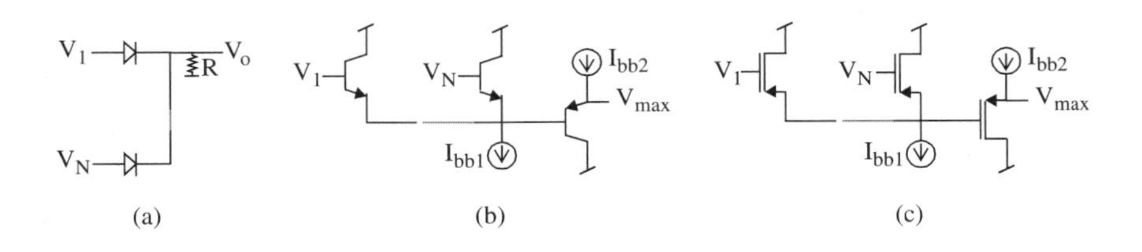

Figure 7.20. Conventional multi-input MAX circuits: (a) with diodes; (b) with bipolar transistors; and (c) with MOS transistors.

7.21 [Graeme, 1977]. This solution reduces the circuit errors by a factor of A, being A the amplifier gain. Since there is a trade-off between resolution and speed, the speed can be increased by a factor of A for the same resolution. Operation at this potentially high speed is not usually possible because this circuit usually requires the inclusion of a dominant pole at very low frequencies to achieve stability. Another drawback of this amplification scheme is that it occupies a large area. These problems can be avoided by adopting simpler amplification schemes, as discussed in the following section.

Current-mode circuits

The simplest amplification circuit in MOS technology is based on a transistor in source-common configuration (transistor M_i in Figure 7.22a). Since this amplifier is not differential, the voltage imposed at node *x* is the voltage required by M_i to operate in the saturation region when conducting a current I_i [Lazzaro, 1989]. If several of these cells are connected, as illustrated in Figure 7.22b, the voltage at node *x* takes the value that transistor M_{max} (the one which receives the maximum current, I_{max}) needs to work in the saturation region. The other M_i transistors operate in the ohmic region, which causes the voltages at their drains to be low and the associated transistors T_i to be off. Only the

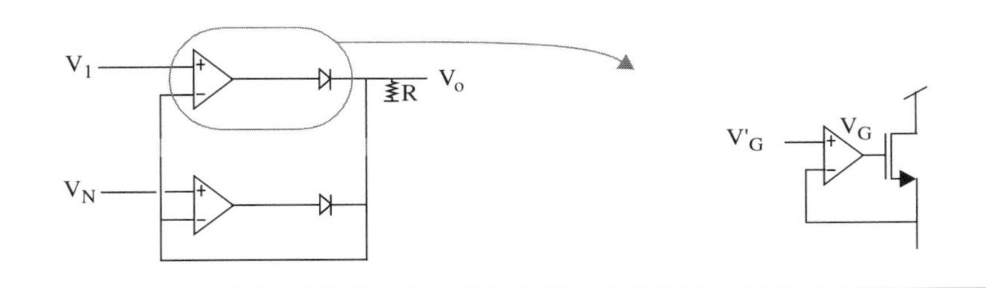

Figure 7.21. Improvement on the conventional diode circuit.

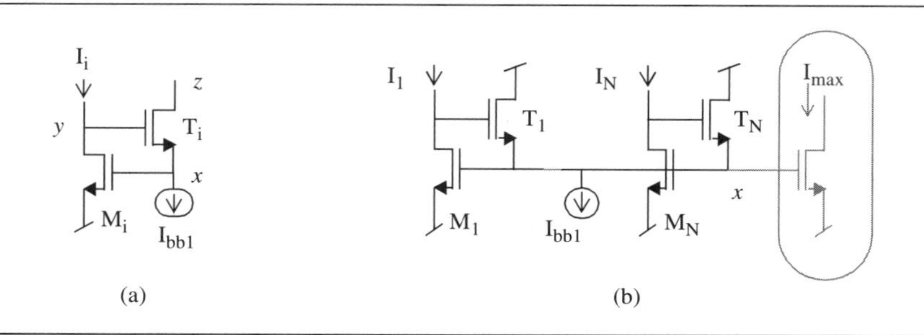

Figure 7.22. (a) Amplification scheme based on a single transistor; (b) Extension of the circuit proposed in [Lazzaro, 1989] to operate as a MAX circuit.

voltage at the input node of I_{max} is high. This is why the current I_{bb1} only flows through the transistor T_i associated to that node. A current of value I_{max} can be obtained from the voltage at the node x by adding a transistor as shown with the discontinuous lines in Figure 7.22b [Sasaki, 1991] (a cascode or regulated output stage can be added instead of a single transistor to obtain greater resolution [Sasaki, 1992a; Vidal-Verdú, 1995]).

The two-transistor cell shown in Figure 7.22a can be also seen as a current-controlled current-conveyor [Andreou, 1991] so that it features two significant functionalities. On the one hand, it can perform as a voltage follower between nodes y and x. On the other, it conducts the current I_{bb1} from a low-impedance node, such as x, to a high-impedance one, such as z. In the realizations mentioned in the above paragraph, only the voltage follower behavior is exploited. However, the other functionality can be also exploited to obtain MIN/MAX circuits, as discussed in [Baturone, 1994]. The proposal in [Baturone, 1994] is to replace the current source I_{bb1} by a diode-connected transistor, thus making I_{bb1} equal to I_i. Combining several of these cells as shown in Figure 7.23a makes the transistor T_i associated to I_{max} replicate this current to a high

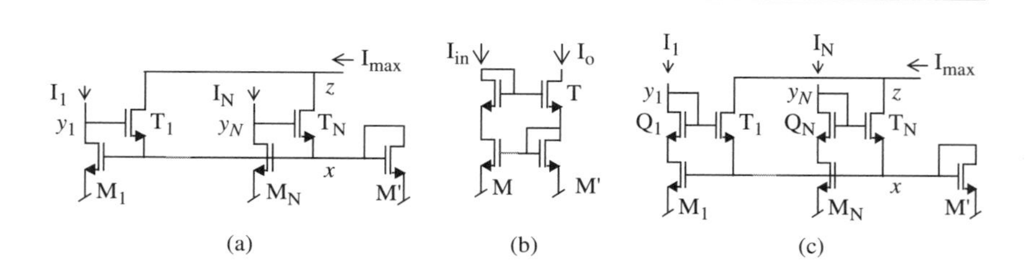

Figure 7.23. (a) Simplified version of the MAX circuit proposed in [Baturone, 1994]; (b) Schematic of an improved Wilson mirror; (c) Multi-input MAX circuit proposed in [Baturone, 1994].

impedance node. As a result, not only the current source I_{bb1} but also the cascode or regulated output stages can be eliminated. The consequence of this simplification is a reduction of the area and power consumption, maintaining resolution and speed performance. The circuit can be seen as the connection of Wilson current mirrors (remember Figure 7.12e) which share their diode-connected output transistor. Since Wilson current mirrors suffer from DC mismatching errors due to their asymmetric biasing [Wang, 1990b], better performance can be obtained with the circuit shown in Figure 7.23c, which results from connecting improved Wilson mirrors (Figure 7.23b). If the circuit operates in weak inversion, resolution in the range of picoamperes can be provided. If it operates in strong inversion, resolution is directly proportional to $(WL/I)^{1/2}$, so that large transistors and low currents are preferred for high resolution.

MAX circuits are usually employed to implement the minimum operations required for the antecedent connection in fuzzy ICs. The way to do this is to apply De Morgan's Laws as follows: $\overline{MFC}$ blocks (which provide the complement of the membership degrees) are followed by a MAX circuit and the output of the MAX circuit is complemented. This was explained in Section 7.2.2 when describing the architecture that shares blocks:

$$\overline{max}[\overline{I\mu}(x), \ldots, \overline{I\mu}(x)] = min[I\mu(x), \ldots, I\mu(x)] \tag{7.20}$$

7.4.2 PRODUCT OPERATORS

Product operators are more difficult to implement with analog circuitry than minimum/maximum operators. The literature on them is extensive because they are required by many applications besides fuzzy systems. This section briefly describes the continuous-time techniques that have been more widely employed, such as: (a) the "log-antilog" as well as translinear bipolar techniques; (b) MOS translinear techniques; and (c) variable-transconductance techniques that use MOS transistors working in saturation and ohmic regions.

- **"Log-antilog" and bipolar translinear techniques**

A direct-active-biased bipolar transistor features an exponential relation ("antilog") between its collector current, I_c, and its base-to-emitter voltage, V_{BE}. This is the cause that motivates the following behavior stated by the bipolar translinear principle [Gilbert, 1990]: A closed loop containing an even number of directly biased junctions placed in such a way that there is the same number of junctions in the clockwise as in the counterclockwise direction features that the product of current densities (I_c/A, where A is directly proportional to the emitter area) flowing in the clockwise direction is the same as the product of current densities flowing counterclockwise. Figure 7.24 shows two typical translinear loops (a cascade and a parallel schemes). These translinear loops permit the implementation of multiplier/divider circuits.

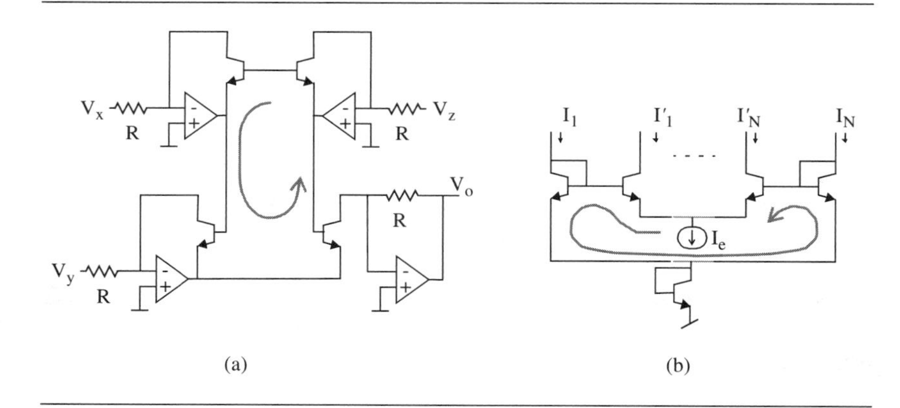

Figure 7.24. (a) Schematic of a "log-antilog" multiplier/divider circuit ($V_o = V_x \bullet V_y / V_z$); (b) Schematic of the normalizer proposed in [Gilbert, 1984].

Figure 7.24a illustrates the schematic of a so-called "log-antilog" multiplier/divider circuit [Sheingold, 1976; Graeme, 1977]. These circuits, which were employed prior to the statement of the bipolar translinear principle, are conventional voltage-mode designs. They place the transistors in the feedback loops of amplifiers.

Bipolar translinear multiplier/divider circuits are current-mode designs that simplify the "log-antilog" circuits because they eliminate the amplifiers as well as the resistors that convert voltages into currents. An example of bipolar translinear multiplier/divider circuits is the normalizer proposed in [Gilbert, 1984] (Figure 7.24b). Applying the bipolar translinear principle to the 4-transistor N-1 loops of this circuit, it follows that:

$$(I_1 \cdot I'_2 = I_2 \cdot I'_1, \ldots, I_1 \cdot I'_N = I_N \cdot I'_1) \Rightarrow$$

$$I_1/I'_1 = I_2/I'_2 = \ldots = I_N/I'_N = K \tag{7.21}$$

where K is a constant. Since $\Sigma_k I'_k = I_e$, it follows that:

$$\frac{I_1}{I'_1} \cdot I'_1 + \ldots + \frac{I_N}{I'_N} \cdot I'_N = K \cdot I_e = \sum_k I_k \Rightarrow$$

$$K = \frac{\sum_k I_k}{I_e} \Rightarrow I'_i = I_e \cdot \frac{I_i}{\sum_k I_k} \tag{7.22}$$

There are three basic approaches for reproducing this multiplying behavior when working with MOS circuitry. One of them is to employ BiCMOS technologies, which allow combining bipolar and MOS transistors on the same chip. This is the approach adopted in [Kettner, 1993b] to design a defuzzification circuit. The second solution is to work with either the vertical or the lateral bipolar transistors that appear in conventional MOS technologies [Vittoz, 1983]. Finally, the third solution is to work in weak inversion where the MOS transistor behavior is similar to that of direct-active-biased bipolar transistors. Working with MOS transistors biased in strong inversion and with conventional MOS technologies it is not possible to reproduce the bipolar translinear behavior and other techniques have to be employed to implement multiplier/divider circuits, as is discussed in the following section.

• **MOS translinear techniques**

While 4-bipolar-transistor translinear loops permit the realization of multiplier/divider circuits, the loops of 4 MOS transistors biased in saturation are suitable for implementing squarer/square-rooter circuits [Bult, 1987; Seevink, 1991; Wiegerink, 1993; Lu, 1994; Liu, 1994; Baturone, 1995]. The two basic structures of 4-MOS-transistor translinear loops are shown in Figure 7.25.

Designing multiplier/divider circuits with MOS translinear techniques can be done in two ways. One way consists of combining two squarer circuits (SQR) to implement the "quarter-square technique":

$$\frac{x \cdot y}{z} = \frac{1}{4}\left[\frac{(x+y)^2}{z} - \frac{(x-y)^2}{z}\right] \tag{7.23}$$

This solution is proposed in [Wiegerink, 1993] for designing a current-mode multiplier/divider circuit, whose block diagram is shown in Figure 7.26a (the schematic of the SQR block is illustrated in Figure 7.26b). The output current provided by this multiplier/divider circuit is:

$$I_o = I_{sqr1} - I_{sqr2} = I_{ab} + \frac{(I_x + I_y)^2}{8I_{ab}} - I_{ab} - \frac{(I_x - I_y)^2}{8I_{ab}} = \frac{I_x \cdot I_y}{2I_{ab}} \tag{7.24}$$

Figure 7.25. Basic structures of 4-MOS-transistor translinear loops.

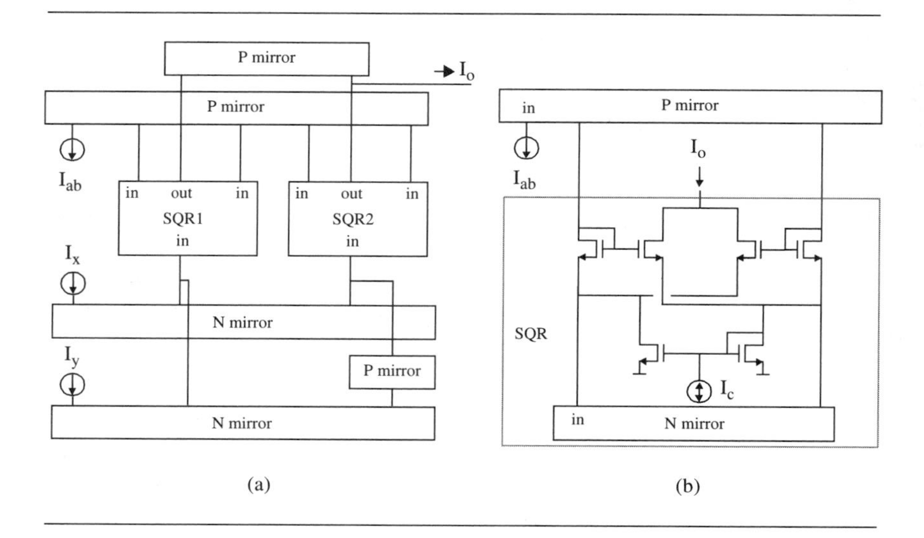

Figure 7.26. (a) Current-mode multiplier/divider circuit obtained by combining two SQR blocks; (b) Schematic of the SQR circuit proposed in [Wiegerink, 1993].

where $(I_x+I_y) \leq 4I_{ab}$.

The other solution for obtaining a multiplier/divider translinear circuit is to combine a square-rooter circuit (SQRT) with a squarer circuit (SQR), as shown in Figure 7.27a [Liu, 1994; Baturone, 1995]. The output current in this case is given by:

$$I_o = I_{sqr} - I_{ab} = \frac{(\sqrt{I_x \cdot I_y})^2}{8I_{ab}} = \frac{I_x \cdot I_y}{8I_{ab}} \qquad (7.25)$$

where $(I_x \cdot I_y)^{1/2} \leq 4I_{ab}$.

Figure 7.27b illustrates the schematic of a circuit that operates as a square-rooter when I_c is null [Baturone, 1995].

Both bipolar and MOS multiplier/divider translinear circuits offer high operation speeds because their nodes are of low impedance and, hence, the voltage swings are also low. They are also independent, in a first approximation, from temperature or fabrication process variations. On the other hand, the main drawback of MOS translinear circuits is that they are very sensitive to deviations from the simple quadratic law behavior (Schichmann–Hodges' model) of the MOS in saturation. These deviations can be caused by second-order effects like length modulation, mobility reduction, body effect, or transistor mismatching [Gregorian, 1986].

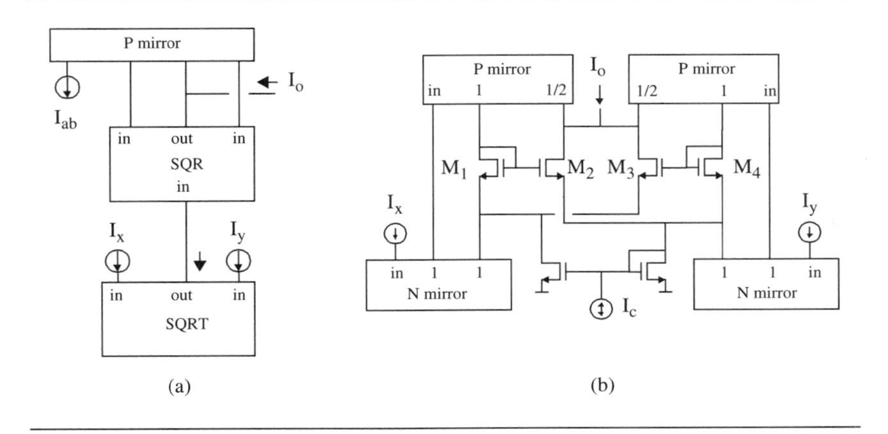

Figure 7.27. (a) Current-mode multiplier/divider circuit obtained by combining a SQR block with an SQRT block; (b) Schematic of the SQR/SQRT circuit proposed in [Baturone, 1995].

• Variable-transconductance techniques

These techniques rely on controlling the transconductance of MOS transistors through an electric variable, usually a voltage. Variable-transconductance circuits are generally based on combining 4 perfectly matched transistors. There are two widely employed approaches; one of them employs MOS transistors biased in saturation while the other makes use of transistors biased in the ohmic region. An example of the first approach is the 4-transistor NMOS cell shown in Figure 7.28a. This structure has been used to design many voltage-mode multipliers [Qin, 1987; Czarnul, 1990; Kim, 1992; Sakurai, 1993; Szczepanski, 1993; Liu, 1993]. According to Figure 7.28a, the following relation is verified:

$$I_n = I_b - I_a = \beta(V_1 - V_2) \cdot (V_a - V_b) \tag{7.26}$$

Since the source terminals of MOS transistors are low-impedance nodes, buffer circuits should be employed to provide the voltages V_a and V_b. These buffers can be eliminated by using CMOS pairs, resulting in the structure shown in Figure 7.28b. This CMOS structure has been employed in the multipliers reported in [Czarnul, 1990; Sakurai, 1993; Liu, 1993; Szczepanski, 1993].

The other approach for implementing variable-transconductance cells is to employ transistors biased in the ohmic region. An example is the 4-transistor cell shown in Figure 7.29a. If the voltages at nodes *a* and *b* are the same, the following relation is verified:

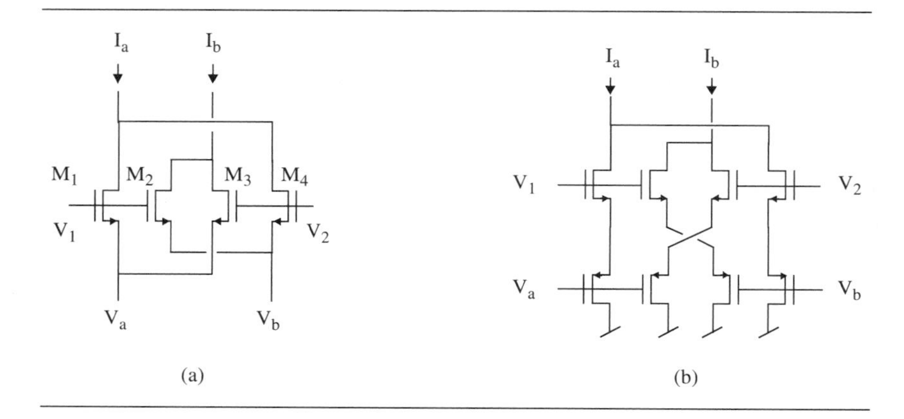

Figure 7.28. Basic cell of a multiplier with (a) NMOS transistors or (b) CMOS pairs biased in the saturation region.

$$I_a - I_b = \beta(V_a - V_b)(V_1 - V_2) \tag{7.27}$$

This equation is verified independent of the voltage values at the drain and sources of the transistors, provided that they are biased in the ohmic region. Several voltage-mode multiplier/divider circuits based on this structure have been proposed in the literature [Czarnul, 1986; Kachab, 1989]. For instance, the circuit in Figure 7.29b provides the following output signal:

$$V_1 - V_2 = (V'_2 - V'_1) \cdot \frac{V'_a - V'_b}{V_a - V_b} \tag{7.28}$$

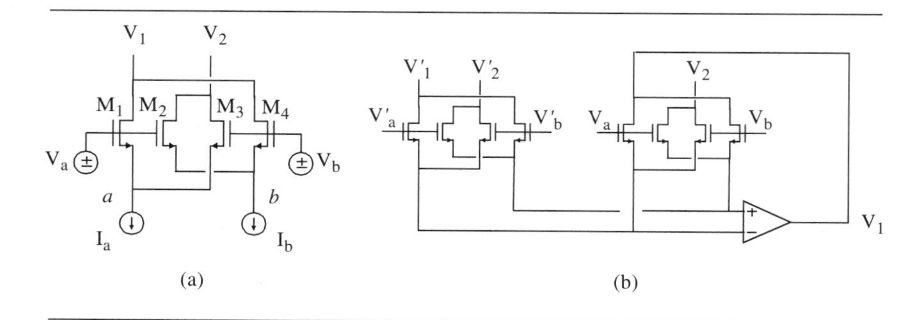

Figure 7.29. (a) Basic cell of a multiplier with MOS transistors biased in the ohmic region; (b) Voltage-mode multiplier/divider circuit proposed in [Czarnul, 1986; Kachab, 1989].

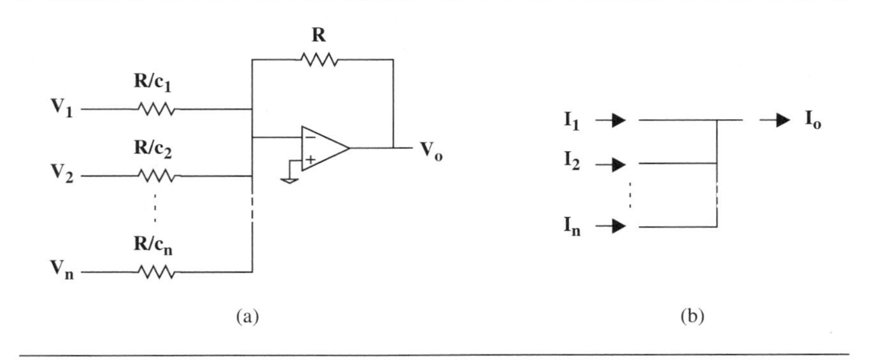

Figure 7.30. (a) Adder/scaler circuit realized with voltage-mode RC-active techniques; (b) Wire connection for implementing a current-mode addition.

7.4.3 SCALING OPERATORS

A product operator can be employed as the antecedent connective to calculate the activation degree of a rule. In this case, the inputs to the multiplier circuit are the outputs of the MFCs. A product operator can also be employed as the implication operator between antecedents and consequents. In this case, one of the inputs of the multiplier circuit represents the activation degree of the rule while the other, which is provided by the MFG, represents the consequent. If the consequents are defined by singleton values, the MFG output is constant, and multipliers can be replaced by scaler circuits, which are simpler. The circuit shown in Figure 7.30a is a conventional adder/scaler circuit realized with voltage-mode RC-active techniques. It scales the voltage values by the resistance values:

$$V_o = -\sum_{i=1}^{n} c_i V_i \tag{7.29}$$

Working in current-mode, we have already seen that a current mirror can act as a scaler circuit whose scaling factor is controlled by the geometries of its constituent transistors. In particular, it can be controlled digitally, as was shown previously in Figure 7.17. In many fuzzy ICs, these scaler circuits that weight the consequent singleton values by their corresponding activation degrees are called MFGs, as was shown in the architectures in Figure 7.5 and Figure 7.6. In general, any D/A converter can be used to implement a digitally controlled scaling operation.

7.4.4 ADDITION OPERATORS

The circuit shown in Figure 7.30a permits the implementation of the addition of several voltages. The resistors convert the voltages into currents that are added at the negative input terminal of the amplifier. Addition and subtraction operations are very easily carried out in current-mode designs. According to Kirchhoff's current law, they can be implemented by simply connecting the corresponding wires to a low-impedance node (Figure 7.30b). This is usually achieved by connecting the output stages of current mirrors to the input stage of another mirror. In order to implement high-accuracy addition or subtraction operations, the current mirrors should be designed so as to reduce replication errors. Among the wide variety of current mirrors reported in the literature (which were summarized in Figure 7.12), simple current mirrors do not offer high accuracy, so that more complex mirrors could be required. The mirrors that usually offer a good performance for many applications are regulated, active, and externally biased cascode mirrors.

7.5 OUTPUT STAGE

The circuitry required in the output stage depends to a large degree on the type of fuzzy inference method implemented. The most significant factor is how the rule consequents are managed, which is very much related to the type of defuzzification method employed. We can distinguish two groups of fuzzy systems from a hardware point of view. One group is formed by the systems that work with fuzzy sets to represent the consequents. They are the most costly to realize as fuzzy ICs. This is the case of Mamdani-type fuzzy systems that use the Center-of-Area defuzzification method. The output obtained in these systems at the end of the rule processing stage is the membership function, $\mu_{B'}$, of the fuzzy set, B', which results from applying the whole rule base. Applying the Center-of-Area calculation, the system output, y_{out}, is given by:

$$y_{out} = \frac{\sum_{i=1}^{n} \mathrm{y_i} \cdot \mu_{B'}(\mathrm{y_i})}{\sum_{i=1}^{n} \mu_{B'}(\mathrm{y_i})} \tag{7.30}$$

where $\mathrm{y_i}$ are the discrete points of the output universe of discourse.

The second group we can distinguish is formed by fuzzy systems that summarize the fuzzy nature of the consequents into several parameters. Mamdani-type systems that employ the minimum as the implication operator, triangular membership functions with the same basis for the consequents and the Center of Sums as the defuzzification method ($\mathrm{CoS_{min}}$) as well as systems that employ the Fuzzy Mean as the defuzzification method (FM) represent each consequent by a single parameter. If the defuzzification method is the Center of Sums with the product as the implication operator (CoS), the Weighted Fuzzy Mean (WFM), the Quality (QM), the Level Grading Method with triangular membership functions (LGM) or Yager's Method (YM), each consequent is defined by two parameters. Finally, three parameters are used to represent the consequents if the defuzzification method is the ξ-quality (ξ-QM). Takagi–Sugeno type systems (TS) also represent the consequents by a set of parameters (as discussed in Sections 5 and 6 of Chapter 3). The output of all these fuzzy systems that work with non-fuzzy representations of the consequents is calculated by evaluating the following generic expression:

$$y_{out} = \frac{\sum_{i=1}^{R} w_i(h_i, B_i) \cdot f_i(B'_i)}{\sum_{i=1}^{R} w_i(h_i, B_i)} \tag{7.31}$$

where R is the number of rules in the knowledge base and w_i is a weighting parameter that depends on the activation degree of the rule, h_i. This weighting parameter summarizes the main features of the consequent fuzzy set, B_i (in the case of Mamdani-type systems). On the other hand, $f_i(B'_i)$ is a parametric representation of the rule conclusion, B'_i. Table 7.2 shows the expressions of w_i and f_i for the above mentioned fuzzy systems that belong to this second group.

Table 7.2. Summary of fuzzy systems with parametric representation of consequents.

	$w_i(h_i, B_i)$	$f_i(B'_i)$
CoS_{min}	$1-(1-h_i)^2$	c_i
FM	h_i	c_i
CoS	$h_i \cdot S_i$ (with $S_i \propto$ area)	c_i
WFM	$h_i \cdot \gamma_i$ (with $\gamma_i \propto$ support)	c_i
QM	$h_i \cdot \gamma_i$ (with $\gamma_i \propto$ 1/support)	c_i
LGM	$h_i \cdot (1-h_i) \cdot \gamma_i$ (with $\gamma_i \propto$ 1/support)	c_i
YM	h_i	$p_{1i}+p_{2i}h_i$
ξ-QM	$h_i \cdot \gamma_i$ (with $\gamma_i \propto 1/\text{support}^{\xi}$)	c_i
TS	h_i	$\Sigma_{k=1}^{u} a_{ik}x_k+a_{i0}$

The circuitry required by these two groups of fuzzy systems is described in the following section.

7.5.1 FUZZY REPRESENTATION OF CONSEQUENTS

The output stage of Mamdani-type fuzzy systems that employ the Center-of-Area defuzzification method has to implement Equation (7.30), so the output universe of discourse has to be swept. It was already shown in Section 7.1 that the first analog defuzzifier chips divided the output space into 25 points and that they provided the 25 values of $\mu_{B'}(y_i)$ in parallel. The circuitry employed consists of two adder/scaler circuits similar to that shown in Figure 7.30a and of a "log-antilog" multiplier/divider circuit (which was explained in Section 7.4.2).

Another continuous-time technique for calculating the center of area has been proposed in [Bouras, 1998]. This structure is more compact than the previous one because it does not work with parallel data. Another difference is that the output space is not discrete, so the centroid is not calculated from the ratio of two summations but from the ratio of two integrals:

$$y_{out} = \frac{\int_{y_{low}}^{y_{high}} y \cdot \mu_{B'}(y)dy}{\int_{y_{low}}^{y_{high}} \mu_{B'}(y)dy} \qquad (7.32)$$

The block diagram of this defuzzifier circuit is shown in Figure 7.31. The input signal V_{in} that appears in this figure represents the value of $\mu_{B'}(y)$. This signal is integrated along a time interval T to obtain the denominator of Equation (7.32) and it is also multiplied by the signal V_{sweep} to obtain the integrand value of the numerator so that:

$$V_{out} = K\frac{\int_0^T V_{sweep} \cdot V_{in}(t)dt}{\int_0^T V_{in}(t)dt} \qquad (7.33)$$

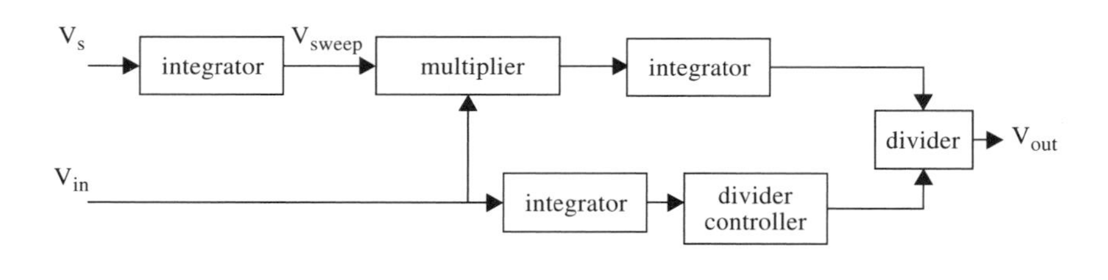

Figure 7.31. Block diagram of the defuzzifier circuit proposed in [Bouras, 1998].

where K is a factor introduced by the gain of the different block, and the voltage V_{sweep} is a saw-shaped signal of period T, which is generated from the DC voltage V_S.

The multiplier, the integrators, and the divider circuits employed in [Bouras, 1998] are fully differential circuits that employ the 4-transistor cell with MOS transistors biased in the ohmic region that was shown in Figure 7.29a. They employ the MOSFET-C design style, which is very adequate for voltage-mode continuous-time ICs. The divider circuit used in this structure employs a multiplier in the feedback loop of an amplifier, which is one of the approaches for designing divider circuits, as will be discussed in the following section.

7.5.2 PARAMETRIC REPRESENTATION OF CONSEQUENTS

Calculating the output of these systems is simpler than in the previously discussed systems. Once the weight parameters, $w_i(h_i, B_i)$, have been calculated and the parametric representations of the conclusions, $f_i(B'_i)$, have been obtained, the equation for obtaining the system output is simply a calculation of a weighted average:

$$y_{out} = \frac{\sum_{i=1}^{R} w_i \cdot f_i}{\sum_{i=1}^{R} w_i} \tag{7.34}$$

where R is the number of rules in the knowledge base.

- **Strategies for implementing a weighted average**

Among the operations required to realize Equation (7.34), division is the most difficult to implement in hardware. Division can be eliminated if the denominator in Equation (7.34) is constant. This happens to Takagi–Sugeno type systems with normalized antecedent fuzzy sets (the sum of the membership degrees for each input is always '1'), which employ the product or the normalized conjunction (Section 7.4) as the antecedent connective, and also implement the whole rule base. When this is not the case and division must be implemented, basically three strategies can be employed, as discussed in the following section.

Normalization locked loops and normalizer circuits

One strategy for implementing the weighted average consists of normalizing the weights by means of a normalizer circuit, placing multiplier circuits at its output to multiply them by the values f_i, and connecting the multipliers to an adder circuit. In this way, the rule processing and output stages are merged. The resulting structure, shown in Figure 7.32a, implements the following equation:

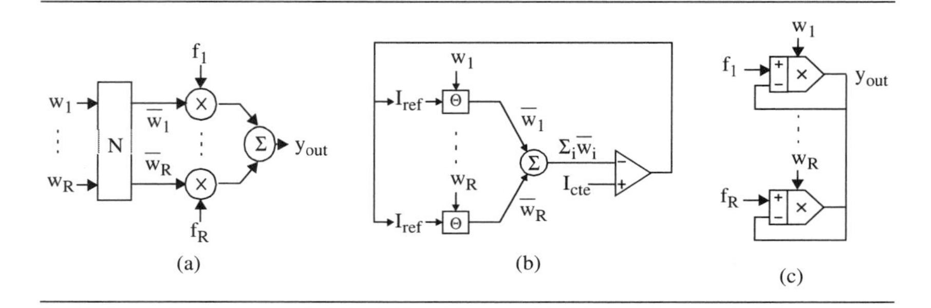

Figure 7.32. (a) Using a normalizer circuit to implement the weighted average; (b) Block diagram of the normalization locked loop proposed in [Sasaki, 1992a] and [Yamakawa, 1993]; (c) Aggregation circuit based on multipliers (represented by "x").

$$y_{out} = \sum_{i=1}^{R} f_i \cdot \frac{w_i}{\sum_{i=1}^{R} w_i} = \sum_{i=1}^{R} f_i \cdot \overline{w_i} \tag{7.35}$$

A normalizer circuit is the bipolar translinear circuit proposed in [Gilbert, 1984], which was already shown in Figure 7.24b. If MOS technologies are employed, one of the techniques used by several authors to approximate the behavior of a true normalizer is to employ the so-called normalization locked loops [Sasaki, 1992a; Yamakawa, 1993]. The purpose of these structures is to maintain the denominator of Equation (7.34) constant via negative feedback, as shown in Figure 7.32b. They approximate the behavior of a true normalizer because the signals $\overline{w_i}$ do not have the value $w_i/\Sigma_i w_i$ but $I_{ref}\,\Theta w_i$, where I_{ref} (the maximum membership degree) is modified by the normalization loop. With this approximation, the importance of an f_i on the global system not only depends on its weight w_i (according to Equation (7.34)) but also on the normalization operation.

Aggregation circuits

A second strategy for implementing a weighted average that has been adopted in several fuzzy analog ICs [Landolt, 1993; Rojas, 1994; Manaresi, 1994; Tsukano, 1995; Peters, 1995] consists in implementing weighting and averaging operations simultaneously with an aggregation circuit such as that reported in [Mead, 1989]. As a consequence, the rule processing and output stages are again merged.

The circuit proposed in [Mead, 1989] is based on transconductance operators (OTAs). As a consequence, it implements an approximation to Equation (7.34) in which the weights are not w_i but directly proportional to $(w_i)^{1/2}$. To implement Equation (7.34), the aggregation circuit should employ true multipliers with outputs represented by currents, as shown in Figure 7.32c. In this way, the following function is obtained after applying Kirchhoff's current law at the output node:

$$\sum_{i=1}^{R} (f_i - y_{out}) w_i = 0 \tag{7.36}$$

which is equivalent to Equation (7.34)

Divider and scaler circuits

The third strategy for implementing the weighted mean operator is to realize the rule processing and the output stages separately. In the rule processing stage, scaler or multiplier circuits (the MFGs) provide the f_i values weighted by their corresponding w_i. It is convenient to work in current-mode so that the sums can be realized by connecting wires. Hence, the output stage only needs a divider to implement the weighted average [Huertas, 1996; Baturone, 1997b]. This is the scheme that was illustrated in Section 7.2 when the architecture that shares circuitry and the active-rule-driven architecture were described. Scaler circuits, which are simpler than multipliers, can be used when the f_i values are single parameters, for instance when they are the singleton values of the consequents in the case of zero-order Takagi–Sugeno or singleton fuzzy systems. In particular, we have seen that current-mode scaler circuits can be realized as digitally programmed current mirrors that do not suffer from offset (that is, the output current is null when the input is null, even considering second-order effects or random errors). Therefore, only the divider circuit should be compensated against offset. Regarding the dynamic behavior, the load capacitance which affects the divider circuit is smaller than that which affects the multipliers in the aggregation circuit, so that this strategy can offer a higher operation speed.

• **Design of divider circuits**

Section 7.4.2 already showed that bipolar and MOS translinear techniques can be used to design multiplier/divider circuits. Other techniques that can be used are described in the following section.

Inverting multipliers

A technique widely employed for the realization of analog divider circuits is to invert the operation of a multiplier. This is usually done by including the multiplier in the feedback loop of an amplifier [Sheingold, 1976], as shown in Figure 7.33a. This technique is employed in the divider circuit with voltage output and current inputs proposed in [Baturone, 1997b] (Figure 7.33b). It is based on a 4-transistor multiplier biased in the ohmic region. The amplifier

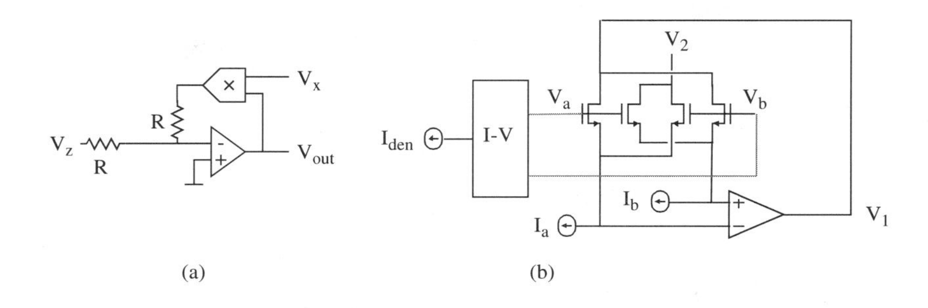

Figure 7.33. (a) Inverting a multiplier with an amplifier; (b) Divider circuit with voltage output and current inputs proposed in [Baturone, 1997b].

makes it possible to feed the multiplier output back into the input voltage node V_1. It provides high impedance nodes to the signals I_a, I_b, and low impedance to the signal V_1. In addition, it fixes the same voltage value at the nodes of I_a and I_b (provided the amplifier operates in the linear region). An I-V converter is added to obtain $(V_a–V_b)$ from the current I_{den}, so that:

$$V_1 - V_2 = \frac{I_a - I_b}{\beta(V_a - V_b)} = K_d \cdot \frac{I_{num}}{I_{den}} \tag{7.37}$$

This divider circuit is stable if V_a is bigger than V_b, so that I_{den} should be unipolar. Once this condition is verified, both static and dynamic performance of the circuit depend widely on the selected amplifier. As usually happens in amplifier-based circuits, high precision is achieved with high-gain amplifiers while high operation speed is obtained with high-gainbandwidth amplifiers (where gainbandwidth, GB, is the product of the gain and the bandwidth). As an example, 6-bit resolution has been experimentally confirmed with an amplifier of 70-dB gain [Baturone, 1997b].

Other types of transconductance multipliers can be more easily inverted without the need for an amplifier, as shown in Figure 7.34a [Sánchez-Sinencio, 1989]. This is the strategy employed by the aggregation circuit proposed in [Mead, 1989]. For instance, the multiplier based on CMOS pairs, which was shown in Figure 7.28b, can support local feedback, as illustrated in Figure 7.34b. To obtain a current-mode divider circuit, as required in many fuzzy ICs, two of these resistor blocks can be combined as shown in Figure 7.34c. From this it follows that:

$$V_o = V_1 - V_2 = \frac{I_{num}}{\beta_{num}(V_a - V_b)} =$$

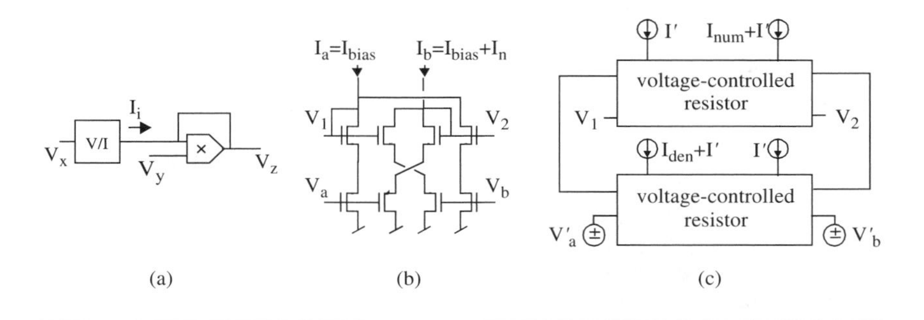

Figure 7.34. (a) Inverting a multiplier via local feedback; (b) Voltage-controlled resistor; (c) Current-mode divider obtained from two resistors like those in (b).

$$= \frac{\beta_{den}(V'_a - V'_b)}{\beta_{num}} \cdot \frac{I_{num}}{I_{den}} = K_d \cdot \frac{I_{num}}{I_{den}} \qquad (7.38)$$

V_b should always be superior to V_a in order to achieve stability. The response time of the circuit is longer as I_{den} is smaller. In particular, the operation speed is reduced by the well-to-substrate capacitance that appears if the PMOS transistors are placed in independent wells. A resolution of around 4 bits has been measured experimentally in prototypes of these circuits [Baturone, 1997b].

Successive-approximation technique based on current-mode continuous-time converters

The use of the successive-approximation technique for implementing a multiplication/division is reported in [Allen, 1984]. This technique can be carried out by A/D and D/A converters. An A/D converter receiving an input voltage x_{num} and a reference voltage x_{den} provides a code $\{b_1, b_2, \ldots, b_N\}$ which represents the binary value of the ratio x_{num}/x_{den}:

$$\sum_{i=1}^{N} 2^{-i} \cdot b_i = Q\left(\frac{x_{num}}{x_{den}}\right) \qquad (7.39)$$

where N is the number of bits, Q(.) is a quantization operator, and the A/D converter has been considered as a divider-type converter (because the reference voltage x_{den} fixes the full scale value).

The output of a D/A converter whose reference signal is x_p or $x_p/2^N$ and whose input binary code is the output of the previous A/D converter is the multiplication/division required:

$$x_{out} = x_p \cdot \sum_{i=1}^{N} 2^{-i} \cdot b_i = x_p \cdot Q\left(\frac{x_{num}}{x_{den}}\right) \tag{7.40}$$

if the D/A converter is a divider-type converter, or:

$$x_{out} = \frac{x_p}{2^N} \cdot \left(\sum_{i=1}^{N} 2^{N-i} \cdot b_i\right) = x_p \cdot Q\left(\frac{x_{num}}{x_{den}}\right) \tag{7.41}$$

if it is a multiplicative-type converter.

If this A/D-D/A structure is employed as the divider block in a fuzzy IC, the fuzzy inference output is provided in both analog and digital formats, which is very advantageous for interacting with analog actuators and digital processing environments.

Algorithmic current-mode A/D and D/A converters are very efficient for implementing these A/D-D/A dividers because they offer good trade-offs between power consumption and speed and between area occupation and resolution. The conversion method implemented by an algorithmic A/D converter is shown in Figure 7.35a, where I_{den} represents the full scale current and $I_{den}/2^N$ is the quantization step. Considering this situation, the input range for I_{num} is $I_{num} \le I_{den}$. This conversion method allows a very modular design: the structure of N-bit resolution converters is based on the cascade of N identical or alternating (odd and even) bit cells, as shown in Figure 7.35b. Each bit cell contains an A/D and a D/A subcell that provide, respectively, one bit of the digital output, $\{b_i\}$, and a contribution to the analog output, I_{out}. The output $\{b_i\}$ is the digital code of the current-mode division I_{num}/I_{den} while I_{out} is the discrete current that represents the multiplication/division $I_p \cdot I_{num}/I_{den}$. A disadvantage of a cascade connection is that the converters are sensitive to mismatching among cells, so that resolution is typically reduced to 9 bits [Pelgrom, 1989], although systematic errors can decrease this potential resolution.

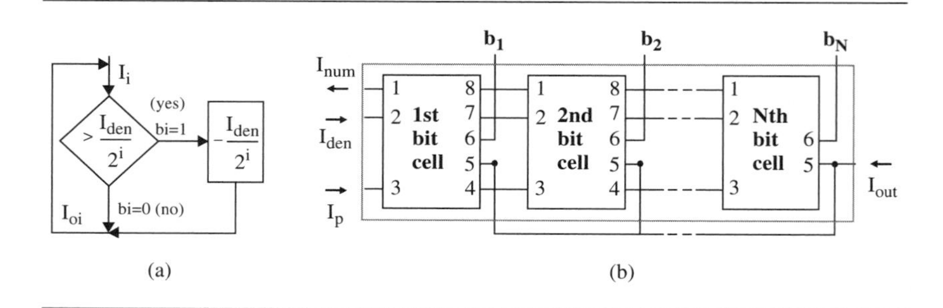

Figure 7.35. (a) Flow chart of an algorithmic A/D conversion method; (b) Modular design of a current-mode successive-approximation multiplier/divider circuit.

This resolution is usually enough for most fuzzy systems.

An advantage of using continuous-time techniques is that control circuitry is not required to govern the signal transmission from one cell to another [Nairn, 1990]. In addition, the response time can be inferior to N times the duration of the slowest operation. The slowest operation depends on the transient response of the current mirrors and the settling time of the current comparators.

As an example of this type of multiplier/divider circuits, the performance of a 5-bit 2.4-μm CMOS prototype is reported in [Baturone, 1998b]. It occupies a silicon area of 0.077 mm^2, operates at 3-V supply, and achieves a response time of a few hundreds of nanoseconds with power consumption below the milliwatt.

7.6 DESIGN OF A CURRENT-MODE CMOS PROTOTYPE

In order to illustrate all that has been discussed previously about the design of continuous-time fuzzy ICs, let us consider the design of a zero-order Takagi–Sugeno or singleton fuzzy system with two inputs, one output, 9 rules with the format "*if A and B then c*", using the minimum as the antecedent connective. The input universes of discourse are covered by a symmetric triangular membership function (*Z*) and by two saturated functions (*N*, *P*), as shown in Figure 7.36a. The output universe of discourse is covered by three singleton labels (*n*, *z*, *p*), as illustrated in Figure 7.36a. The rule base selected for this system is an antisymmetric base (Figure 7.36b), which is widely used in control applications. The approximate knowledge that is codified by this rule base can be summarized linguistically as: "*If the plant to control deviates from its goal state, Z, the control action should counteract deviation*".

Since there are input fuzzy sets and output singleton values that repeat in different rules, an efficient architecture should share MFCs and MFGs, as shown in Figure 7.37. This structure, except for the divider circuit (DIV), has been integrated in a 2.4-μm CMOS technology. The resulting IC, which will

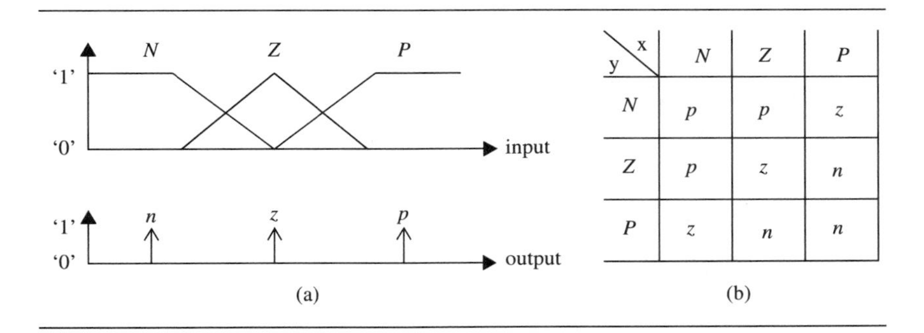

Figure 7.36. (a) Covering of input and output spaces; (b) Rule base.

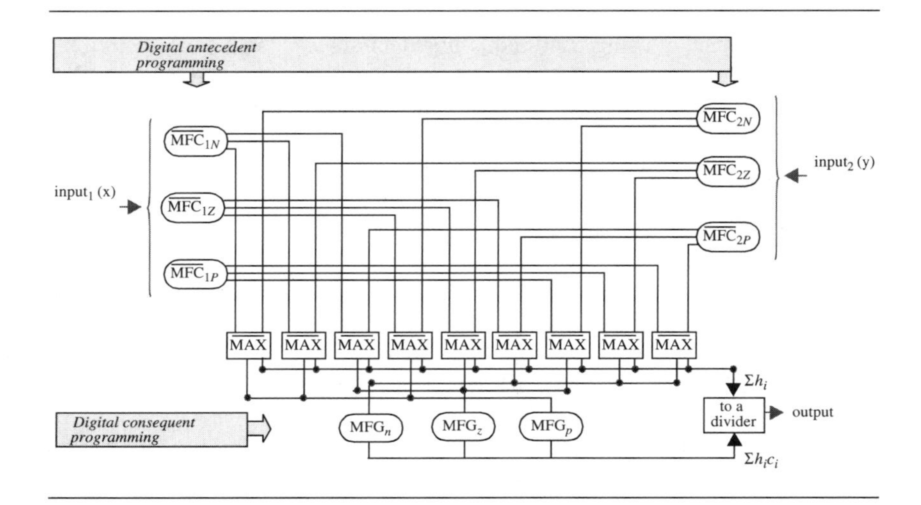

Figure 7.37. Block diagram of the rule chip.
(©1998 IEICE. Reprinted with permission from [Baturone, 1998c]).

be called the rule chip, provides two current outputs, one that represents the weighted sum of the partial conclusions of all the rules, $\Sigma h_i c_i$, and the other that represents the sum of all the activation degrees, Σh_i. These are the two signals whose ratio corresponds to the fuzzy system output, $output = \Sigma h_i c_i / \Sigma h_i$. Figure 7.38a shows a microphotograph of the rule chip, while Figure 7.38b depicts the different blocks so as to identify their relative area occupation. The rule chip occupies an active area of 0.95 mm^2. It can be combined with different divider circuits, integrated in other chips, to implement a complete fuzzy system.

The $\overline{\text{MFC}}$s included in this rule chip basically correspond to the third structure described in Section 7.3.2. Their schematics are shown in Figure 7.39, where the blocks 5-D/A and 3-D/A represent current-mode converters of 5 and 3 bits, respectively. The implemented equations are:

$$\overline{I_{\mu Z}} = m_b \cdot (|I_{in} - I_{aux}|) \tag{7.42}$$

$$\overline{I_{\mu N}} = m_b \cdot (I_{in} \Theta I_{aux}) \tag{7.43}$$

$$\overline{I_{\mu P}} = m_b \cdot (I_{aux} \Theta I_{in}) \tag{7.44}$$

where I_{aux} and m_b are given by:

$$I_{aux} = (a_0 + 2 \cdot a_1 + 4 \cdot (a_2 + 2 \cdot a_3 + 4 \cdot a_4))I_u \tag{7.45}$$

$$m_b = (b_0 + 2 \cdot b_1 + 4 \cdot b_2)I_u \tag{7.46}$$

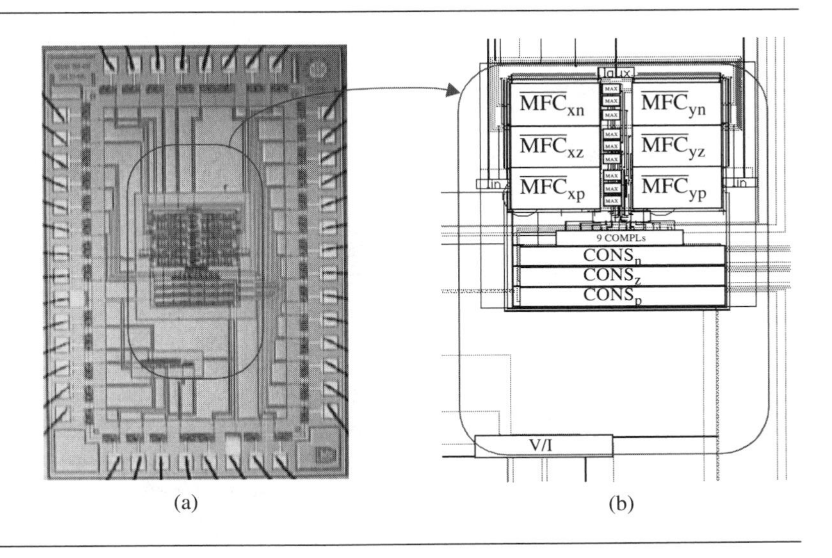

Figure 7.38. (a) Rule chip photograph; (b) Identification of the functional blocks. (©1996 IEEE. Reprinted with permission from [Huertas, 1996]).

with a_i, b_i the programming bits and I_u a reference current.

Figure 7.40 shows experimental results corresponding to the fuzzification stage. They illustrate how the slope of the label Z and the location of the labels N and P can be programmed so as to obtain different overlapping. The inputs to the rule chip can be voltages or currents because it includes V/I converters, like those shown in Figure 7.41a [Bult, 1986].

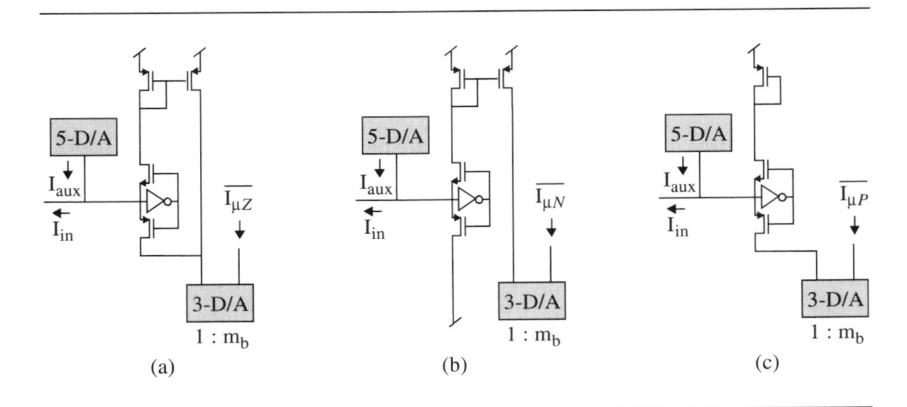

Figure 7.39. Schematics of (a) $\overline{\mathrm{MFC}_Z}$; (b) $\overline{\mathrm{MFC}_N}$; (c) $\overline{\mathrm{MFC}_P}$.

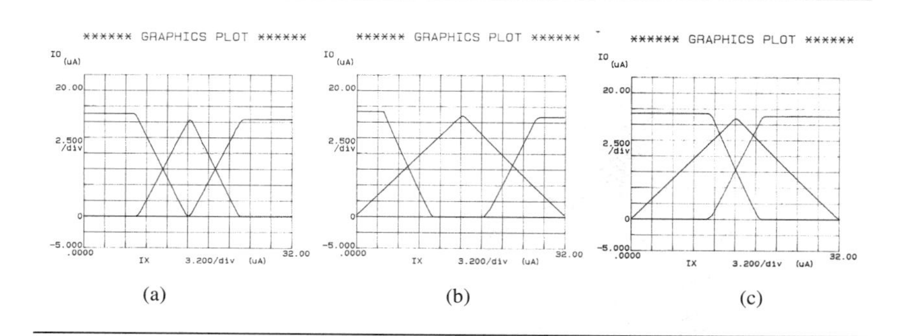

Figure 7.40. Experimental results corresponding to the fuzzification stage. (©1998 IEICE. Reprinted with permission from [Baturone, 1998c]).

The output current of the previous $\overline{\text{MFC}}$ is replicated three times to realize all the possible combinations between antecedents. These currents, which represent the complements of the membership degrees, enter into a current-mode MAX circuit, whose output is complemented with I_{ref}, as shown in Figure 7.41b. The result is the rule activation degree, represented by the current h_i. Since there are 9 MAX circuits in the rule chip, the current I_{ref} is replicated 9 times. Each current h_i is replicated twice, one current that is summed with the other h_i so as to provide the value Σh_i (the sum of the rule activation degrees) as one output of the chip, and the other one that is summed with the h_i currents which should weight the same consequent (n, z, or p). Figure 7.42 illustrates experimental results of the antecedent connection. They correspond to rules like "*if* V_x *is P and* V_y *is Z then* ...".

The MFG blocks that implement the implication operator (weighting of singleton values by their activation degrees) are 4-bit D/A converters, whose

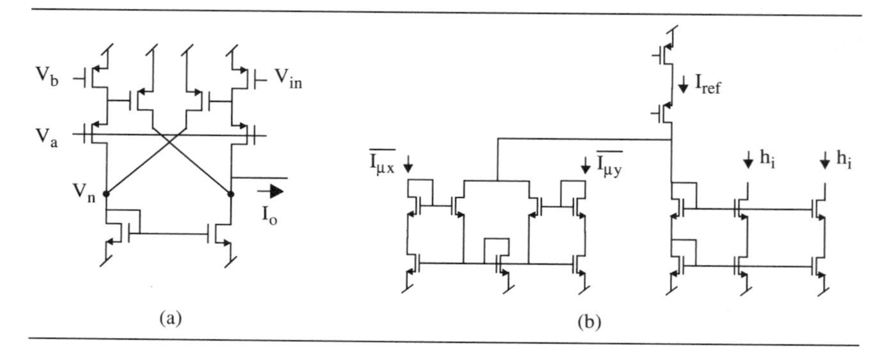

Figure 7.41. (a) Linear V/I converter proposed in [Bult, 1986]; (b) Circuit to obtain the activation degree of a rule.

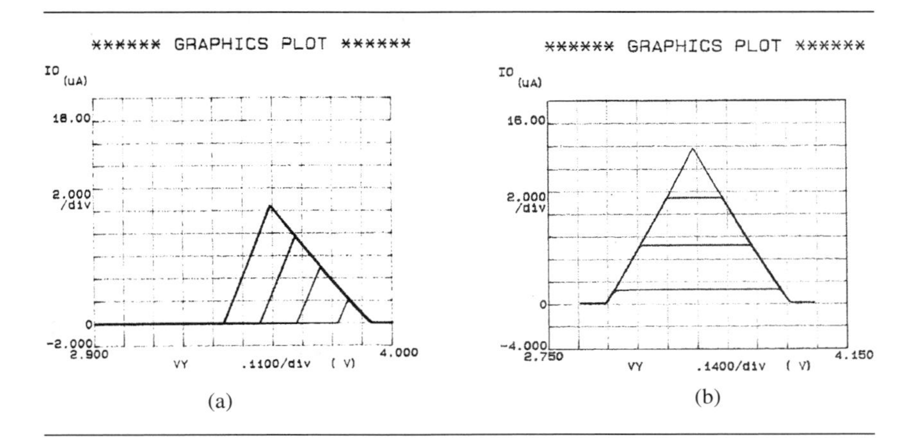

Figure 7.42. Experimental results of the minimum operation: (a) Between a label *Z* and different labels *P* (with V_x=V_y); (b) Between a label *Z* and a label *P* (for different values of V_x). (©1998 IEICE. Reprinted with permission from [Baturone, 1998c]).

schematic is shown in Figure 7.43. They implement the following equation:

$$I_o = \left(\frac{1}{2} \cdot d_1 + \frac{1}{4} \cdot d_2 + \frac{1}{8} \cdot d_3 + \frac{1}{16} \cdot d_4\right) I_v \qquad (7.47)$$

The value of the current I_v that enters into each MFG is given by the sum of those h_i which should weight the same consequent. According to the rule base in Figure 7.36b, these I_v take the following values:

$$I_{vp} = h_{NN} + h_{NZ} + h_{ZN} \qquad (7.48)$$

$$I_{vz} = h_{NP} + h_{ZZ} + h_{PN} \qquad (7.49)$$

$$I_{vn} = h_{ZP} + h_{PZ} + h_{PP} \qquad (7.50)$$

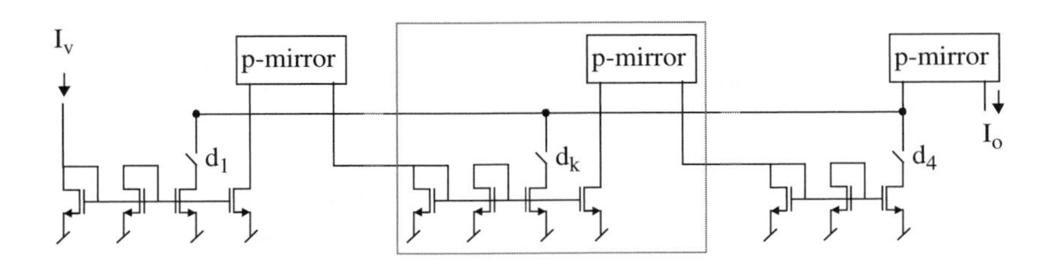

Figure 7.43. Schematic of an MFG.

All the MFG outputs are connected to an output pad. Hence, this pad provides the following output current:

$$I_o = \Sigma h_i c_i = (h_{NN} + h_{NZ} + h_{ZN})c_p + \\ (h_{NP} + h_{ZZ} + h_{PN})c_z + (h_{ZP} + h_{PZ} + h_{PP})c_n \qquad (7.51)$$

which is the weighted sum of the conclusions of the entire rule base.

The response time of the rule chip is shorter than 240 ns. In order to illustrate the behavior of a complete fuzzy system, the rule chip was connected to another chip containing a divider circuit working in the ohmic region like the one shown in Figure 7.33. The output surfaces obtained for different coverages of the input universes of discourse are shown in Figure 7.44.

Hspice simulations considering all the parasitic capacitances of the layouts were made to analyze the dynamic behavior of a fuzzy IC which includes the circuitry of the rule chip together with the circuitry of a divider working in the ohmic region. Response times shorter than 500 ns were obtained for different input pulses and 5-pF capacitance at the output node of the divider. In these cases, the average power consumption measured by simulation was inferior to 7mW [Huertas, 1996].

The fuzzy system response time is obviously shorter if the divider circuit is integrated on the same chip, thus avoiding delays of the external connections. As an example, the complete scheme of Figure 7.37 was integrated on the same chip. In this case, a 5-bit successive-approximation divider of the type described in Section 7.5.2 was included. The bit cells of this divider have a schematic like that in Figure 7.45 [Baturone, 1998b]. When signal H takes a high value ('1'), the digital word that represents the quotient I_{num}/I_{den} is calculated. When S is '1', all the bits $\{b_i\}$ that control the D/A part are simultaneously updated. This synchronization reduces the presence of spurious transient signals known as glitches so as to ensure a good dynamic behavior of the D/A part. In addition, all the D/A bit cells are identical in order to reduce asymmetric switching. Switching speed increases whenever there is a conducting path for the current. This is the purpose of transistor M_a in Figure 7.45. The bits $\{b_i\}$ do not change when H is '1', so that the analog (I_{out}) and digital ($\{b_i\}$) outputs are maintained while performing a new division operation.

The microphotograph of the fuzzy IC in which this divider is included is shown in Figure 7.46. Its active area is 1.1 mm^2. The chip output represents the fuzzy system output in digital (a 5-bit resolution word) or analog format (a discrete current with a maximum of 32 levels).

Figure 7.47 shows several output surfaces measured from this fuzzy IC. They represent the output current (analog format of the output) for different coverages of the input and output universes of discourse. The singleton values $\{n, z, p\}$ were programmed to take {0.9375, 0, 0.9375} for obtaining the surface at the middle of the figure and {0, 0.4375, 0.875} for obtaining the other

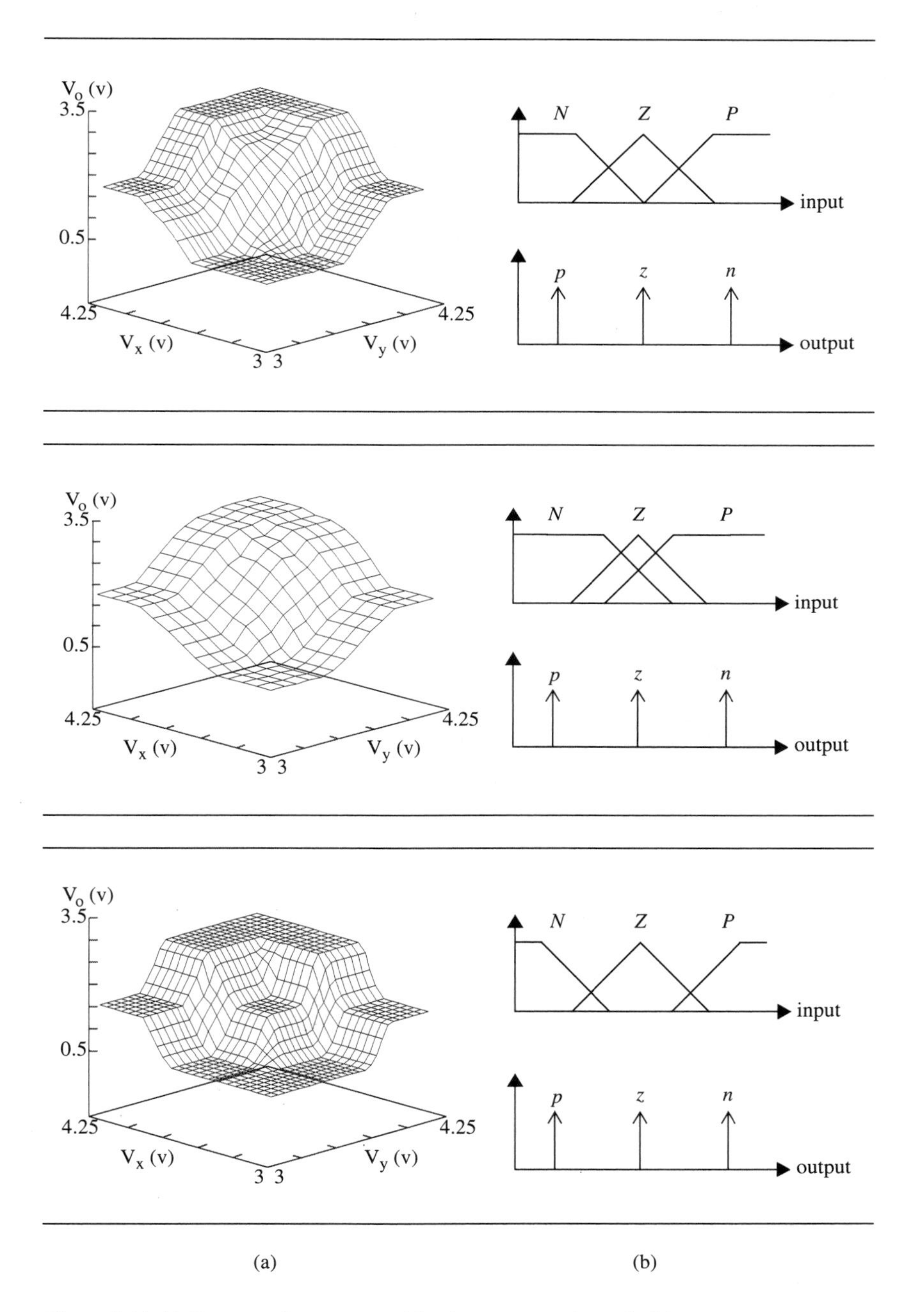

Figure 7.44. (a) Output surfaces measured for the coverages shown in (b).

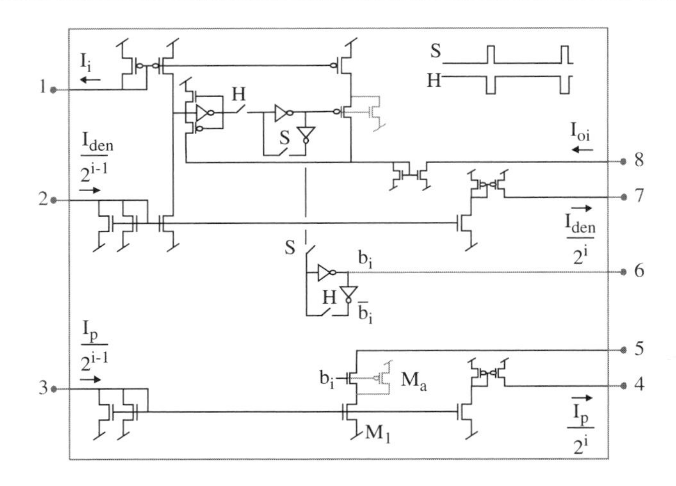

Figure 7.45. Schematic of the A/D-D/A bit cell proposed in [Baturone, 1998b].

two surfaces. The static power consumption of the chip ranges from 21 mW (in the surface at the middle of the figure) to less than 9 mW (in the other two cases) with a 5-V power supply and a current of 16 μA to represent the maximum

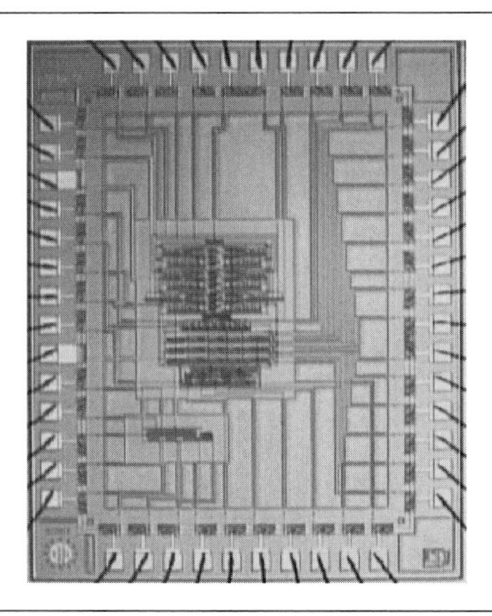

Figure 7.46. Microphotograph of a continuous-time fuzzy IC.
(©1998 IEICE. Reprinted with permission from [Baturone, 1998c]).

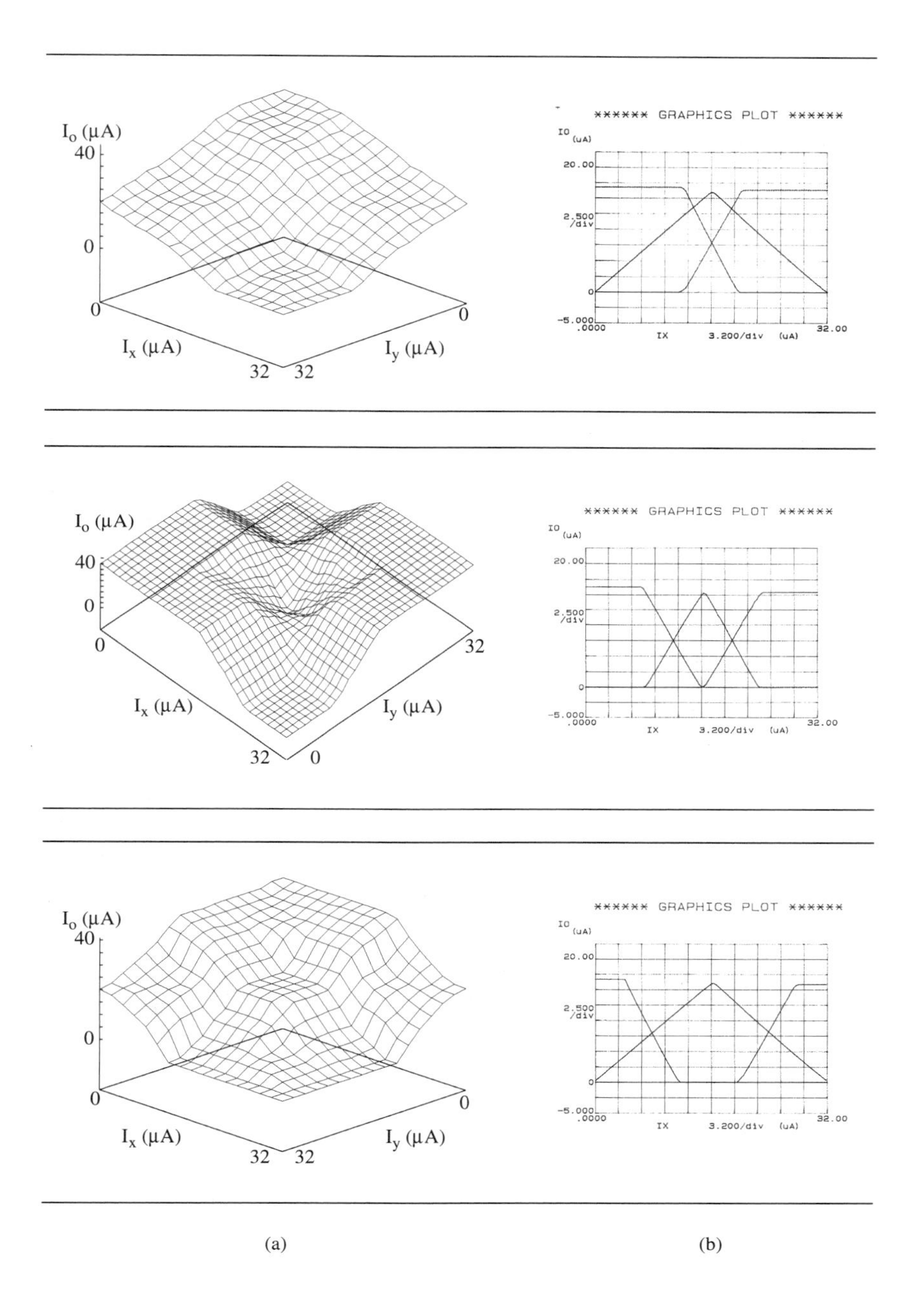

Figure 7.47. (a) Output surfaces measured for the input coverages shown in (b). (©1997 IEEE. Reprinted with permission from [Baturone, 1997a]).

membership degree. The measured response time was shorter than 2 μs. The divider is the circuit that limits the inference speed of this chip. Response time below the microsecond can be accomplished if the design is optimized and submicron technologies are employed.

REFERENCES

1. **Allen, P. E., Sánchez-Sinencio, E.**, *Switched Capacitor Circuits*, Van Nostrand Reinhold Co., 1984.
2. **Andreou, A. G., Boahen, K. A., Pouliquen, P. O., Pavasovic, A., Jenkins, R. E., Strohbehn, K.**, Current-mode subthreshold MOS circuits for analog VLSI neural systems, *IEEE Transactions on Neural Networks,* Vol. 2, N. 2, pp. 205-213, 1991.
3. **Babanezhad, J. N., Gregorian, R.**, A programmable gain/loss circuit, *IEEE Int. Journal of Solid-State Circuits*, Vol. 22, N. 6, pp. 1082-1089, 1987.
4. **Baturone, I., Barriga, A., Sánchez-Solano, S., Huertas, J. L.**, Una implementación hardware de circuitos básicos de lógica difusa, in *Proc. VII Congreso de Diseño de Circuitos Integrados,* pp. 407-412, Toledo, 1992.
5. **Baturone, I., Huertas, J. L., Barriga, A., Sánchez-Solano, S.**, Current-mode multiple-input maximum circuit, *Electronics Letters*, Vol. 30, N. 9, pp. 678-680, 1994.
6. **Baturone, I., Huertas, J. L., Barriga, A., Sánchez-Solano, S.**, Extending the functionality of a flexible current-mode CMOS circuit, *Electronics Letters*, Vol. 31, N. 15, pp. 1231-1232, 1995.
7. **Baturone, I., Sánchez-Solano, S., Barriga, A., Huertas, J. L.**, Flexible fuzzy controllers using mixed-signal current-mode techniques, in *Proc. IEEE Int. Conference on Fuzzy Systems,* pp. 875-880, Barcelona, 1997a.
8. **Baturone, I., Sánchez-Solano, S., Barriga, A., Huertas, J. L.**, Implementation of CMOS fuzzy controllers as mixed-signal IC's, *IEEE Transactions on Fuzzy Systems,* Vol. 5, N. 1, pp. 1-19, 1997b.
9. **Baturone, I., Barriga, A., Sánchez-Solano, S., Huertas, J. L.**, Mixed-signal design of a fully parallel fuzzy processor, *Electronics Letters*, Vol. 34, N. 5, pp. 437-438, 1998a.
10. **Baturone, I., Sánchez-Solano, S., Huertas, J. L.**, A CMOS current-mode multiplier/divider circuit, in *Proc. IEEE Int. Symposium on Circuits and Systems*, pp. WPA8-3, Monterey, 1998b.
11. **Baturone, I., Sánchez-Solano, S., Huertas, J. L.**, Towards the IC implementation of adaptive fuzzy systems, *IEICE Transactions on Fundamentals*, Vol. E81-A, N. 9, pp. 1877-1885, 1998c.
12. **Bouras, S., Kotronakis, M., Suyama, K., Tsividis, Y.**, Mixed analog-digital fuzzy logic controller with continuous-amplitude fuzzy inferences and defuzzification, *IEEE Transactions on Fuzzy Systems,* Vol. 6, N. 2, pp. 205-215, 1998.
13. **Bult, K., Wallinga, H.**, A CMOS four-quadrant analog multiplier, *IEEE Int. Journal of Solid-State Circuits*, Vol. 21, N. 3, pp. 430-435, 1986.

14. **Bult, K., Wallinga, H.**, A class of analog CMOS circuits based on the square-law characteristic of an MOS transistor in saturation, *IEEE Int. Journal of Solid-State Circuits*, Vol. 22, N. 3, pp. 357-365, 1987.

15. **Chen, J. J., Chen, C. C., Tsao, H. W.**, Tunable membership function circuit for fuzzy control systems using CMOS technology, *Electronics Letters*, Vol. 28, N. 22, pp. 2101-2103, 1992.

16. **Crawley, P. J., Roberts, G. W.**, High-swing MOS current mirror with arbitrarily high output resistance, *Electronics Letters*, Vol. 28, N. 4, pp. 361-363, 1992.

17. **Czarnul, Z.**, Novel MOS resistive circuit for synthesis of fully integrated continuous-time filters, *IEEE Transactions on Circuits and Systems,* Vol. 33, N. 7, pp. 718-721, 1986.

18. **Czarnul, Z., Takagi, S.**, Design of linear tunable CMOS differential transconductor cells, *Electronics Letters*, Vol. 26, N. 21, pp. 1809-1811, 1990.

19. **Domínguez-Castro, R., Rodríguez-Vázquez, A., Medeiro, F., Huertas, J. L.**, High resolution CMOS current comparators, in *Proc. ESSCIRC'92,* pp. 242-245 Copenhagen, 1992.

20. **Franchi, E., Manaresi, N., Rovatti, R., Bellini, A., Baccarani, G.**, Analog synthesis of nonlinear functions based on fuzzy logic, *IEEE Int. Journal of Solid-State Circuits*, Vol. 33, N. 6, pp. 885-895, 1998.

21. **Gilbert, B.**, A monolithic 16-channel analog array normalizer, *IEEE Int. Journal of Solid-State Circuits*, Vol. 19, N. 6, pp. 956-963, 1984.

22. **Gilbert, B.**, Current-mode circuits from a translinear point of view: a tutorial, in Toumazou, C., Lidgey, F. J., Haigh, D. G., Eds., *Analogue IC design: The Current-Mode Approach*, Peter Peregrinus Ltd., 1990.

23. **Graeme, J. G.**, *Designing with Operational Amplifier: Applications Alternatives,* McGraw Hill, 1977.

24. **Gregorian, R., Temes, G. C.**, *Analog MOS Integrated Circuits for Signal Processing*, Wiley-Interscience Pub., 1986.

25. **Huertas, J. L., Sánchez-Solano, S., Barriga, A., Baturone, I.**, Serial architecture for fuzzy controllers: hardware implementation using analog/ digital VLSI techniques, in *Proc. 2nd Int. Conference on Fuzzy Logic and Neural Networks*, pp. 535-538, Iizuka, 1992.

26. **Huertas, J. L., Sánchez-Solano, S., Barriga, A., Baturone, I.**, A hardware implementation of fuzzy controllers using analog/digital techniques, *Int. Journal of Computers and Electrical Engineering*, Vol. 20, N. 5, pp. 409-419, 1994.

27. **Huertas, J. L., Sánchez-Solano, S., Baturone, I., Barriga, A.**, Integrated circuit implementation of fuzzy controllers, *IEEE Int. Journal of Solid-State Circuits*, Vol. 31, N. 7, pp. 1051-1058, 1996.

28. **Inoue, T., Ueno, F., Motomura, T.**, Analysis and design of analog CMOS building blocks for integrated fuzzy inference circuits, in *Proc. IEEE Int. Symposium on Circuits and Systems*, pp. 2024-2027, Singapore, 1991.

29. **Kachab, N. I., Ismail, M.**, MOS multiplier/divider cell for analogue VLSI, *Electronics Letters*, Vol. 25, N. 23, pp. 1550-1552, 1989.

30. **Kettner, T., Heite, C., Schumacher, K.**, Analog CMOS realization of fuzzy logic membership functions, *IEEE Int. Journal of Solid-State Circuits,* Vol. 28, N. 7, pp. 857-861, 1993a.

31. **Kettner, T., Schumacher, K., Goser, K.**, Realization of a monolithic fuzzy logic controller, in *Proc. ESSCIRC'93*, pp. 66-69, Seville, 1993b.

32. **Kim, Y. H., Park, S. B.**, Four-quadrant CMOS analog multiplier, *Electronics Letters*, Vol. 28, N. 7, pp. 649-650, 1992.

33. **Landolt, O.**, Efficient analog CMOS implementation of fuzzy rules by direct synthesis of multidimensional fuzzy subspaces, in *Proc. IEEE Int. Conference on Fuzzy Systems*, pp. 453-458, San Francisco, 1993.

34. **Landolt, O.**, Low power analog fuzzy rule implementation based on a linear MOS transistor network, *Proc. MicroNeuro*, pp. 86-93, Lausanne, 1996.

35. **Lazzaro, J., Ryckebusch, S., Mahowald, M. A., Mead, C. A.**, Winner-take-all network of O(n) complexity, in Touretzky, D., Ed., *Advances in neural information processing systems*, pp. 703-711, Morgan Kaufmann, 1989.

36. **Lemaitre, L., Patyra, M. J., Mlynek, D.**, Analysis and design of CMOS fuzzy logic controller in current mode, *IEEE Int. Journal of Solid-State Circuits*, Vol. 29, N. 3, pp. 317-322, 1994.

37. **Liu, B. D., Huang, C. Y., Wu, H. Y.**, Modular current-mode defuzzification circuit for fuzzy logic controllers, *Electronics Letters,* Vol. 30, N. 16, pp. 1287-1288, 1994.

38. **Liu, S. Y., Hwang, Y. S.**, CMOS four-quadrant multiplier using bias offset crosscoupled pairs, *Electronics Letters*, Vol. 29, N. 20, pp. 1737-1738, 1993.

39. **Lu, L. H., Wu, C. Y.**, The design of the CMOS current-mode general-purpose analog processor, in *Proc. IEEE Int. Symposium on Circuits and Systems*, pp. 549-552, London, 1994.

40. **Manaresi, N., Franchi, E., Guerrieri R., Baccarani, G.**, A modular analog architecture for fuzzy controllers, in *Proc. ESSCIRC'94*, pp. 288-291, Ulm, 1994.

41. **Manaresi, N., Rovatti, R., Franchi, E., Baccarani, G.,** Analog implementation of a piecewise-linear fuzzy system, in *Proc. Int. Symposium on Nonlinear Theory and its Applications,* pp. 145-148, Honolulu, 1997.

42. **Mead, C.**, *Analog VLSI and Neural Systems*, Addison-Wesley, 1989.

43. **Miki, T., Matsumoto, H., Ohto, K., Yamakawa, T.**, Silicon implementation for a novel high-speed fuzzy inference engine: mega-flips analog fuzzy processor, *Journal of Intelligent and Fuzzy Systems*, Vol. 1, N. 1, pp. 27-42, 1993.

44. **Nairn, D. G., Salama, C. A. T.**, Current-mode algorithmic analog-to-digital converters, *IEEE Int. Journal of Solid-State Circuits,* Vol. 25, N. 4, pp. 997-1004, 1990.

45. **Nairn, D. G.**, High-speed switched-current circuits using trans-impedance amplifiers, *Microelectronics Journal*, Vol. 26, N. 1, pp. 35-41, 1995.

46. **Pelgrom, M. J., Duinmaijer A. C. J., Welbers, A. P. G.**, Matching properties of MOS transistors, *IEEE Int. Journal of Solid-State Circuits*, Vol. 24, N. 5, pp. 1433-1440, 1989.

47. **Peters L., Guo, S.**, A high-speed, reconfigurable fuzzy logic controller, *IEEE Micro*, Vol. 15, N. 6, pp. 65, 1995.

48. **Qin, S. C., Geiger, R. L.**, A ±5v CMOS analog multiplier, *IEEE Int. Journal of Solid-State Circuits*, Vol. 22, N. 6, pp. 1143-1146, 1987.

49. **Rojas, I., Pelayo, F. J., Anguita, M., Prieto, A.**, Continuous-time analog defuzzifier for product-sum based implementations, in *Proc. Microneuro,* pp. 324-330, 1994.

50. **Rovatti, R.**, Fuzzy piecewise multilinear and piecewise linear systems as universal approximators in Sobolev norms, *IEEE Transactions on Fuzzy Systems*, Vol. 6, N. 2, pp. 235-249, 1998.

51. **Sackinger, E., Guggenbuhl, W.**, A high-swing, high-impedance MOS cascode circuit, *IEEE Int. Journal of Solid-State Circuits*, Vol. 25, N. 1, pp. 289-298, 1990.

52. **Sakurai, S., Ismail, M.**, High frequency wide range CMOS analog multiplier, *Electronics Letters,* Vol. 28, N. 24, pp. 2228-2229, 1993.

53. **Sánchez-Sinencio, E., Ramírez-Angulo, J., Linares-Barranco, B., Rodríguez-Vázquez, A.**, Operational transconductance amplifier-based nonlinear function syntheses, *IEEE Int. Journal of Solid-State Circuits*, Vol. 24, N. 6, pp. 1576-1586, 1989.

54. **Sánchez-Solano, S., Baturone, I., Barriga, A., Huertas, J. L.**, Estudio comparativo de circuitos básicos en lógica difusa, in Proc. *II Congreso Español sobre Tecnologías y Lógica Fuzzy*, pp. 301-312, Madrid, 1992.

55. **Sasaki, M., Inoue, T., Shirai, Y., Ueno, F.**, Fuzzy multiple-input maximum and minimum circuits in current mode and their analysis using bounded difference equations, *IEEE Transactions on Computer*, Vol. 39, N. 6, pp. 764-774, 1990.

56. **Sasaki, M., Ueno, F.**, A fuzzy logic function generator (FLUG) implemented with current-mode CMOS circuits, in *Proc. Int. Symposium on Multiple Valued Logic*, pp. 356-362, Victoria, 1991.

57. **Sasaki, M., Ishikawa, N., Ueno, F., Inoue, T.**, Current mode analog fuzzy hardware with voltage input interface and normalization locked loop, *IEICE Transactions on Fundamentals*, Vol. E75-A, N. 6, pp. 650-654, 1992a.

58. **Sasaki, M., Ishikawa, N., Ueno, F., Inoue, T.**, Current-mode analog fuzzy hardware with voltage input interface and normalization locked loop, in *Proc. IEEE Int. Conference on Fuzzy Systems*, pp. 451-457, San Diego, 1992b.

59. **Sasaki, M., Ueno, F.**, A novel implementation of fuzzy logic controller using a new meet operation, in *Proc. FUZZ-IEEE'94*, pp. 1676-1681, Orlando, 1994.

60. **Seevinck, E., Wiegerink, R. J.**, Generalized translinear circuit principle, *IEEE Int. Journal of Solid-State Circuits,* Vol. 26, N. 8, pp. 1098-1102, 1991.

61. **Sheingold, D. H.**, *Nonlinear Circuit Handbook*, Analog Devices Inc., 1976.

62. **Szczepanski, S., Wyszynski, A., Schaumann, R.**, Highly linear voltage-controlled CMOS transconductors, *IEEE Transactions on Circuits and Systems*, Vol. 40, N. 4, pp. 258-262, 1993.

63. **Tsukano, K., Inoue, T.**, Synthesis of operational transconductance amplifier-based analog fuzzy functional blocks and its application, *IEEE Transactions on Fuzzy Systems*, Vol. 3, N. 1, pp. 61-68, 1995.

64. **Turchetti, C., Conti, M.**, A general approach to nonlinear synthesis with MOS analog circuits, *IEEE Transactions on Circuits and Systems*, Vol. 40, N. 9, pp. 608-612, 1993.

65. **Vidal-Verdú, F., Rodríguez-Vázquez, A.**, Using building blocks to design analog neuro-fuzzy controllers, *IEEE Micro*, Vol. 15, N. 4, pp. 49-57, 1995.

66. **Vidal-Verdú, F.**, Una aproximación al diseño de controladores neuro-difusos usando circuitos integrados analógicos, Ph.D. dissertation, Univ. Málaga, 1996 (in Spanish).

67. **Vittoz, E. A.**, MOS transistors operated in the lateral bipolar mode and their application in CMOS technology, *IEEE Int. Journal of Solid-State Circuits,* Vol. 18, N. 3, pp. 273-279, 1983.

68. **Wang, Z.**, *Current-mode Analog Integrated Circuits and Linearization Techniques in CMOS Technologies*, Hartung-Gorre Verlag, 1990a.

69. **Wang, Z.**, Analytical determination of output resistance and DC matching errors in MOS current mirrors, *IEE Proceedings, Pt. G*, Vol. 137, N. 5, pp. 397-404, 1990b.

70. **Watkins, S. S.**, A radial basis function neurocomputer implemented with VLSI circuits, in *Proc. Int. Joint Conference on Neural Networks*, Vol. II, pp. 607-612, Baltimore, 1992.

71. **Wiegerink, R. J.**, *Analysis and Synthesis of MOS Translinear Circuits*, Kluwer Academic Pub., 1993.

72. **Yamakawa, T., Miki, T., Ueno, F.**, The design and fabrication of the current mode fuzzy logic semi-custom IC in the standard CMOS IC technology, in *Proc. 15th Int. Symposium on Multiple-Valued Logic,* pp. 76-82, 1985.
73. **Yamakawa, T.**, A simple fuzzy computer hardware system employing Min-Max operations— A challenge to 6th generation computer, in *Proc. 2nd IFSA World Congress*, pp. 827-830, Tokyo, 1987.
74. **Yamakawa, T.**, High-speed fuzzy controller hardware system: the Mega FLIPS machine, *Elsevier Science Pub. Comp. Inc., Information Sciences*, Vol. 45, pp. 113-128, 1988.
75. **Yamakawa, T.**, A fuzzy inference engine in nonlinear analog mode and its application to a fuzzy logic control, *IEEE Transactions on Neural Networks*, Vol. 4, N. 3, pp. 496-522, 1993.
76. **Zhijian, L., Hong, J.**, A CMOS current-mode high speed fuzzy logic microprocessor for a real-time expert system, in *Proc. Int. Symposium on Multiple-Valued Logic*, pp. 394-400, Charlotte, 1990.

Chapter 8

DISCRETE-TIME ANALOG TECHNIQUES FOR DESIGNING FUZZY INTEGRATED CIRCUITS

As discussed in Chapter 6, discrete-time analog techniques work with sampled analog signals, i.e., sampled signals that can take a continuous range of values. Two discrete-time techniques are widely known, switched-capacitor (SC) and switched-current (SI), which are voltage- and current-mode techniques, respectively.

SC techniques can be seen as an intermediate solution between the reliability, flexibility, and highly automatized design process typical of digital realizations and the low area occupation, low power consumption, and highly manual design process typical of continuous-time analog implementations. The basic elements of SC designs are switches, which are controlled by clock signals, and linear capacitors, which are usually fabricated with two polysilicon layers. Other elements usually employed are operational amplifiers and comparators.

Implementation of linear capacitors with two polysilicon layers requires special fabrication processes since standard digital fabrication processes do not usually contain those two layers. Another limitation of SC designs is that they are not suitable for working at low voltage supplies, like 3.3 V or inferior, towards which integrated circuitry is moving at present. These limitations prompted the development of SI techniques. SI designs also employ switches but their other basic elements are simpler. On the one hand, the capacitors they employ do not need to be linear but those that appear at the transistor gate nodes are exploited. On the other hand, MOS transistors are used instead of amplifiers. These simpler basic elements make SI designs more compact. As drawbacks of SI techniques, we can cite the lack of CAD tools to automatize their design and their low precision compared to SC designs.

While continuous-time techniques are adequate for parallel signal processing, discrete-time techniques are very well suited to sequential signal processing because data can be stored in some blocks and transmitted to other blocks according to control signals. Section 8.1 describes sequential architectures of fuzzy ICs. Sequential architecture can affect the rule processing, the data processing within each rule, or both. The architectures described are generic so that they can be realized with discrete-time analog or digital techniques. Advantages of using analog techniques are that the signal flows through a single wire instead of an N-wire bus (if the signal is represented by N bits) and that operations like addition, scaling, etc. can be carried out in a single clock cycle. The next sections of this chapter focus on the design of the basic building blocks in a fuzzy IC with discrete-time techniques. Section 8.2 describes the MFC blocks (SC designs) that have been employed in analog fuzzy ICs to generate piecewise linear membership functions. Section 8.3 describes the circuits dedicated to rule processing: minimum, maximum, scaler, and adder circuits as well as SC and SI circuits that act as memory cells and are fundamental to sequential processing. Section 8.4 focuses on the circuits required in the output stage. SC and SI designs reported in literature are described. As a summary of this chapter, timing considerations of complete systems are discussed in Section 8.5 and the design of a programmable SC fuzzy IC prototype is detailed and illustrated with experimental results in Section 8.6.

8.1 SEQUENTIAL ARCHITECTURES

Sequential architectures usually employ less circuitry than fully parallel architectures because the same blocks can implement several required operations in different time intervals and, hence, they are usually more effective in terms of area and power consumption. On the other hand, the time invested in calculating the output is greater because processing is not performed in one step. As a consequence, selecting a fully parallel, fully sequential, or a mixed architecture depends on the trade-off area/speed required. When selecting it, the designer should consider if a fully parallel architecture can be implemented on the same chip because, if it requires connecting several chips, the delays introduced can result in a sequential architecture implemented on the same chip being more effective.

We can distinguish two sequential architectures that have been employed in fuzzy ICs. One architecture sequentially processes both rules and data within each rule. The other architecture only processes the data in series within each rule because the rules are processed in parallel. The first architecture has not been employed in analog fuzzy ICs but in digital ones, so it will be described in the next chapter. This section focuses on the second architecture, which has been employed in analog fuzzy ICs that implement Mamdani-type fuzzy systems with the Center-of-Area defuzzification method and in analog fuzzy ICs that implement zero-order Takagi–Sugeno or singleton fuzzy systems. Since this architecture processes the rules in parallel, it can employs one of the schemes that were described in Section 7.2 of Chapter 7, that is, a rule by rule scheme, a scheme that shares circuitry, or an active rule-driven scheme. Discrete-time building blocks are fundamental in this architecture to store intermediate results although some circuitry can be continuous-time, as will be explained in the following.

8.1.1 SEQUENTIAL ARCHITECTURE FOR MAMDANI-TYPE FUZZY SYSTEMS

Sequential architectures for Mamdani-type fuzzy systems that use the Center-of-Area defuzzification method have been described in [Huertas, 1992; 1993; Fattaruso, 1994; Bouras, 1998]. All these proposals follow a rule by rule scheme except for the architecture in [Fattaruso, 1994], which shares the antecedent blocks, the MFCs. Some of them employ SI techniques [Huertas, 1992; Bouras, 1998] while others are SC designs [Huertas, 1993; Fattaruso, 1994]. Another difference is that some of them rely on discrete-time functional blocks [Huertas, 1993; Fattaruso, 1994] while others employ mainly continuous-time blocks to implement the required operations, reducing the discrete-time circuitry to the memory cells that store intermediate results [Huertas, 1992; Bouras, 1998]. On the other hand, the architecture reported in [Fattaruso, 1994] differs from the others in that the inputs to the chip are digital instead of analog.

The block diagram of this type of architecture is shown in Figure 8.1a. The operation cycle is divided into three stages. The first stage is devoted to computing the activation degrees of the rules by means of the MFCs and the circuits that implement the antecedent connective, like minimum circuits, or MINs. The second stage calculates the membership functions of the fuzzy sets, which represent the partial conclusions of the rules. For this purpose, the MFG provides (every clock cycle) a membership degree of the consequent fuzzy set to

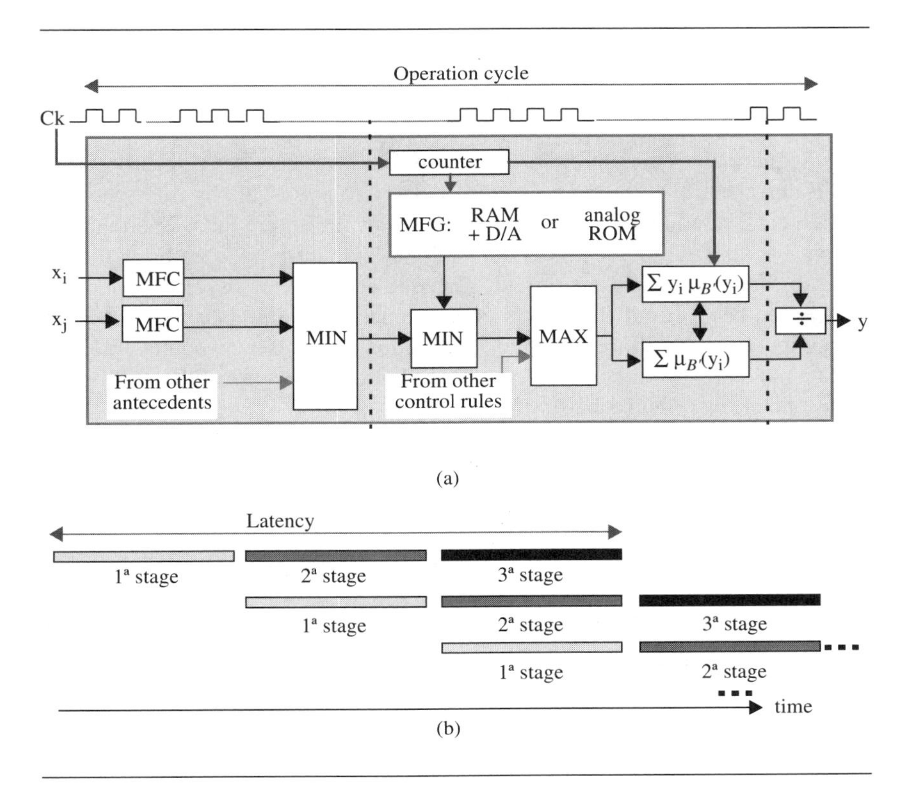

Figure 8.1. (a) Sequential architecture for Mamdani-type fuzzy systems; (b) Pipeline scheme.

be truncated by the activation degree of the rule through another MIN operator. Thus, the bus of q lines employed in a fully parallel architecture to represent the consequent membership function is replaced by a single line that provides q successive samples. The output of each rule is processed by a MAX operator. The output of the MAX enters two blocks, which will be called accumulators if they are discrete-time blocks, or integrators if they are continuous-time blocks. The third stage implements a division to compute the fuzzy system output.

The blocks that implement the required operations within each stage (membership function generation, minimum, maximum, addition, and division) can be continuous-time blocks. However discrete-time blocks are fundamental at the end of the stages to store partial results, so that pipeline schemes can be applied, as shown in Figure 8.1b. Pipeline means that the three stages can be performed at the same time, thus increasing the throughput (data per time unit) and hence the FLIPS achieved by the resulting fuzzy IC. Throughput is mainly limited by the second stage because the number of clock cycles required in this stage is directly proportional to the number of discrete elements in the output space.

8.1.2 SEQUENTIAL ARCHITECTURE FOR SINGLETON FUZZY SYSTEMS

Sequential architectures for singleton fuzzy systems have been described in [Baturone, 1994; Çilingiroglu, 1997]. They follow a rule by rule scheme in which each rule has its own circuitry, as shown in Figure 8.2a. The main difference with the previous architecture for Mamdani-type systems is that the MFG is now a simple scaler circuit since the necessity of sweeping the output space has been eliminated. As a consequence, the inference speed of these chips increases by a factor similar to the number of discrete points employed by a Mamdani-type fuzzy system to represent the membership function of the consequent fuzzy set (a factor of around 25).

Four groups of operators can be distinguished: (a) the MFCs, which calculate the membership degrees of the inputs; (b) the MINs, which compute the activation degrees of the rules through a minimum connective; (c) a scaler/adder circuit and an adder circuit that prepare, respectively, the numerator and denominator for the weighted average; and (d) a divider, which provides the output. As in the previous architecture, synchronization of the system is governed by a clock signal. As an example, in the architecture reported in [Baturone, 1994], a clock signal with two non-overlapping phases, ϕ_1 and ϕ_2, is employed, as shown at the top of Figure 8.2a. The switches can be directly controlled by this clock signal or by control signals derived from it.

There are operations that require more clock cycles than others depending on the circuits employed. For instance, division is usually the most time consuming operation if it is performed by a serial divider circuit. In this case, a

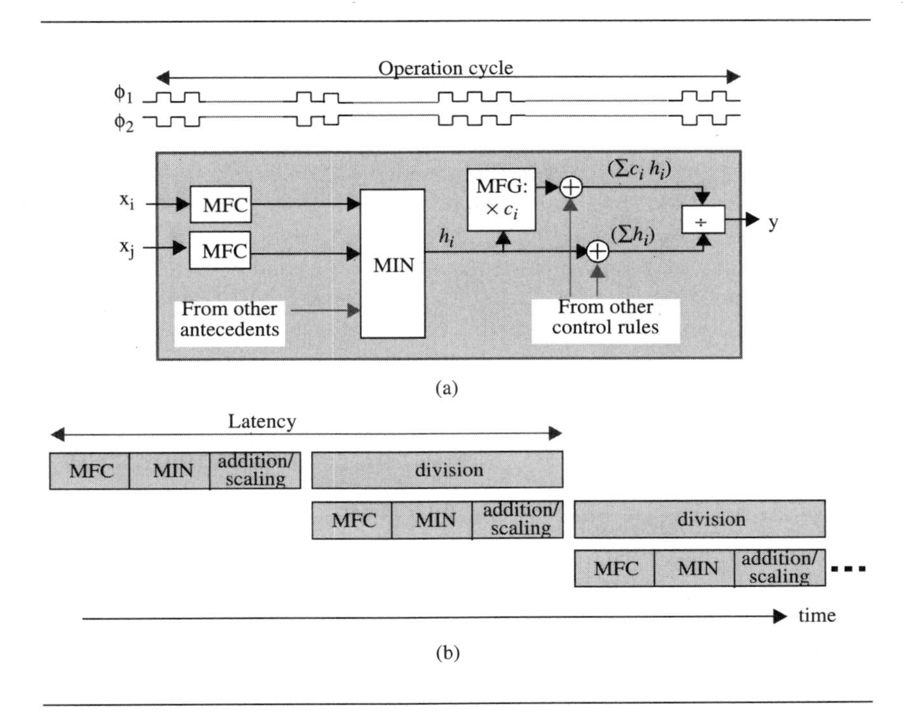

Figure 8.2. (a) Sequential architecture for singleton fuzzy systems; (b) Pipeline scheme.

pipeline scheme like that shown in Figure 8.2b can be employed. Discrete-time blocks which act as memory blocks must be included at least preceding the divider. This architecture is described in more detail in Sections 8.5 and 8.6.

8.2 FUZZIFICATION STAGE

The basic operator in the first stage of the fuzzy IC realizes a non-linear transformation on the input variables to obtain the membership degrees to the antecedent fuzzy sets. The non-linear transformations that are easily implemented with discrete-time techniques are piecewise linear transformations. Among them, those which are usually employed as membership functions are trapezoidal and triangular functions. Two types of discrete-time MFCs reported in literature that generate trapezoidal/triangular membership functions are described in the following. One of them processes the input variable in series while the other one works in parallel.

8.2.1 SERIAL MFC

The circuit illustrated in Figure 8.3a is a serial MFC that implements the symmetric trapezoidal function shown in Figure 8.3b [Huertas, 1993]. The input variable X is sampled and sequentially compared with some reference voltages that represent the break points of the membership function. These voltages, V_1, V_2, V_3, V_4, are provided by the reference circuit *ref*-1. A finite-state machine (FSM) takes the result of this comparison and controls the different switches according to one of the possible five linear pieces. The other reference circuit, *ref*-2, provides the coefficients required by each transformation. Therefore, the number of clock cycles invested is the number of break points plus an additional cycle to perform the transformation. A way to implement reference circuit *ref*-1 is shown in Figure 8.3c while Figure 8.3d shows the state diagram of the FSM (considered as a Moore's machine). Table 8.1. shows the output values of the MFC and the values that provide the reference circuit *ref*-2 for each state of the FSM.

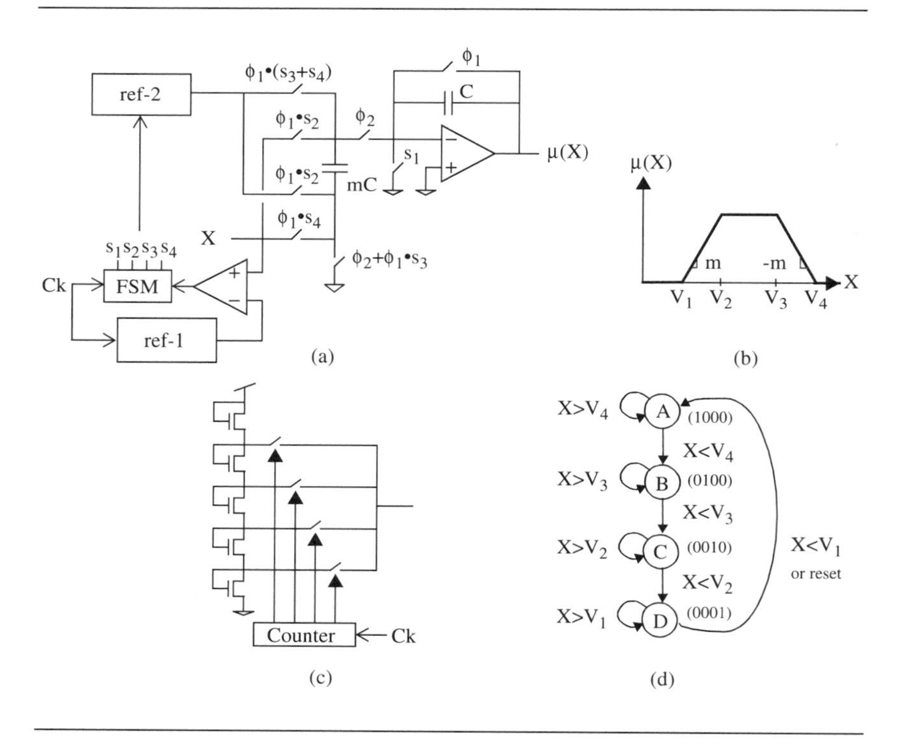

Figure 8.3. (a) Serial MFC reported in [Huertas, 1993]; (b) Symmetric trapezoid implemented by the circuit in (a); (c) Possible realization of the circuit *ref*-1; (d) State diagram of the FSM.

Table 8.1. Output values of the MFC, $\mu(X)$, and circuit *ref-2* depending on the FSM state.

(state) s_1 s_2 s_3 s_4	μ(X)	ref-2
(A) 1000	0	---
(B) 0100	$m(V_4-X)$	V_4
(C) 0010	$m(V_2-V_1) = m(V_4-V_3)$	$V_1-V_2 = V_3-V_4$
(D) 0001	$m(X-V_1)$	V_1

8.2.2 PARALLEL MFC

An MFC designed with SC techniques that processes the input variable in parallel has been reported in [Baturone, 1994]. Its schematic is shown in Figure 8.4a while Figure 8.4b indicates the parameters that define the membership function generated. This MFC is based on the adder/scaler SC structure proposed in [Krummenacher, 1982; Gregorian, 1986], which is depicted within the discontinuous rectangle in Figure 8.4a and will be described in Section 8.3.2. Once the operation cycle has begun and the input variable has been sampled, the capacitor of capacitance C/m is precharged to the voltage V_{top} while the comparator checks if the input X is bigger or smaller than V_{aux}. The latch after the comparator maintains the comparison result during the first clock cycle to control the input switches adequately. During the second phase of the

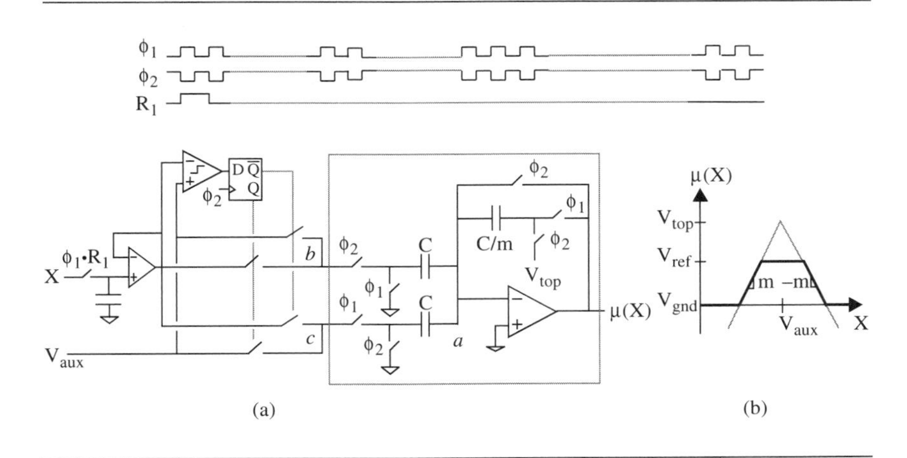

Figure 8.4. (a) Parallel MFC reported in [Baturone, 1994]; (b) Symmetric trapezoid generated by the circuit in (a) when it is combined with a minimum circuit.

first clock cycle, when ϕ_2 is high ('1'), the switch controlled by ϕ_2 which is placed in the feedback loop of the amplifier, makes node *a* be virtual ground. If the charge conservation principle is applied at node *a* and the behavior of the switches and the amplifier is considered ideal, the output of this MFC, which is available at the first phase, ϕ_1, of the second clock cycle is: $V_{top}+m\bullet[V(b)-V(c)]$. If X is bigger than V_{aux}, $V(b)$ is V_{aux} and $V(c)$ is X and the negative-slope piece, $V_{top}+m\bullet(V_{aux}-X)$, is generated. On the other hand, if X is smaller than V_{aux}, $V(b)$ is X and $V(c)$ is V_{aux} and the positive-slope piece, $V_{top}+m\bullet(X-V_{aux})$, is provided. These pieces are illustrated with discontinuous lines in Figure 8.4b.

In order to obtain a symmetric trapezoidal function, the pieces generated should be truncated by the values corresponding to the null ('0') and maximum ('1') membership degrees, which are denoted, respectively, by V_{gnd} and V_{ref} in Figure 8.4b. This can be done by implementing a minimum operation, $\min(\mu(X), V_{ref})$, and a bounded-difference operation, $\mu(X) \ominus V_{gnd}$. The value V_{gnd} is chosen to be ground so as to verify that non-active rules have null activation degrees.

Another MFC similar to this one because it is based on the same adder/scaler scheme has been described in [Çilingiroglu, 1997]. It also implements trapezoidal/triangular membership functions but in this case the horizontal pieces are generated by the saturation of the amplifier employed.

The MFCs mentioned thus far offer the advantage of being easily programmed by controlling their capacitance ratios with switches. This is a general advantage of SC circuits. The typical structures employed in SC circuits to include digital programmability are arrays of parallel capacitors with binary-weighted capacitance values, as shown in Figure 8.5. [Allen, 1984]. The equiv-

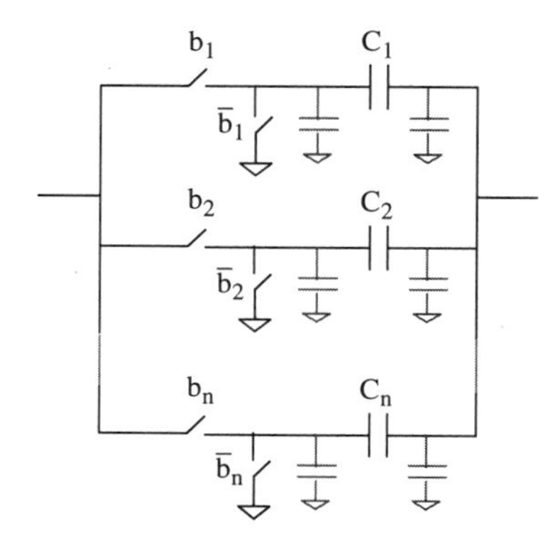

Figure 8.5. Programmable array of parallel capacitors.

alent capacitance of this structure is given by:

$$C_{eq} = b_1 \cdot C_1 + b_2 \cdot C_2 + \ldots + b_n \cdot C_n \tag{8.1}$$

In the described MFCs these structures can be used to program the slope, m, of the membership function. The effect of parasitic capacitances (shown with gray lines in Figure 8.5) is negligible because the capacitor terminals are connected to low-impedance nodes.

Offset of the comparator does not usually require to be cancelled. For instance, if the MFC works with a full scale of 4 V, and 6-bit resolution is employed for the input spaces, the comparator in Figure 8.4a may have an error of 30 mV (and the input-referred DC offset of MOS amplifier/comparators does not usually exceed 10 mV [Gregorian, 1986]).

8.3 RULE PROCESSING STAGE

The discrete-time fuzzy ICs that have been reported in the literature implement Mamdani-type fuzzy systems with Center-of-Area defuzzification method and singleton fuzzy systems. The operators needed to implement these systems are the minimum, maximum, addition, and scaling. This section describes discrete-time techniques that carry out these operations. Since fuzzy ICs with sequential architecture and pipeline schemes require memory cells to store intermediate results, SC and SI memory cells are also described in this section. Finally, there are fuzzy systems such as those implementing the Product-Sum inference method or Takagi–Sugeno methods of first or superior order that require product operators. Multiplier circuits that implement these operators will be described in Section 8.4.2 together with divider circuits.

8.3.1 MINIMUM AND MAXIMUM OPERATORS

Once the membership degrees have been calculated, the next operation to be performed in a fuzzy IC is the connection of the rule antecedents to compute the rule activation degree, h_i. If an "and" connective is considered, the associated operator is usually the minimum and the circuit which implements it is a multi-input MIN circuit. Two types of MIN/MAX circuits are described in the following. As with the above mentioned MFCs, one of them processes the inputs in series while the other one operates in parallel.

- **Serial MIN/MAX circuits**

A multi-input MIN circuit that processes the input in series is shown in Figure 8.6 with black lines, for the particular case of two inputs. A MAX circuit has the same schematic but interchanging the comparator terminals. The additional circuitry shown with gray lines is employed to realize the horizontal pieces of the membership function that was shown in Figure 8.4b, that is, to perform the minimum operation min(μ(X), V_{ref}) and the bounded difference

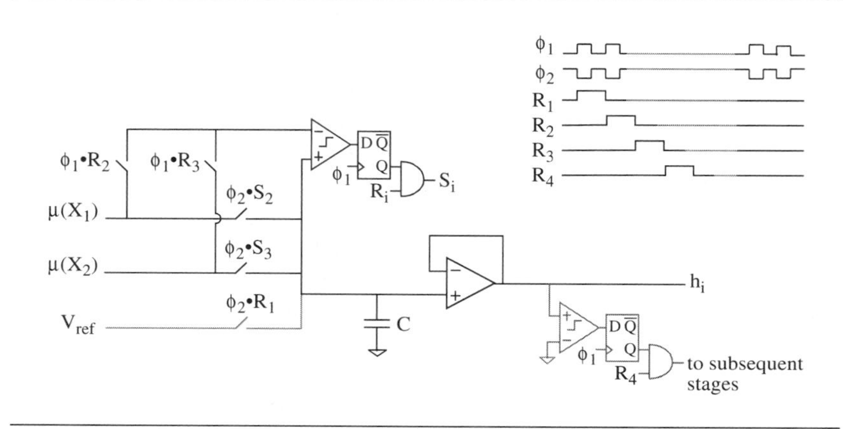

Figure 8.6. Two-input MIN circuit.

operation, $\mu(X)\ \Theta\ V_{gnd}$. The latter operation is carried out by the comparator placed at the bottom of Figure 8.6. The output of this comparator controls subsequent blocks of the fuzzy IC to avoid considering negative activation degrees.

Let us consider that the MIN circuit in Figure 8.6 is placed after the MFC shown in Figure 8.4a. When signals ϕ_2 and R_1 are high (the MFC is calculating the membership degree), the capacitor precharges to the value V_{ref} (V_{ref} is V_{top} if the membership function is a triangle). During the next clock cycles (R_2 is high, R_3 is high, and so on) and when phase ϕ_1 is high, the input voltages are sequentially compared with the voltage stored in the capacitor. If the input voltage is smaller than the stored voltage, the associated switch is closed and the input voltage is stored when phase ϕ_2 is high. Otherwise, the stored voltage is not modified. The comparator is again followed by a latch so that comparison result is available during a complete clock cycle. Thus, the first phase of the clock cycle is employed to perform the comparison while the evaluation is carried out during the second one. This is more secure than performing comparison and evaluation in just one phase because comparison delays may make the capacitor charge erroneously to a voltage superior to that stored.

The time invested by this circuit in computing the output is as many clock cycles as inputs. Considering the circuit in Figure 8.6, the operation is performed in two cycles and the final result, h_i, can be transferred to the subsequent block when ϕ_2 and R_3 are high. The response time of this circuit is low because the only constant times involved are those of capacitor charge and discharge. Hence the clock frequency can be very high. As a matter of fact, the clock frequency of the fuzzy IC is not limited by this circuit.

• **Parallel MIN/MAX circuits**

A multi-input MIN/MAX circuit that processes the inputs in parallel is described in [Çilingiroglu, 1997] and illustrated in Figure 8.7 for the case of three inputs. The circuit consists of U identical cells, where U is the number of inputs, and contains a number of transistors of order U^2 (this is different from the serial MIN/MAX circuit described above which is of order-U complexity). Each cell contains a competition node, n_i, which is also an input/output node connected to all the other cells. There are two basic transistors within each cell, a grounded NMOS transistor, M_{di}, and a PMOS transistor, M_{ui}, connected to the positive supply. The NMOS transistor is called the *pull-down* transistor because it tends to fix the voltage at node n_i to a low value when signal ϕ_2 is high ('1'). The PMOS transistor is called the *pull-up* transistor because it tends to fix that voltage to a high value when ϕ_2 is high ('1') and provided that some of the intermediate PMOS transistors, M_{ij}, of the cell are conducting. Once the competition nodes take the voltages of the input signals, V_i, the switches controlled by ϕ_1 are open and signal ϕ_2 rises so that the minimum input selection phase starts. During this phase, the circuit converges to a stable state in which only one node n_i, the one associated to the minimum input, takes a low value

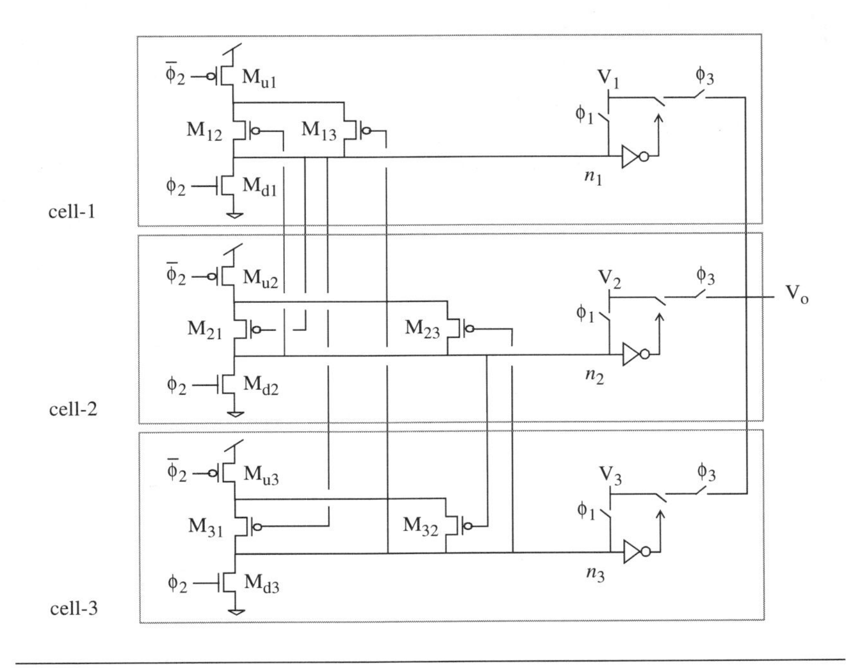

Figure 8.7. Three-input parallel MIN circuit proposed in [Çilingiroglu, 1997].

(ground) and all the others take a high value (V_{DD}-$|V_{TP}|$). All the intermediate PMOS transistors of the i-th cell, M_{ij}, are off. Among the intermediate PMOS transistors of the other cells, only the one whose gate is connected to n_i, M_{-i}, is on. In the output phase, the switches controlled by ϕ_3 are closed, thus transferring the result, V_o=min{V_1, ..., V_U} to the subsequent processing block. The circuit invests two clock cycles independent of the number of inputs; ϕ_1 and ϕ_2 are non-overlapping phases, and ϕ_2 and ϕ_3 are overlapping phases.

A multi-input MAX circuit can be obtained with the same structure by replacing the intermediate PMOS transistors with NMOS transistors and by taking the drains of the pull-up transistors instead of the drains of the pull-down transistors as both competition and input/output nodes [Çilingiroglu, 1993].

8.3.2 ADDITION AND SCALING OPERATORS

If the fuzzy IC works in current mode, addition and scaling operations are usually realized in continuous time by connecting wires and employing current mirrors. If the results of these operations have to be stored, SI memory cells like those which will be described in Section 8.3.3 must be employed. On the other hand, if the fuzzy IC works in voltage mode, SC adder/scaler circuits are usually employed.

The circuit shown in Figure 8.8a represents a conventional continuous-time voltage-mode adder/scaler circuit (realized with RC-active techniques). The SC voltage-mode adder/scaler circuit shown in Figure 8.8b [Krummenacher, 1982; Gregorian, 1986] results from replacing the resistors of the former circuit with switches and capacitors (this is the circuit that was commented on when describing the parallel MFCs in Section 8.2.2). Considering that the amplifier and the switches are ideal, the output of this circuit during the *k*-th clock cycle of period T is:

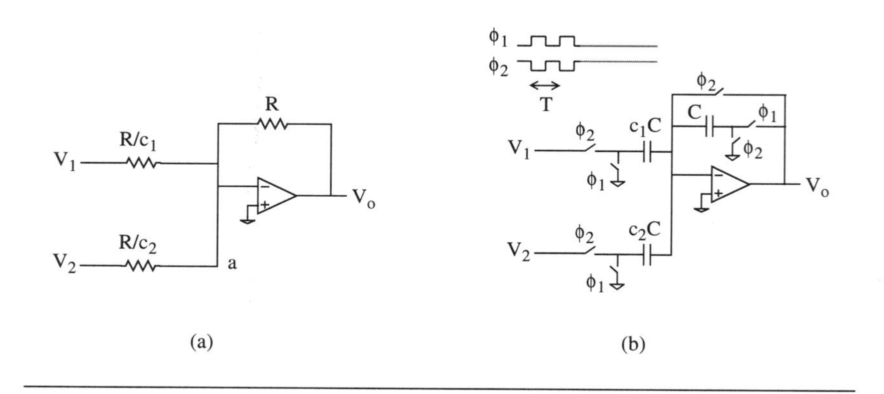

Figure 8.8. Adder/scaler circuits realized with (a) RC-active and (b) SC techniques.

$$V_o(kT) = c_1 V_1\left(kT-\frac{T}{2}\right)+c_2 V_2\left(kT-\frac{T}{2}\right) = \sum_{i=1}^{2} c_i V_i\left(kT-\frac{T}{2}\right) \tag{8.2}$$

A subtraction operation is implemented by simply interchanging the switches associated to the input signal that has to be subtracted (as was also seen when describing the parallel MFCs).

Several errors appear in SC circuits that modify the ideal behavior. On the one hand, the switches are not ideal because their resistance is not null when they are conducting and they have parasitic capacitances that produce a charge injection effect known as *feedthrough*. These errors can be reduced by employing CMOS switches of minimum geometries (provided they allow a full charge transference in the specified time). On the other hand, the capacitance values are affected by systematic errors due to parasitic capacitances and random errors in the fabrication process. The first cause of errors is avoided by employing structures like that in Figure 8.8b, provided that the amplifier gain is high. The other cause is reduced if unit capacitors are used when designing the layout of the circuit. Regarding the amplifier non-idealities, we can cite saturation, offset (V_{os}), and a finite gain (A_o). This means that the transfer function of the amplifier working in the linear region (not saturated) is given by:

$$V_o = A_o(V_{plus} - V_{minus} + V_{os}) \tag{8.3}$$

where V_{plus} and V_{minus} are, respectively, the voltages at the negative and positive input terminals.

If the amplifier gain is high, the SC structure of Figure 8.8b is not only insensitive to parasitic capacitances but also to the amplifier offset.

Another non-ideality of real amplifiers is that their output resistance is not null and hence the voltage at the output node can be affected by the load. However, this is not an important problem in SC designs because the amplifiers only drive capacitive loads. The maximum clock frequency at which an SC circuit can operate is usually limited by the *slew-rate* of the amplifier (which is defined as the maximum variation rate of the voltage at the output node), especially if working with SC structures where the amplifier has to commutate between virtual ground and the corresponding nominal value. As a consequence, perhaps the most important consideration when designing a high-frequency SC circuit is the selection of an amplifier that has a very short settling time and ensures enough gain to achieve the required resolution.

Adder/scaler circuits like that in Figure 8.8b can be employed to generate the numerator signal, $\Sigma h_i c_i$, of the weighted average, which is required by singleton fuzzy systems. The consequent singleton values can be digitally programmed by replacing the capacitors $c_i C$ with a binary-weighted capacitor array. On the other hand, switches ϕ_2 and ϕ_1 can always be controlled (as was discussed in Section 8.3.1) so that they do not consider negative rule activation degrees, h_i.

8.3.3 MEMORY CELLS

Memory cells are particular to discrete-time techniques. They are fundamental for storing intermediate results of data sequential processing, especially data that is transferred between different pipeline stages.

The adder/scaler circuit described in the previous section (Figure 8.8b) can be employed as a memory cell that stores the result V_o while the switch controlled by ϕ_1 is closed and the one controlled by ϕ_2 is open. It is reset and takes a new sample when the switch controlled by ϕ_2 is closed and the one controlled by ϕ_1 is open.

Memory cells that work in current mode are simpler, as can be deduced from Figure 8.9. The input transistor of these blocks translates the input current into a voltage that is stored in the gate-to-source parasitic capacitance and that can be later recovered and translated into a current. The block in Figure 8.9a represents a first-generation memory cell that invests a single clock cycle and employs two transistors to store and recover the data. The block in Figure 8.9b is a second-generation memory cell that invests two clock cycles but employs only one transistor. The advantage of the latter cell is that it is insensitive to mismatching because the same transistor is used to store and to recover the data [Hughes, 1990]. However, it remains sensitive to errors due to the channel modulation effect. To avoid them, cascode or regulated memory cells can be employed (remember cascode and regulated current mirrors discussed in Section 7.3.2 of Chapter 7). The other important cause of errors is the switch feedthrough that produces variations in the charge stored. A more detailed discussion about these types of errors and how to reduce them can be seen in [Toumazou, 1990; 1993] and the references cited there.

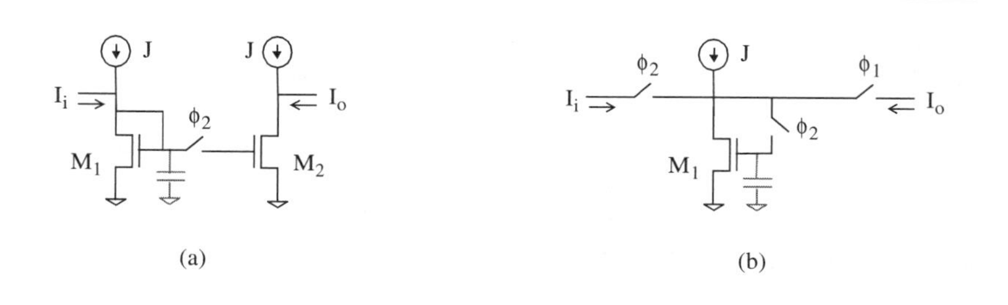

Figure 8.9. (a) First-generation and (b) second-generation SI memory cells.

8.4 OUTPUT STAGE

As was explained in Section 7.5 of Chapter 7, the circuitry employed in the output stage of a fuzzy IC mainly depends on how the fuzzy system implemented represents the rule consequents. This section is devoted to explaining the discrete-time circuitry that can be employed to carry out the Center-of-Area calculation in systems that deal with fuzzy representation of consequents and to perform the weighted average calculation required by systems with parametric representation of consequents.

8.4.1 FUZZY REPRESENTATION OF CONSEQUENTS: ACCUMULATOR CIRCUITS

The operation that has to be implemented to calculate the center of area is given by:

$$y_{out} = \frac{\sum_{y_i=1}^{n} y_i \cdot \mu_{B'}(y_i)}{\sum_{y_i=1}^{n} \mu_{B'}(y_i)} \tag{8.4}$$

The values $\mu_{B'}(y_i)$ represent the membership function of the fuzzy set resulting from processing all the rules. These values are provided by a MAX circuit as the Mamdani Min-Max method is implemented. The values y_i represent the points of the discrete output space. They are given by a sequential MFG, one after the other every time step. This means that the sums of numerator and denominator in (8.4) have to be implemented by accumulators. These accumulators are also called integrators but as they are discrete-time circuits their operation is different from that of continuous-time integrators, which were discussed in the previous chapter. An SC adder/scaler block, like that shown in Figure 8.8b, can act as an accumulator if: (1) it has a single input; (2) it receives the data sequentially; and (3) the switch that resets the capacitor in the feedback loop of the amplifier closes every integration cycle.

Regarding SI techniques, a simplified version of an accumulator is shown in Figure 8.10a [Toumazou, 1990; 1993]. A current-mode circuit to calculate the center of area that employs SI accumulators like that shown in Figure 8.10a has been proposed in [Pelayo, 1993]. Its block diagram is illustrated in Figure 8.10b. The cascade connection of two accumulators makes it possible to obtain the numerator and denominator of (8.4) without using a multiplier circuit that multiplies y_i and $\mu_{B'}(y_i)$. The idea is to exploit that:

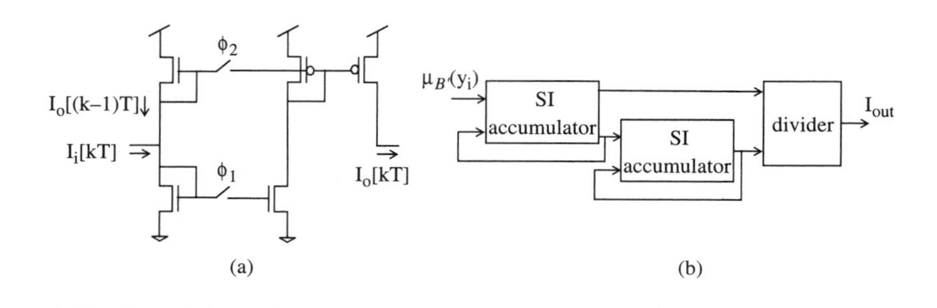

Figure 8.10. (a) Simplified SI accumulator (T represents the clock period); (b) Block diagram of the defuzzifier circuit proposed in [Pelayo, 1993].

$$\sum_{y_i=1}^{n} y_i \cdot \mu_{B'}(y_i) = \mu_{B'}(n) + [\mu_{B'}(n) + \mu_{B'}(n-1)] + \ldots$$

$$\ldots + [\mu_{B'}(n) + \mu_{B'}(n-1) + \ldots + \mu_{B'}(1)] \qquad (8.5)$$

which is the idea used in [Watanabe, 1990] to implement (8.4) with a digital circuit.

Once the numerator and denominator have been computed, the next step is to perform division. The divider circuit employed in [Pelayo, 1993] is a MOS translinear circuit (therefore, a continuous-time circuit) like those described in the previous chapter. Discrete-time divider circuits can also be employed. Their design is addressed in the following section.

8.4.2 PARAMETRIC REPRESENTATION OF CONSEQUENTS: DISCRETE-TIME MULTIPLIER/DIVIDER CIRCUITS

As explained in Chapter 7, the operation that has to be performed in the output stage of fuzzy systems that represent the rule consequents by several parameters is the weighted average:

$$y_{out} = \frac{\sum_{i=1}^{R} w_i \cdot f_i}{\sum_{i=1}^{R} w_i} \qquad (8.6)$$

where f_i is the parametric representation of the i-th rule conclusion and w_i is its weight.

The sums of numerator and denominator can be realized by the circuits described in Section 8.3.2, and the scaling circuits shown in that section can also be employed if f_i is a constant value, like a singleton value. Multiplier circuits are required whenever f_i is a variable that depends, for instance, on the activation degree of the rule or on the input signal values. In addition, a divider circuit is needed. The most significant discrete-time techniques for designing multiplier/divider circuits are discussed in the following.

- **Time division techniques**

The divider circuits based on this technique (one of the first reported techniques) generally consist of a commutator that generates a pulse train and a circuit that modulates the width of the pulses (a pulse-width-modulation or PWM circuit). The output is the mean value of these pulses, exploiting that this mean value is directly proportional to the product between the pulse amplitude and its duration (duty-cycle). Computing this average makes these circuit operate slowly. As an example, the circuits reported in [Cichoki, 1986; De Cock, 1992] operate at frequencies that range from 130 to 300 KHz.

- **Successive-approximation techniques**

It was already described in the previous chapter that given three input signals x_p, x_{num}, and x_{den}, the multiplier/division operation $x_{out}=x_p \cdot x_{num}/x_{den}$ can be implemented by combining an A/D with a D/A converter. The A/D block receives x_{num} as input and x_{den} as the reference signal, and provides a digital code $\{b_i\}$ that represents the discrete value of the ratio x_{num}/x_{den}, which can be denoted as $Q(x_{num}/x_{den})$. The D/A block receives $\{b_i\}$ as input and x_p as the reference signal and provides the following output:

$$x_{out} = x_p \cdot \sum_{i=1}^{N} 2^{-i} \cdot b_i = x_p \cdot Q\left(\frac{x_{num}}{x_{den}}\right) \tag{8.7}$$

if the D/A is a divider-type converter or, equivalently:

$$x_{out} = \frac{x_p}{2^N} \cdot \left(\sum_{i=1}^{N} 2^{N-i} \cdot b_i\right) = x_p \cdot Q\left(\frac{x_{num}}{x_{den}}\right) \tag{8.8}$$

if the D/A is a multiplicative-type converter. The quantization error is given by $\pm x_{den} / 2^{(N+1)}$.

A fuzzy IC with this A/D-D/A scheme as a divider circuit can interact directly with analog actuators and digital processing environments since the output is provided in both analog and digital formats. The previous chapter describes the design of continuous-time A/D-D/A schemes. The design of discrete-time A/D-D/A schemes with SC techniques is discussed in the following.

Open-loop converters

Numerous types of SC A/D and D/A converters with different features regarding accuracy, operation speed, and silicon area have been reported in the literature [Allen, 1984]. Flash (or parallel) open loop A/D converters are the fastest ones because their speed depends on the response of combinatorial circuitry and comparators. Conversion is usually performed in just one clock cycle. On the other hand, they occupy a large silicon area because they employ 2^N-1 comparators for an N-bit resolution, a resistor array to divide the reference voltage signal into 2^N parts, and a digital decoder network. The trade-off area/speed also appears in D/A converters. Those that consist of a single stage, for instance, those based on binary-weighted capacitor arrays are the fastest ones but they occupy larger areas than the sequential converters that rely on charge redistribution or on cyclic or iterative algorithmic structures.

Figure 8.11 illustrates an SC multiplier/divider circuit that contains parallel A/D and D/A converters. Once the numerator and denominator of the weighted average are available (the values Σw_i and $\Sigma w_i f_i$), the digital and analog signals that represent their quotient are obtained during the second phase, ϕ_2, of the first clock cycle. The divider circuit of the fuzzy IC reported in [Çilingiroglu, 1997] employs a similar structure but the A/D block does not contain digital decoder circuitry, so the output is provided only in analog format.

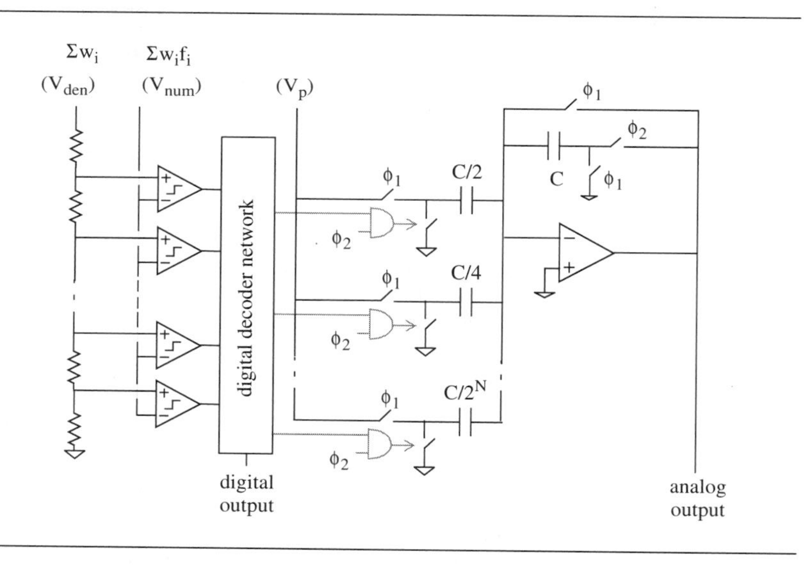

Figure 8.11. A/D-D/A divider based on parallel converters.

Closed-loop converters

Compared with open-loop converters, the response time of sequential closed-loop converters is longer since it is directly proportional to their resolution (number of bits) but their area occupation is smaller because it does not increase exponentially with resolution. Among this type of converters we can cite the *serial converters*, such as the single-, constant-, or double-slope converters, which are typically slow (conversion frequency lower than approximately 100 Hz) but can achieve high resolution (superior to 12 bits) [Allen, 1984]. Another type is the group of *successive-approximation converters*, such as those based on charge redistribution [McCreary, 1975] and the algorithmic converters that can be realized by replicating blocks or by iterating operations in time (also known as cyclic converters) [McCharles, 1978]. The constant times involved in converters based on charge redistribution among capacitors are related to capacitor charges and discharges while the response time of cyclic converters is limited by the settling time of the amplifiers. This is why the first converters are faster than the latter converters. On the other hand, cyclic converters occupy the least space and can achieve resolutions of approximately 12 and 13 bits [Shih, 1986; Ohara, 1987; Onodera, 1988], because many techniques have been reported to obtain structures capable of self-calibration, independent of capacitor ratios, and insensitive to amplifier/comparator offset as well as to parasitic capacitances.

The conversion mechanisms employed by successive-approximation A/D converters are shown in the flow charts of Figure 8.12 [Nairn, 1990].

Figure 8.13 shows a multiplier/divider circuit based on the redistribution-charge A/D converter reported in [McCreary, 1975]. During the sampling mode, which can be performed during the first phase, ϕ_1 is '1', of the first clock

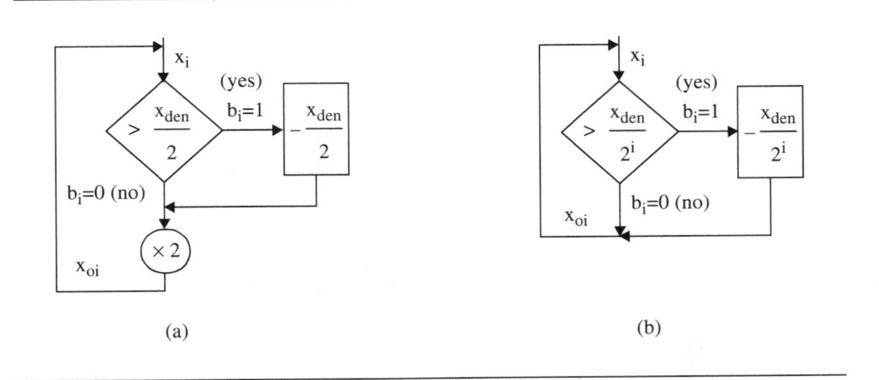

Figure 8.12. Flow charts of successive-approximation A/D converters (x_{den} represents the full scale): (a) multiplicative-type and (b) divider-type.

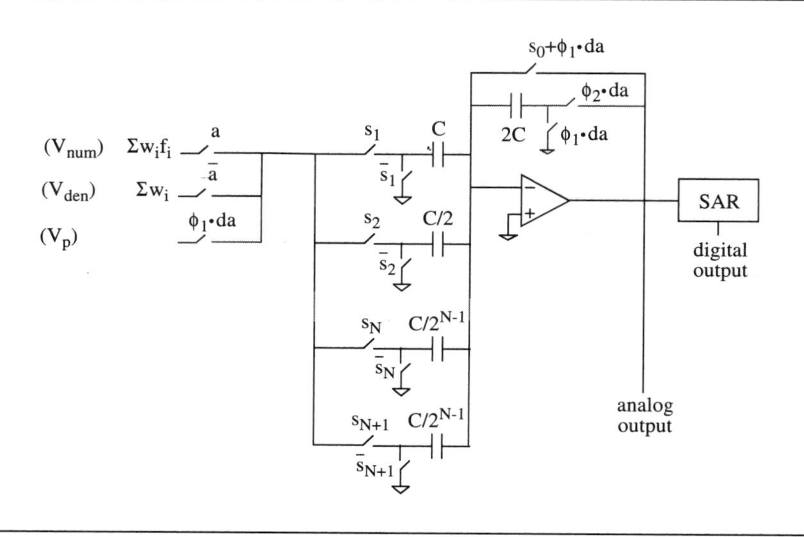

Figure 8.13. A/D-D/A divider circuit based on a charge-redistribution A/D converter.

cycle ($a=s_i=\phi_1$), the switch controlled by s_o is closed so as to configure the amplifier as a voltage follower to eliminate the potential influence of amplifier offset [Gregorian, 1986]. During the next phase, ϕ_2 is '1', the switches controlled by a and s_o are open and those controlled by $\bar{s}_i$ are closed, thus holding a voltage $-V_{num}$ at the negative input terminal of the amplifier. The successive-approximation register (SAR) contains all the circuitry required to control the switches. After the sampling and holding modes, the amplifier, which acts as a comparator when operating in open-loop configuration, provides a bit in each successive phase, so that the N-bit digital code is available after 1+N/2 clock cycles. The response time of this converter depends on charging/discharging constant times that are shorter than amplifier settling times. Therefore, the SAR can generate control signals with periods smaller than the period of the main clock.

Once the digital code has been calculated, it can be converted to analog format during the next clock cycle, taking advantage of the binary-weighted capacitor array (except for the capacitor $C/2^{N-1}$, which is repeated) and employing the circuitry shown with gray lines in the amplifier feedback path. Operation speed is inferior to that obtained by the circuit in Figure 8.11 but the area occupation is smaller. The fuzzy IC described in [Fattaruso, 1994] employs an A/D converter of this type to obtain the global system output (only provided in digital format).

A multiplier/divider circuit based on cyclic or iterative A/D-D/A converters is reported in [Piedade, 1990]. The output is provided in analog format after two operation stages. The first stage consists of N steps to obtain the N-bit digital code. This code is translated to analog format during the next stage, which again consists of N steps. Another realization based on this type of converters is illustrated in Figure 8.14. The two constituent converters show similar structures, thus easing the layout design. This circuit performs division in only one stage so that higher operation speed is achieved. As the bits, b_i, which are sequentially obtained by the divider-type A/D converter, enter the multiplicative-type D/A converter, the analog output is computed. Thus, analog and digital N-bit resolution outputs are simultaneously obtained after N+1 clock cycles. The response time is somewhat superior to the divider circuit based on charge-redistribution converters, but the area occupation is smaller and, in addition, this structure offers the flexibility of providing different resolutions if the fuzzy IC works with different operation cycles.

The possible offset of the comparator in the A/D block of the divider circuit has to be reduced by using offset cancellation techniques whenever high resolution is required (superior to 6 bits).

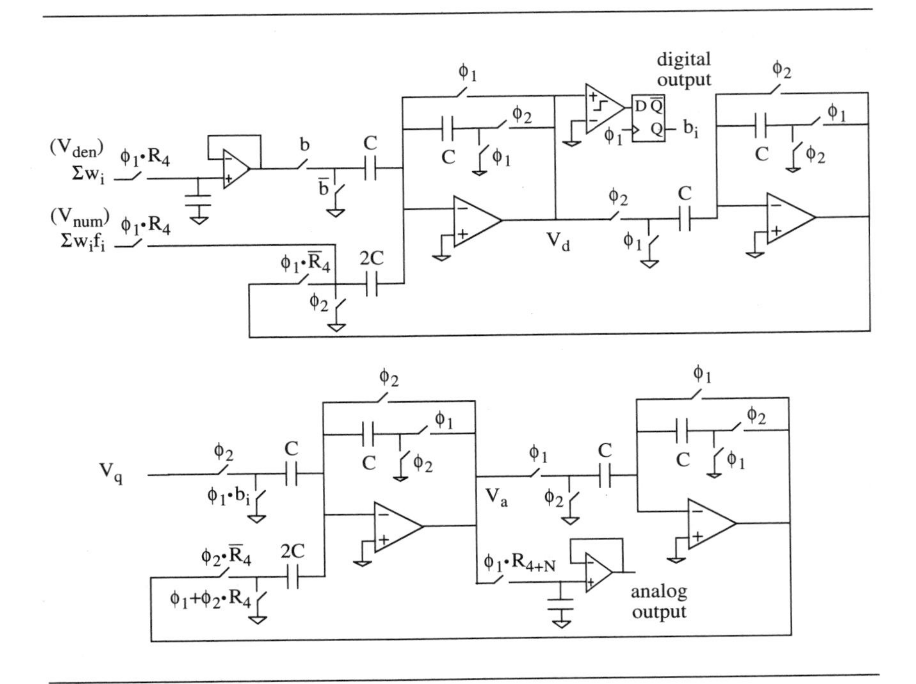

Figure 8.14. A/D-D/A divider circuit based on cyclic converters.

8.5 TIMING AND INFERENCE SPEED OF DISCRETE-TIME SINGLETON FUZZY ICS

Let us focus on the design of discrete-time fuzzy ICs that implement a singleton fuzzy system using the minimum as the antecedent connective. The inference speed depends primarily on the divider circuit employed. Since SC dividers have been discussed, let us consider an SC fuzzy IC with u-antecedent rules, with a serial MIN circuit (controlled by the main clock signal whose period is T), and which provides an N-bit output signal. The latency is given by:

- $(2+u)$T, if the divider circuit is based on parallel converters, or
- $(3+u+N/2)$T, if the A/D part of the divider circuit relies on charge redistribution, or
- $(2+u+N)$T, if the divider circuit consists of cyclic converters.

The SC fuzzy ICs that contain divider circuits with parallel converters are the fastest ones and can operate with latency under 1 microsecond, although they are the most area consuming. The fuzzy ICs whose divider circuit employs cyclic converters occupy the least space, and they are the most versatile (since the number of bits of the output can be modified by changing the operation cycle), but they are the slowest (latency of tens of microseconds for resolutions of 5 and 6 bits).

However, since discrete-time functional blocks are employed, pipeline schemes can be used, so that new input data can be sampled before providing the output corresponding to the old samples, thus increasing the throughput. As an example, Figure 8.15a shows the timing of the different functional blocks of a fuzzy IC with 2 antecedents per rule, a MIN circuit governed by the main clock signal, and a divider circuit based on parallel converters. The throughput is one output every 3 clock cycles (those that are invested by the MFCs and MIN blocks). If the divider circuit with parallel converters is replaced by a 5-bit divider circuit based on cyclic converters, the fuzzy IC provides one output every 5 clock cycles and also samples the input every 5 clock signals (the latency is 9 clock cycles in this case). This is illustrated in Figure 8.15b.

From the above, it can be seen that SC fuzzy ICs employ very few clock cycles to carry out fuzzy inferences (4 or 9 in the examples above). In particular, they usually invest fewer clock cycles than digital fuzzy ICs, however the frequency of their clock signals is also much lower. As discussed previously, clock frequency is limited by the slew-rate and the bandwidth of the operational amplifier selected. For instance, considering a two-stage Miller amplifier, like the one shown in the following section, working with a clock period of 0.8 μs, the fuzzy IC achieves 417 KFLIPS in the case of Figure 8.15a, and 250 KFLIPS in the case of Figure 8.15b. Taking into account that single-stage amplifiers, such as folded-cascode ones, offer a higher speed [Makris, 1990;

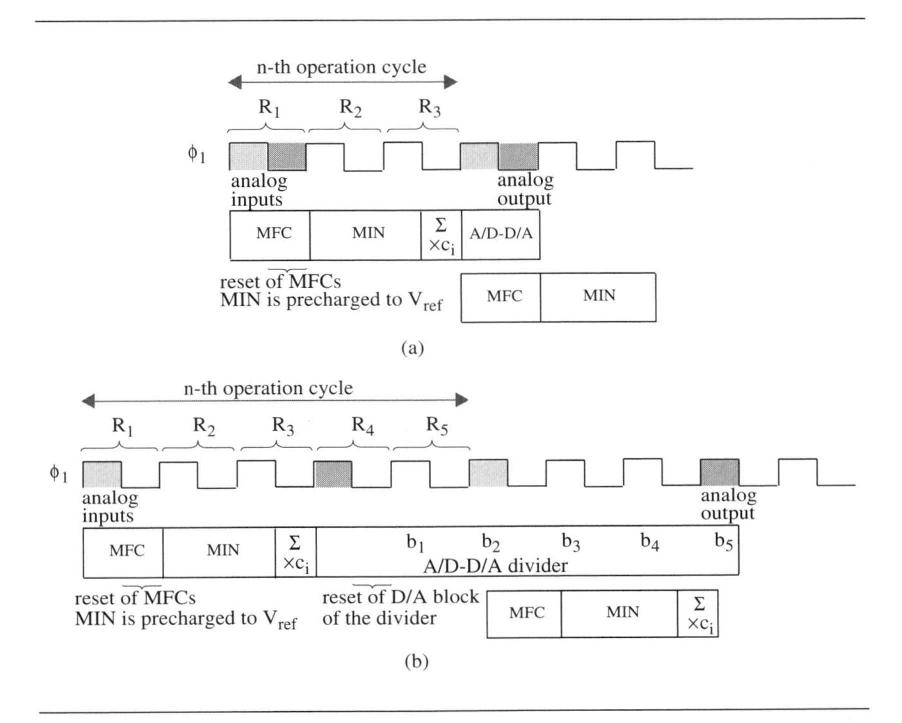

Figure 8.15. Pipeline schemes of a fuzzy IC that contains a divider circuit based on: (a) parallel converters; (b) cyclic converters.

Robertini, 1992] and that, in the case of Figure 8.15a, the self-sequentialization of the MIN circuit increase the throughput, the inference speed of the chip may reach the range of MFLIPS.

8.6 DESIGN OF AN SC CMOS PROTOTYPE

In order to illustrate what has been discussed in this chapter, let us consider a singleton fuzzy system with one input, one output, and 4 rules. It can be implemented with a fuzzy IC whose block diagram is shown in Figure 8.16a. The system is able to approximate different unidimensional functions. In particular, let us consider the approximation of the sinusoidal function shown in Figure 8.16b, by using symmetric triangular membership functions like those illustrated in Figure 8.16c.

The circuitry that implements one rule (an MFC and a scaler circuit) has been integrated in a 2.4-μm CMOS technology. The MFC follows the serial scheme that was shown in Figure 8.4a while the scaler circuit has the schematic that was shown in Figure 8.8b. The input capacitors to the MFC are realized

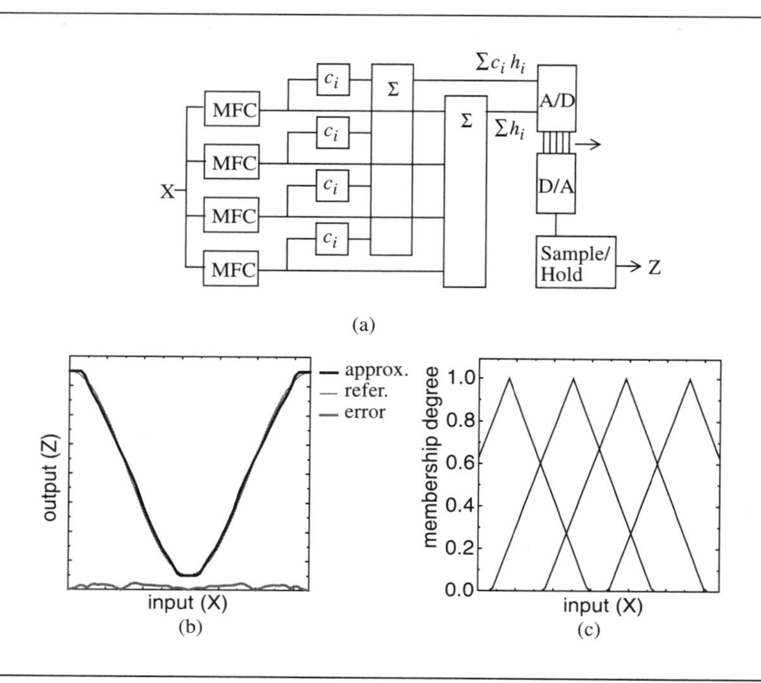

Figure 8.16. (a) Block diagram of the fuzzy IC prototype; (b) Approximation selected as illustrative example; (c) Covering of the input universe to achieve the approximation in (b).

with 3 unit capacitors of 0.25 pF, and the feedback capacitor consists of 4 ones, so that the membership function slope takes the required value of 0.75=3/4. The feedback capacitor of the scaler circuit consists of 16 unit capacitors of 0.128 pF while the input capacitor contains 15 unit capacitors of 0.128 pF plus one of 0.032 pF, thus implementing the required singleton value. The scaler circuit has the same structure as a 4-bit parallel D/A converter with binary-weighted capacitors. A two-stage Miller amplifier, whose schematic is shown in Figure 8.17a, was employed and a similar structure (Figure 8.17b) was selected to realize the comparator. The latter has the novelty that the second inverting stage is a dynamic structure that holds the comparison result when ϕ is high.

The microphotograph of Figure 8.18a shows a layout corresponding to one MFC and one comparator, which controls that the output values transmitted to the subsequent block are negative. The layout of the consequent block is shown in the microphotograph of Figure 8.18b. Figure 8.19 shows experimental results measured with this rule chip working with a 5-V power supply, V_{gnd} = 2.5 V, V_{top} = 1.5 V and a 1-MHz clock frequency compared to simulated results

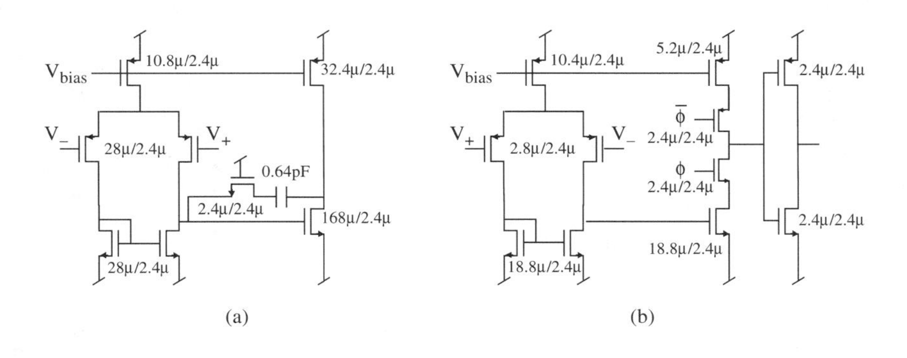

Figure 8.17. Schematics of (a) the amplifier and (b) the comparator used in the prototype.

obtained with *Switcap2*. Figure 8.19a illustrates the MFC output (during the first phase of the clock signal the output follows the input because a sample-and-hold circuit was not included at the input of the MFC). Figure 8.19b shows the signal already rectified and weighted by the singleton value at the output of the consequent block. The maximum absolute error measured in these signals was 6 mV (0.6% for an output voltage range of 1V, that is, 7 bits of equivalent resolution). The average power consumption of this rule chip (without considering the output buffer) is 2.38 mW.

The operation of the whole fuzzy IC implementing 4-rule circuits and a divider circuit based on cyclic converters like those shown in Figure 8.14 is il-

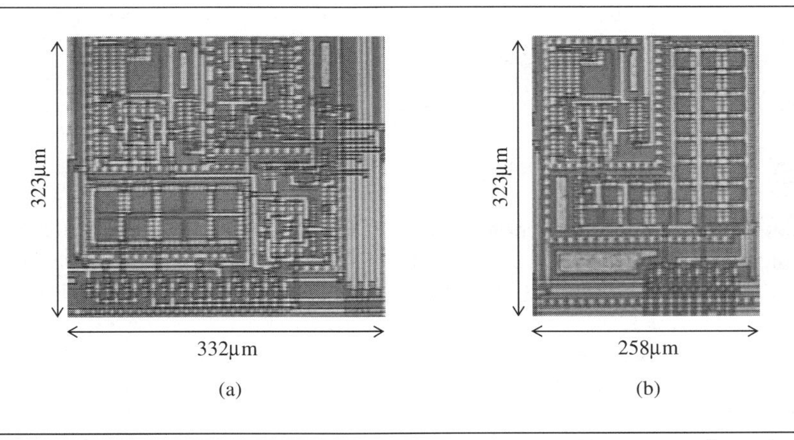

Figure 8.18. Microphotographs of the rule circuit that illustrate (a) the MFC (0.107 mm^2) and (b) the scaler block (0.083 mm^2).

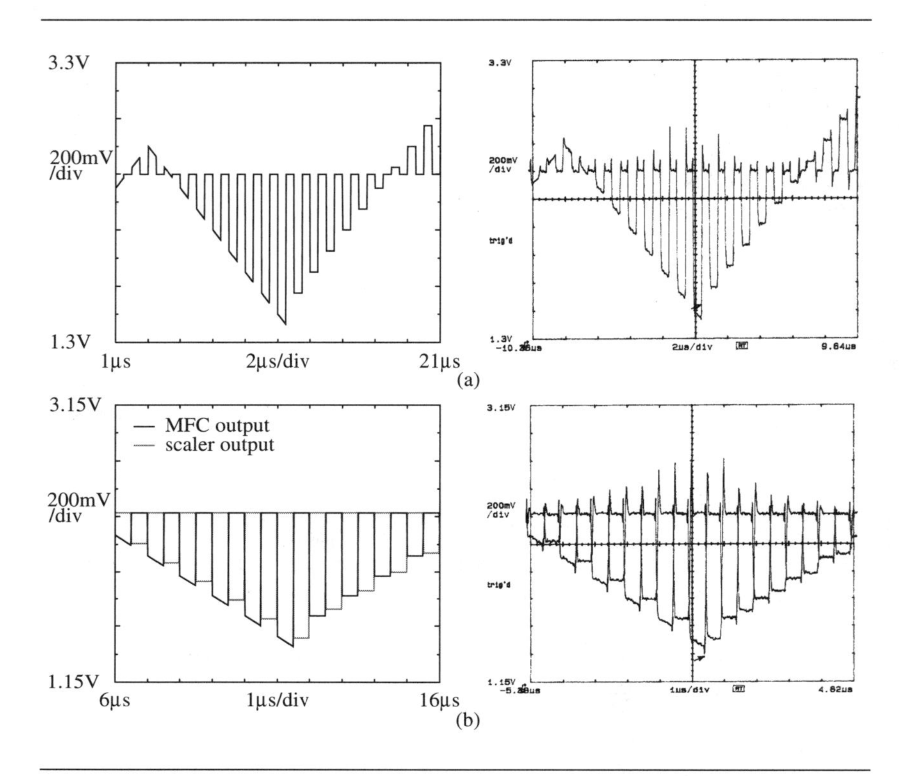

Figure 8.19. (a) Simulated (on the left) and experimental results (on the right) measured at the MFC output; (b) Simulated (on the left) and experimental results (on the right) measured at the scaler output.

lustrated in Figure 8.20 through *Switcap2* simulations. Supposing 4-bit resolution scaler circuits, the singleton values of the 4 rules were fixed to 15.25/16, 1.25/16, 1.25/16, and 15.25/16. If the operation cycle is 5 times the clock period (the digital output has 5 bits) the discrete output may take 29 values (from the code 00010, which corresponds to $2/32 < \Sigma c_i h_i < 3/32$, to the code 11110, which corresponds to $30/32 < \Sigma c_i h_i < 31/32$), as shown in Figure 8.20a. If the operation cycle is 6 times the clock period, the fuzzy IC now provides a 6-bit resolution output, as can be seen in Figure 8.20b. This figure illustrates the fact that increasing the operation cycle also increases the output resolution and the divider circuits approximates the continuous value $V_{num}/V_{den} = \Sigma c_i h_i / \Sigma h_i$ with less quantization error.

Working at a clock frequency of 1.25 MHz, the fuzzy IC achieves 250 KFLIPS when the output is provided with 5 bits (Figure 8.20a) and 208

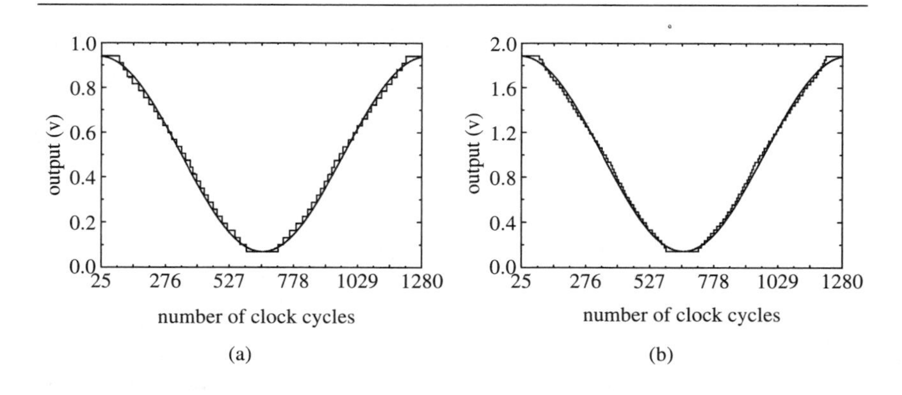

Figure 8.20. *Switcap2* simulation of an SC fuzzy IC implementing the fuzzy system in Figure 8.16a providing an output of (a) 5 and (b) 6 bits. The gray line shows the sinusoidal function that is approximated.

KFLIPS if it is provided with 6 bits (Figure 8.20b). The silicon area occupation in a 2.4-μm CMOS technology is estimated to be 2.9 mm^2 and the average power consumption is 15.69 mW.

REFERENCES

1. **Allen, P. E., Sánchez-Sinencio, E.**, *Switched Capacitor Circuits*, Van Nostrand Reinhold, 1984.
2. **Baturone, I., Sánchez-Solano, S., Barriga, A., Huertas, J. L.**, A switched capacitor based singleton fuzzy controller, in *Proc. 3rd. IEEE Int. Conference on Fuzzy Systems*, pp. 1303-1307, Orlando, 1994.
3. **Bouras, S., Kotronakis, M., Suyama, K., Tsividis, Y.**, Mixed analog-digital fuzzy logic controller with continuous-amplitude fuzzy inferences and defuzzification, *IEEE Transactions on Fuzzy Systems,* Vol. 6, N. 2, pp. 205-215, 1998.
4. **Cichoki, A., Unbehauen, R.**, A novel switched-capacitor four-quadrant analog multiplier-divider and some of its applications, *IEEE Transactions on Instrumentation and Measurement*, Vol. 35, N. 2, pp. 156-162, 1986.
5. **Çilingiroglu, U.**, A charge-based neural Hamming classifier, *IEEE Int. Journal of Solid-State Circuits,* Vol. 28, N. 1, pp. 59-67, 1993.
6. **Çilingiroglu, U., Pamir, B., Günay, Z. S., Dülger, F.**, Sampled-analog implementation of application-specific fuzzy controllers, *IEEE Transactions on Fuzzy Systems*, Vol. 5, N. 3, pp. 431-442, 1997.

7. **De Cock, B. A., Maurissens, D., Cornelis, J.**, An analog CMOS multiplier based on the Tomota-Sugiyama-Yamaguchi principle, in *Proc. ESSCIRC'92,* pp. 327-330, Copenhagen, 1992.

8. **Fattaruso, J. W., Mahant-Shetti, S. S., Barton, J. B.**, A fuzzy logic inference processor, *IEEE Int. Journal of Solid-State Circuits*, Vol. 29, N. 4, pp. 397-402, 1994.

9. **Gregorian, R., Temes, G.C.**, *Analog MOS Integrated Circuits for Signal Processing,* Wiley-Interscience Pub., 1986.

10. **Huertas, J. L., Sánchez-Solano, S., Barriga, A., Baturone, I.**, Serial architecture for fuzzy controllers: Hardware implementation using analog/digital VLSI techniques, in *Proc. 2nd. Int. Conference on Fuzzy Logic and Neural Networks*, pp. 535-538, Iizuka, 1992.

11. **Huertas, J. L., Sánchez-Solano, S., Barriga, A., Baturone, I.**, A fuzzy controller using switched-capacitor techniques, in *Proc. IEEE Int. Conference on Fuzzy Systems*, pp. 516-520, San Francisco, 1993.

12. **Hughes, J. B., Macbeth, I. C., Pattullo, D. M.**, Second generation switched-current signal processing, in *Proc. IEEE Int. Symposium on Circuits and Systems*, pp. 2805-2808, New Orleans, 1990.

13. **Krummenacher, F.**, Micropower switched capacitor biquadratic cell, *IEEE Int. Journal of Solid-State Circuits,* Vol. 17, pp. 507-512, 1982.

14. **Makris, C. A., Toumazou, C.**, Operational amplifier modelling for high speed sampled data applications, *IEE Proceedings, Part. G*, Vol. 137, N. 5, pp. 333-339, 1990.

15. **McCreary, J. L., Gray, P. R.**, All-MOS charge redistribution analog-to-digital conversion techniques-Part I, *IEEE Int. Journal of Solid-State Circuits*, Vol. 10, N. 6, pp. 371-379, 1975.

16. **McCharles, R. H.**, Charge circuits for analog LSI, *IEEE Transactions on Circuits and Systems*, Vol. 25, pp. 490-497, 1978.

17. **Nairn, D. G., Salama, C. A. T.**, Current-mode algorithmic analog-to-digital converters, *IEEE Int. Journal of Solid-State Circuits,* Vol. 25, N. 4, pp. 997-1004, 1990.

18. **Ohara, H., Ngo, H. X., Armstrong, M. J., Rahim, C. F., Gray, P. R.**, A CMOS programmable self-calibrating 13-bit eight-channel data acquisition peripheral, *IEEE Int. Journal of Solid-State Circuits*, Vol. 22, N. 6, pp. 930-938, 1987.

19. **Onodera, H., Tateishi, T., Tamaru, K.**, A cyclic A/D converter that does not require ratio-matched components, *IEEE Int. Journal of Solid-State Circuits*, Vol. 23, N. 1, pp. 152-158, 1988.

20. **Pelayo, F. J., Rojas, I., Ortega, J., Prieto, A.**, Current-mode analog defuzzifier, *Electronics Letters*, Vol. 29, N. 9, pp. 743-744, 1993.

21. **Piedade, M. S., Pinto, A.**, A new multiplier-divider circuit based on switched capacitor data converters, in *Proc. IEEE Int. Symposium on Circuits and Systems*, pp. 2224-2227, New Orleans, 1990.

22. **Robertini, A., Guggenbühl, W.**, Modelling and settling times of amplifiers in SC circuits, *IEE Proceedings, Part.-G*, Vol. 139, N. 1, pp. 131-136, 1992.

23. **Shih, C. C., Gray, P. R.**, Reference refreshing cyclic analog-to-digital and digital-to-analog converters, *IEEE Int. Journal of Solid-State Circuits*, Vol. 21, pp. 544-554, 1986.

24. **Toumazou, C., Lidgey, F. J., Haigh, D. G.**, Eds., *Analogue IC Design: The Current-Mode Approach*, Peter Peregrinus Ltd., 1990.

25. **Toumazou, C., Hughes, J. B., Battersby, N. V.**, Eds., *Switched-Currents: An Analogue Technique for Digital Technology*, IEE Circuits and Systems Series 5, Peter Peregrinus Ltd. 1993.

26. **Watanabe, H., Dettloff, W., Yount, K.E.**, A VLSI fuzzy logic controller with reconfigurable, cascadable architecture, *IEEE Int. Journal of Solid-State Circuits*, Vol. 25, N. 2, pp. 376-382, 1990.

Chapter 9

DIGITAL TECHNIQUES FOR DESIGNING FUZZY INTEGRATED CIRCUITS

As was shown previously, the development of specific integrated circuits to implement fuzzy systems is the most efficient way to design fuzzy hardware in terms of performance and consumption of resources. In particular, this chapter focuses on digital realizations whose design process is considerably simplified, thanks to the aid of CAD tools that automate several design steps. Many proposals have been reported in the literature since Togai, Watanabe, and Dettloff proposed the first digital fuzzy ICs by the mid-1980s. In order to systematize the study of these approaches, they are summarized in this chapter according to the way of evaluating the rule base.

After describing the first realizations of fuzzy systems with digital ICs, the types of architectures employed and the different ways of implementing the operations involved in the calculus of a fuzzy inference are analyzed. Finally, the realization of an active rule-driven architecture that employs simplified defuzzification methods is presented. This description is accompanied by a set of analyses concerning temporal behavior and area occupation of the resulting integrated circuits. In addition, two CMOS prototypes designed with this architecture are presented.

9.1 THE FIRST DIGITAL FUZZY INTEGRATED CIRCUITS

The first hardware realizations of fuzzy systems employ architectures that provide a data path for each rule. This results in a high inference speed since the rules are processed in parallel. The problem is that their size increases proportionally to the number of rules. This is the case of the architecture proposed by Togai and Watanabe [Togai, 1985] whose block diagram is shown in Figure 9.1 for the particular case of two inputs, one output, and two rules. Since this architecture admits fuzzy inputs (something not reported in the analog domain of fuzzy ICs), blocks that perform operations on all the points of the universe of discourse have to be included (those represented with bold lines in the figure). As a consequence, the calculation of the membership degrees is more complex and the circuitry required by each rule increases notably.

The first prototype designed with this type of architecture was made up of a rule memory (where the consequents were stored), an inference unit, and a control circuit, but no defuzzification block was included. The fuzzy inference process implemented is the Min-Max, with maximum and minimum operations performed sequentially so that the area occupied by these operators is

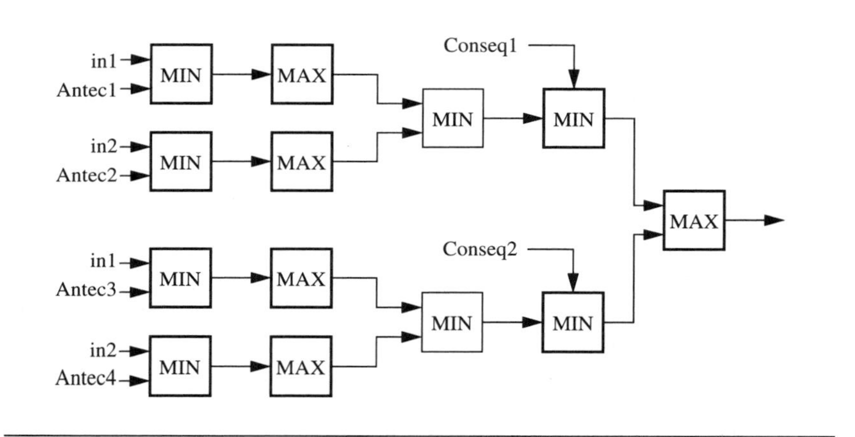

Figure 9.1. Block diagram of an architecture with a data path for each rule.

small. This prototype implements 16 rules with one antecedent and one consequent and represents each fuzzy label by 31 elements with 16 possible values of membership. Since its maximum operation frequency is 20.8 MHz and 256 clock cycles are needed to accomplish each fuzzy inference, its maximum inference speed is 80 KFLIPS.

Some years later, the works developed by Watanabe and Dettloff [Watanabe, 1989; 1990] presented the implementation of the three components of a fuzzy system (fuzzifier, inference engine, and defuzzifier) within a single integrated circuit. The inputs have 6-bit resolution and admit 16 values of membership, so that each membership function is stored in a memory with 64 words of 4 bits. The circuit allows two configurations. One configuration admits 51 rules with 4 antecedents and 2 consequents, while the other configuration is able to process up to 102 rules with 2 antecedents and 1 consequent. Defuzzification is performed via the center of area calculation. The circuit, designed in a 1-μm CMOS technology, contains about 688.000 transistors, occupies a silicon area of 70 mm^2 and is able to provide 625 KFLIPS operating at 40 MHz.

Togai and Chiu also proposed an accelerator of fuzzy inferences in [Togai, 1987]. The main differences compared to the previous Togai's prototype were the handling of non-fuzzy inputs and the implementation of 16 rules of 4 antecedents and 2 consequents. The rule base is stored into a static RAM so that it can be programmed. Since input universes of discourse are represented by 16 elements with 16 levels of membership and fuzzy sets are associated with the rules instead of being related to the universes of discourse, the size of the memory unit is of 6 Kbits (64×4×16 = 4 Kbits for the antecedents and 64×2×16 = 2

Kbits for the consequents). The calculation of the input membership degrees is simplified considerably due to the non-fuzzy nature of the inputs, which means that only one memory location has to be addressed. The content of each memory location is divided into 16 zones (one for each rule), so that the word length is 64 bits. What is stored into the n-th zone is the membership degree of the corresponding input to the fuzzy set associated with the n-th rule. The consequent memory is read sequentially and the minimum between this value and that provided from the connection of the antecedents is calculated. The outputs of these minimum units enter a maximum unit whose output is connected in turn to the defuzzifier. The chip contains several registers so as to establish two pipeline stages. Applying this pipeline, the inference is carried out in 64 clock cycles, which means an inference speed of 250 KFLIPS considering a clock frequency of 16 MHz.

9.2 ARCHITECTURES OF DIGITAL FUZZY INTEGRATED CIRCUITS

After the first realizations of digital fuzzy ICs discussed above, many other approaches have been reported in the literature, which feature a wide variety of strategies to carry out the fuzzy inferences. Their architectures can be classified into three main groups according to their way of evaluating the rules: (a) architectures with parallel rule processing; (b) architectures with sequential rule processing; and (c) active rule-driven architectures. As a result, they cope with parallelism by using different solutions: replicating the required hardware, performing the operations sequentially, or trying to optimize the number of operations. On the other hand, the three groups of architectures share common features like the introduction of pipeline stages that separate independent operations and allow improvement of the inference speed.

9.2.1 PARALLEL RULE PROCESSING ARCHITECTURES

The main feature of these architectures is that they provide a data path for each rule (Figure 9.2a). Apart from being the architectures employed by the first digital realizations, they are also usual in analog fuzzy ICs, as was shown in Chapter 7. The membership functions of antecedents and consequents are associated with the rules, so that calculation of the activation degrees of all the rules is carried out in parallel. However, while the consequent processing is performed in parallel in many analog realizations, this is very costly in the digital domain, thus prompting the introduction of sequential operation. Even so, the system has to be provided with multi-input operators dedicated to aggregate the rules. Since these operators are usually designed by the cascade connection of two-input operators, system operation is considerably slowed down. In addition, the memory size required to store the membership functions of antecedents and consequents is proportional to the number of rules, which may impose severe limitations on the realization of systems with a high number of rules.

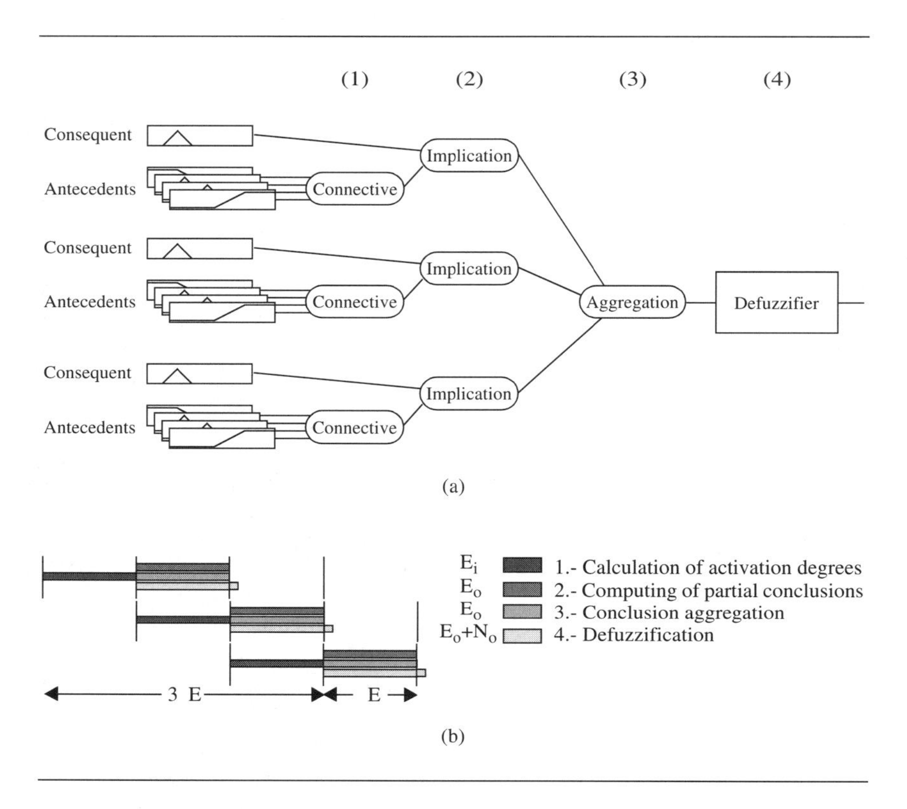

Figure 9.2. Parallel rule processing: (a) architecture; (b) timing scheme.

The processing operations to be performed by a system with fuzzy inputs are: (1) calculation of the activation degrees of the rules; (2) computing of partial conclusions; (3) aggregation of partial conclusions; and (4) defuzzification. Figure 9.2b shows a timing scheme of their realization. The number of elements in the input and output universes of discourse are denoted, respectively, by E_i and E_o while $N_o = \log_2 E_o$ is the number of resolution bits necessary to represent the inferred final result. Once operation (1) has finished, operations (2) and (3) together with the calculation of the numerator and the denominator required to perform the defuzzification can be carried out simultaneously. Once the numerator and denominator values are available, the digital implementation of the division operation (typically following a successive approximation algorithm) invests one clock cycle per bit of the quotient (which means N_o cycles). If segmentation techniques with three pipeline stages are applied, the number of clock cycles required to conclude an inference is inferior to 3E (E being the maximum number of elements in the input and output universes of discourse), but the system is able to provide a new output every E cycles.

When the inputs to the system are non-fuzzy, the operations needed to calculate the activation degrees of the rules are reduced to one scalar operation and, consequently, one of the pipeline stages can be eliminated. In this case, the number of clock cycles invested to produce an inference is reduced to less than 2E while the system goes on providing a new output every E cycles.

9.2.2 SEQUENTIAL RULE PROCESSING ARCHITECTURES

This second type of architecture, widely employed in digital designs, organizes the knowledge base into two clearly different zones. On the one hand, the rule base is stored into a memory in a symbolic way, that is, the rules are represented by the fuzzy labels of their antecedents and consequents. On the other hand, the generation of the antecedent and consequent membership functions is performed by blocks that are shared by all the rules. Hence, the membership functions are obtained from a fuzzy partition of the corresponding universe of discourse (Figure 9.3a).

The operations to carry out are the same as in the previous parallel architectures: (1) calculation of the activation degrees of the rules; (2) computing of partial conclusions; (3) aggregation of partial conclusions; and (4) defuzzification. The use of a memory imposes that only one rule can be addressed every

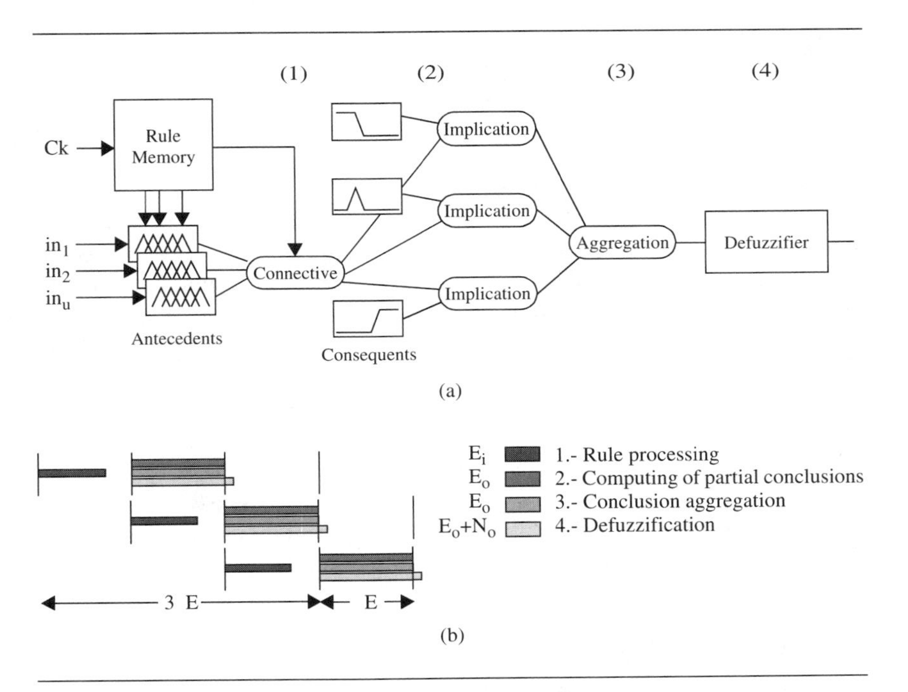

Figure 9.3. Sequential rule processing: (a) architecture; (b) timing scheme.

clock cycle. Therefore, the processing of the rule base must be sequential and the number of clock cycles required to carry out operation (1), considering non-fuzzy inputs, is equal to the number of rules. The activation degrees of those rules sharing the same consequent can be adequately combined and stored into memory elements [Ungering, 1993; Katashiro, 1993], so that the implementation cost of operation (3) depends on the number of the consequent membership functions instead of on the number of rules. The response time remains the same as in the former case (an inference every E cycles) provided that the number of rules is inferior to the number of elements in the output universe of discourse (Figure 9.3b).

Several authors have proposed other solutions to increase the number of rules without degrading the response time of the system. As an example, the works in [Sasaki, 1993a; 1993b] propose a rule rearrangement in which the original rule base is partitioned into several rule sub-bases associated with each different consequent. This approach results in a structure made of many processing elements, where each element carries out the inference over each rule sub-base.

A strategy that improves considerably the inference speed is to process only the active rules [Ikeda, 1991a]. The architecture employed in [Ikeda, 1991b] finds which rules are activated (storing the results into registers) after calculating the membership degrees of the antecedents. In this first version, the rule base consists of 64 rules stored into a ROM, so that it is not programmable because the design was aimed at solving a particular control problem. Later, the architecture was extended to offer a programmable knowledge base [Ikeda, 1992] by adding a scheme of programmable multiplexors connecting the different antecedents and consequents.

Another architecture that combines rule rearrangement with active rule evaluation was presented in [Ascia, 1997]. A basic element in this architecture determines which antecedents are activated taking into account the input values and the support of the membership functions, and stores those antecedents into a register. Since fuzzy rules usually share the same antecedents, the shared antecedents are stored into one memory while another memory is used to store the different ones. This achieves a remarkable reduction not only of the rule memory size but also of the processing time because if one of the membership degrees of the shared antecedents is null, the rest of the rules using it does not need to be processed. Otherwise, if the activation degrees of the terms shared by a group of rules are not null, the minimum of them is calculated and stored into a register. The advantage is that this minimum is then employed by all the rules of the group.

9.2.3 ACTIVE RULE-DRIVEN ARCHITECTURES

This third type of architecture is based on the use of an active rule-driven scheme, where the conventional memory that stores the symbolic rule base is replaced by an associative memory in which the rule antecedents address their consequents. The antecedent membership functions of fuzzy systems implemented with this architecture have a maximum overlapping degree in order to fix the maximum number of rules that can be activated by any combination of the input values. Given a set of input values, the blocks that generate the membership functions provide as many pairs (label, membership-degrees) as the overlapping degree selected for the system. A set of multiplexors controlled by a counter makes it possible to sweep the potentially active rule combinations sequentially. Every counter cycle, the membership degrees are processed through the connective operator to calculate the rule activation degree, while the antecedent labels address the memory locations that contain the corresponding consequents. Finally, defuzzification is performed.

An architecture of this kind was proposed in [Chiueh, 1992]. In this proposal, the maximum overlapping degree among the antecedent membership functions is limited to two, so that each input activates two fuzzy labels and two membership degrees have to be calculated. In order to improve the inference speed, the use of one rule memory per active rule is proposed so that all the active rules can be processed in parallel. The same scheme to evaluate the active rules was described in [Eichfeld, 1992]. However, this architecture only employs a single rule memory, which is sequentially addressed by the active rules. Compared to the rule memory replication employed in [Chiueh, 1992], this solution has the drawback of requiring as many clock cycles as active rules and the advantage of saving area, which can be very important when implementing systems with many rules.

The advantages of evaluating only the active rules are not fully exploited if conventional defuzzification methods are employed because the number of clock cycles invested in the calculation of the antecedent labels and membership degrees and in the computing of the activation degrees of the active rules is much lower than the number of clock cycles required to perform the defuzzification. Conversely, if simplified defuzzification methods are implemented (for instance, the Fuzzy-Mean method illustrated in Figure 9.4a), the number of clock cycles required to perform the accumulation operations of the defuzzification is of the same order as the number of clock cycles needed to calculate the activation degrees of the active rules (Figure 9.4b). This has been the solution proposed in [Jiménez, 1994; 1995]. More recently, [Gabrielli, 1996; 1997] have presented a modification of this idea in order to allow either antecedent combinations that do not form rules or rules that do not contain all the possible inputs. For this purpose, one data column per input is added to the rule memory. It contains a logic '1' if that input contributes to the output or a '0', otherwise.

In addition to the different possibilities to evaluate the rules depending on the selected architecture, digital implementations also allow many ways of performing the rest of the operations involved in a fuzzy inference. This is explained in the following sections by distinguishing two main groups of operations: oper-

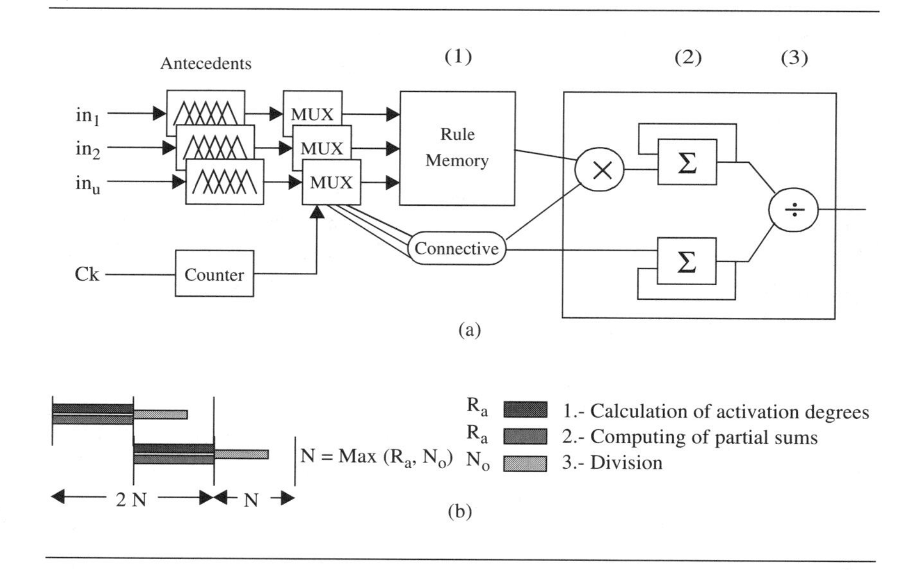

Figure 9.4. Active rule processing: (a) architecture; (b) timing scheme.

ations carried out in the fuzzification stage and operations of the inference and defuzzification stages.

9.3 FUZZIFICATION STAGE

The first operations that a fuzzy system has to perform when receiving new input values are the calculations of the membership degrees of those inputs to the antecedent fuzzy sets. Although some authors have reported hardware solutions for the processing of fuzzy inputs [Togai, 1985; Ascia, 1995; 1997], most hardware realizations work with non-fuzzy inputs since this is the usual way in which hardware interacts with the external world. This means a remarkable reduction of the operations to be performed and, consequently, of the hardware resources required. Hence, we will focus on non-fuzzy inputs in the following.

The approaches proposed in the literature follow two basic strategies to calculate the membership degrees: the use of memories addressed by the inputs and the employment of algorithms. Each strategy offers advantages and drawbacks and different options can be employed for their implementation.

9.3.1 MEMORY-BASED APPROACH

This solution stores the membership degrees of every possible input value into a memory. The input value acts as the address of the memory to retrieve the fuzzy labels that are activated and the corresponding membership degrees. The only limitation imposed on the membership function shape is given by the resolution bits of the memory, and obviously this shape does not influence in any way the computational load. On the other hand, the memory size required to implement high-resolution systems can be very large because the number of rows of the memory increases exponentially with the bit number of the input. Considering N membership functions with P resolution bits for the input and J resolution bits for the membership degree, the memory size required by each input is given by:

$$T = N \cdot J \cdot 2^P \tag{9.1}$$

This way of processing the antecedents is employed, for example, in [Hung, 1994]. In this work, the low resolution of both the input and the membership degrees (4 bits both) permits us not only to use a RAM for each of the 3 membership functions associated to the 2 inputs, but also to duplicate them so as to reprogram the antecedents as the circuit is operating.

To reduce memory requirements, several modifications to this memory-based approach have been reported. In the architecture presented in [Ikeda, 1991a] the proposal is to store certain types of membership functions into several memories which are sequentially accessed by the inputs. To finely adjust these membership functions, the authors use what they name "virtual paging" which consists in displacing the membership function by adding or subtracting a specific value for each input [Arikawa, 1989].

Other authors focus on limiting the maximum degree of overlapping to 2. In the solutions proposed in [Chiueh, 1992; Gandolfi, 1994] the membership functions are stored into two memories for the even and odd functions, respectively. Each row of these memories stores the membership degrees together with their associated fuzzy label (Figure 9.5a). The alternative proposed in [Eichfeld, 1992] further reduces the memory requirements. For this purpose, the maximum overlapping degree is limited to 2, the membership degrees of the two activated fuzzy sets are stored into each row of the memory and only one of the two fuzzy labels is registered. The other fuzzy label can be obtained by adding one to the stored value since the activated labels are represented by consecutive values. This means that each memory row is divided into three fields, two of them dedicated to storing membership degrees while the other one stores the fuzzy label. As shown in Figure 9.5b, a membership function changes from the part MEM-1 to the part MEM-0 of the memory as the value stored for the fuzzy label varies. These two solutions, although proposed for a maximum overlapping degree of 2, can be generalized for any other degree.

9.3.2 ALGORITHM-BASED APPROACH

This approach evaluates the membership degrees by using specific algorithms that restrict the shapes of membership functions to be piecewise linear.

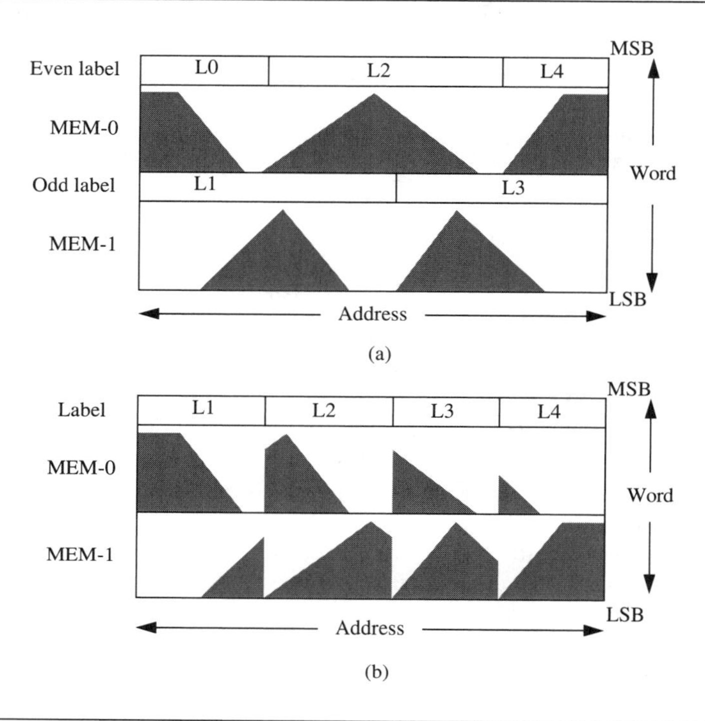

Figure 9.5. Options for storing the antecedents into memories: (a) storing two fuzzy labels or (b) only one fuzzy label.

The antecedent memory only stores the parameters required to carry out this calculation, usually the break points and the slopes. This means that, contrary to the memory-based approach, the membership function shape now influences the computational load but the memory size is smaller and the way of programming the antecedents is easier. From a hardware point of view, the drawback is that membership degrees are obtained performing operations sequentially.

[Katashiro, 1993] proposes an algorithm to implement four shapes of membership functions: trapezoids, triangles, Z, and S (Figure 9.6a). The parameters to define the trapezoidal function (which is the most general) are four: a_n, b_n, m_n, and p_n, where a_n and b_n represent the break points, and m_n and p_n represent the slopes. Given an input value x, the membership degree to the fuzzy set results from evaluating the following expression:

$$\begin{aligned} \mu(x) &= m_n(x-a_n) \qquad \textit{if} \ \ a_n \le x < b_n \\ &= (1+p_n(x-b_n)) \ \ \textit{if} \ \ b_n \le x \end{aligned} \tag{9.2}$$

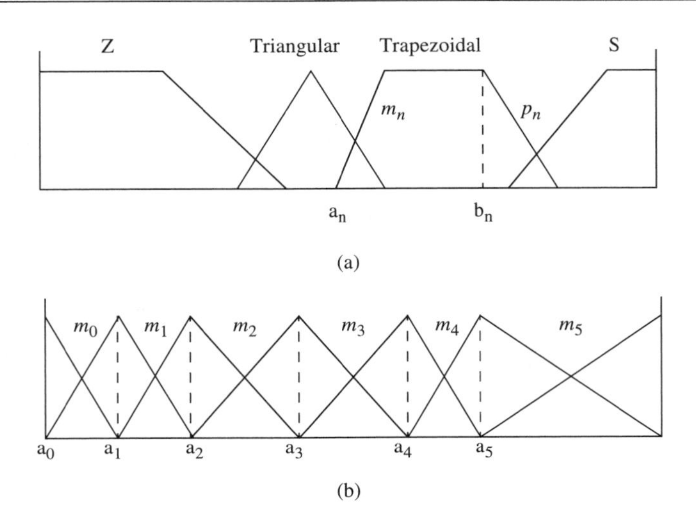

Figure 9.6. Usual shapes of membership functions used in arithmetic-based approaches. (a) Parameters for storing a generic trapezoidal function; (b) Normalized triangular functions.

The sequence the circuit must follow to implement the above expression consists of three stages. During the first one the calculus of $(x - a_n)$ and $(x - b_n)$ is carried out. The expressions $m_n(x - a_n)$ and $1 + p_n(x - b_n)$ are computed during the second stage with floating-point operations. Finally, the result is adapted to the allowed range, interval [0, 1]. Since the proposed architecture permits a maximum overlapping degree of 2, two membership function circuits (MFCs) operate in parallel, one dealing with the even function and the other with the odd one. Each of them compares sequentially the input value x with a_n (or with b_n). The operation concludes and the adequate fuzzy label is selected when the comparison result changes from being positive to negative. This is the same computing algorithm described in [Sasaki, 1993a], with the difference that fixed instead of floating-point arithmetic is used for that task.

A simpler representation of membership functions, which is used in several applications, is shown in Figure 9.6b. From a hardware point of view, this representation has two advantages. On the one hand, the parameters stored for each function are a point and a slope. On the other hand, only one MFC is required to calculate the two membership degrees per input because the functions are normalized so that having calculated one of the degrees, μ_n, the other is obtained as $1 - \mu_n$. The flow chart of this kind of computing is illustrated in Figure 9.7. Given a new input value, x, it is compared with all the break points, a_i. If

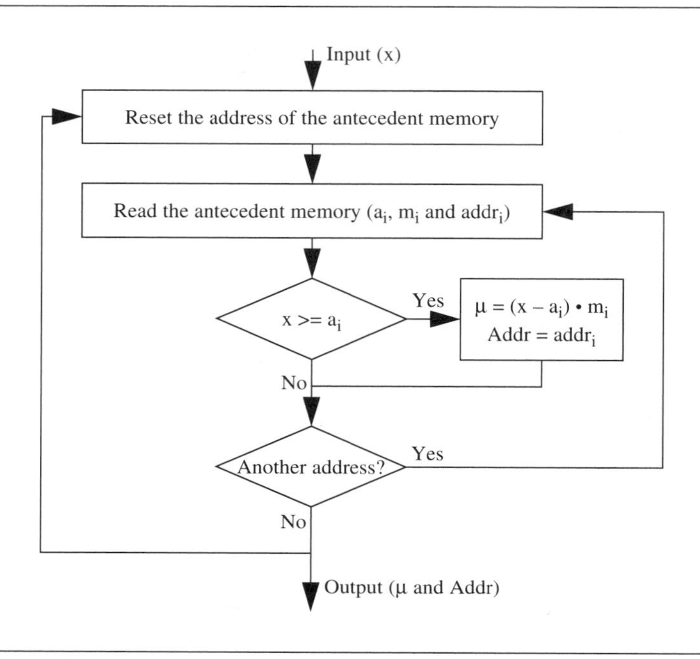

Figure 9.7. Flow chart of the membership degree computing in the arithmetic-based approach.

the value of x is greater or equal to a_i, the membership degree is calculated and the label is selected. Otherwise, no operation is performed. The membership degrees calculated and the selected fuzzy label are provided as outputs and another cycle begins with a new input value. The circuits required to implement this algorithm are a multiplier, a subtracter, and registers to store intermediate results (Figure 9.8), as described in [Jiménez, 1995].

9.4 RULE PROCESSING AND DEFUZZIFICATION STAGES

The hardware required to implement the operations related to the rule processing stage depends on the type of fuzzy inference selected. As discussed for analog realizations, these operations are usually minimum, maximum, product, addition, and subtraction. The realization of these operators with digital circuits is generally accomplished by using library cells (logic gates, flip-flops, etc.) and well-known design techniques in which CAD tools play a relevant role. In particular, automatic synthesis tools that start from a high level description of the system via hardware description languages are

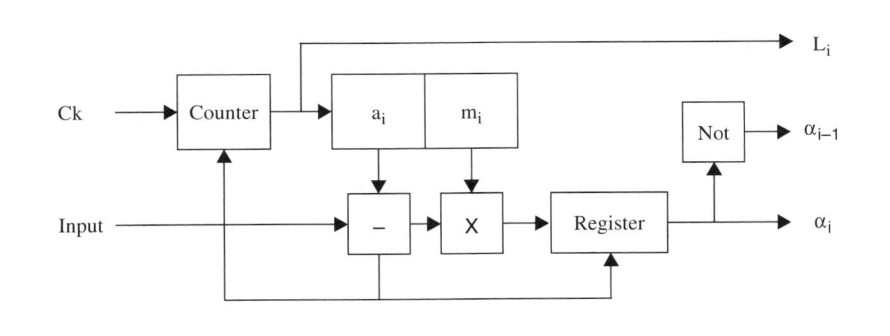

Figure 9.8. Block diagram of an arithmetic MFC.

usually employed, as was discussed in Chapter 6 and which will be detailed in Chapter 10.

Concerning the defuzzification stage, the hardware components required are multipliers, adders, subtracters, accumulators, and a divider. These blocks are designed with conventional techniques, with parallelism as the more relevant factor from the point of view of digital design. As was described in Chapter 3, and also in Chapters 7 and 8, fuzzy systems can be classified into two groups: those that deal with fuzzy consequents and those that represent them by several parameters. There are great differences between these two groups when considering their hardware implementation. The systems that employ fuzzy consequents must previously compute the membership function that results from the inference. Both this calculation and the defuzzification method perform operations on all the points of the output universe of discourse. On the other hand, the systems that deal with non-fuzzy consequents do not have to compute any membership function after the inference and the simplified defuzzification method employed only operates on the rules. Hence, their parallelism is lower than in the former ones. In both cases, the operations to obtain the system output are performed sequentially.

Among the defuzzification methods employed by digital fuzzy systems that represent consequents as fuzzy sets, the most usual one is the Center-of-Area method. Several proposals have been reported in the literature to optimize its hardware implementation and performance either by reducing the required operators or precomputing some values. [Watanabe, 1990] proposes a method of calculating the Center of Area without multipliers but using a cascade of accumulators, as briefly discussed in Chapter 8. Given the equation of the Center-of-Area calculation as:

$$\text{CoA} \rightarrow \hat{y} = \frac{\sum_{y_i=1}^{n} y_i \cdot \mu_{B'}(y_i)}{\sum_{y_i=1}^{n} \mu_{B'}(y_i)} \tag{9.3}$$

the use of recurrent equations to calculate the numerator and denominator is proposed. A first accumulator, S0, sequentially aggregates the values $\mu_{B'}(y_i)$ beginning with $y_i = n$ as follows:

$$\begin{gathered} S0(y_i) = S0(y_{i+1}) + \mu_{B'}(y_i) \quad ; y_i = n, n-1, \ldots, 1 \\ S0(n+1) = 0 \end{gathered} \tag{9.4}$$

thus obtaining the denominator value:

$$S0(1) = \mu_{B'}(n) + \ldots + \mu_{B'}(1) = \sum_{y_i=1}^{n} \mu_{B'}(y_i) \tag{9.5}$$

Simultaneously, the numerator in (9.3) can be calculated by a second accumulator, S1, which receives the results from the previous one:

$$\begin{gathered} S1(y_i) = S1(y_{i+1}) + S0(y_i) \quad ; y_i = n, n-1, \ldots, 1 \\ S1(n+1) = 0 \end{gathered} \tag{9.6}$$

so that S1(1) equals the numerator value.

Another approach to optimize the computing of the Center of Area has been proposed in [Ruiz, 1995]. A feature of this solution is that it does not employ an explicit divider. The idea is to consider the membership degree $\mu_{B'}(y_i)$ as a weight applied to point y_i. Hence, the product $y_i \cdot \mu_{B'}(y_i)$ is equivalent to the momentum of the weight with respect to the origin of the coordinates. If the origin is displaced to the center of area, the momentum of the two sides of the membership function must be equal, so that:

$$\sum_{left} y_i \cdot \mu_{B'}(y_i) = \sum_{right} y_i \cdot \mu_{B'}(y_i) \tag{9.7}$$

The problem is finding where the above equation is verified, which requires the use of two accumulators. The strategy first divides the y-axis into elemental units. Then, each accumulator calculates the momentum referred to one of the extremes of the output universe of discourse and goes on calculating the momentum but referred to the following point closer to the center of the y-axis. Since each elemental unit has its own weight, each displacement to the y-axis center means a change in the reference point and, hence, an increase in the weights of the units already considered. This scheme can be implemented with two adders and two accumulators, like the scheme in [Watanabe, 1990]. After each accumulation, the accumulated values at the right and left parts have to be compared. The least of them will be displaced and accumulated with the next momentum value. This process continues until the left and right reference points meet together. The meeting point is the Center-of-Area value.

Evaluation of the simplified defuzzification methods is much easier because no fuzzy consequent has to be calculated and the number of accumulations is restricted to the rule number. Figure 9.9 shows block diagrams for implementing the Fuzzy-Mean or Height defuzzification method (3.50), the Weighted-Fuzzy-Mean method (3.51), Yager's method (3.59), and Center-of-Sums-with-minimum method (3.55). The Fuzzy-Mean method is the most easily implemented: it requires a multiplier, two accumulators, and a divider (Figure 9.9a). The rest of the methods require more circuitry: an additional multiplier in the case of the Weighted-Fuzzy-Mean method (Figure 9.9b), an additional multiplier and an adder in the case of Yager's method (Figure 9.9c), and a multiplier and two bit inverters in the Center-of-Sums-with-minimum method (Figure 9.9d).

9.5 ACTIVE RULE-DRIVEN ARCHITECTURE WITH SIMPLIFIED DEFUZZIFICATION METHODS

Among all the architectures previously described, those that perform inferences more efficiently are the active rule-driven architectures because the evaluation of non-active rules not only consumes computing time but also

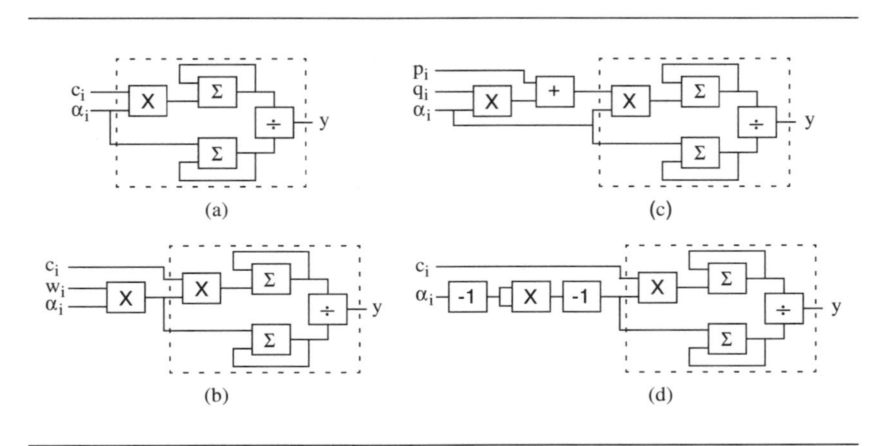

Figure 9.9. Block diagrams for implementing the following defuzzification methods: (a) Fuzzy Mean; (b) Yager's; (c) Weighted Fuzzy Mean; and (d) Center of Sums with minimum.

hardware resources. Among the active rule-driven architectures, the one that offers the best performance in terms of inference speed and cost of specific digital circuits is that whose rule memory is addressed by the antecedents. The number of active rules in this architecture is limited by the maximum overlapping degree of the antecedents. The use of simplified defuzzification methods that represent the consequents by several parameters causes the operations required to calculate the system output to be realized only on the data produced by the active rules.

The dedicated hardware realization of fuzzy systems with this architecture may follow different strategies. Jiménez et al. have automated the design process of digital fuzzy ICs and FPGAs by using a set of functional blocks that are described with generic parameters [Jiménez, 1995; Sánchez-Solano 1997]. Figure 9.10 shows the general structure of the architecture proposed in those works. It consists of three fundamental stages: the fuzzification stage that calculates the pairs (label, membership degree) for each of the inputs; the rule processing stage, in which the activation degrees of the rules are computed; and the output stage, which provides the system output. We will focus on this architecture in the following.

9.5.1 IMPLEMENTATION OPTIONS

Once this active rule-driven architecture with simplified defuzzification method has been selected, there are different implementation options. An issue to be decided is either to provide the system with programmability or to store a fixed knowledge base, which translates into the use of either RAMs or ROMs, PLAs, etc. Another issue concerns the method to store and calculate the membership degrees to the antecedents, either by using a memory- or an arithmetic-based approach. In addition, a block can be added to reduce the number of possible consequents, by assigning the same memory location to several rules. Finally, the designer may choose among four simplified defuzzification

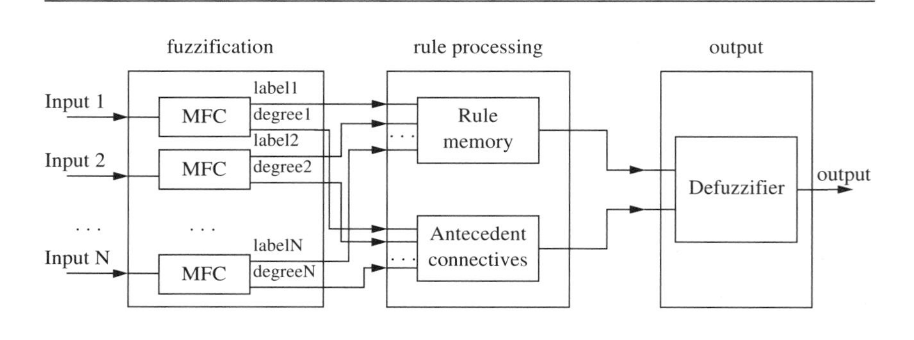

Figure 9.10. Block diagram of the architecture proposed in [Jiménez, 1995].

methods. Apart from the previous choices, the designer has to define the parameters of the different functional blocks, thus fixing options like the number of inputs, resolution bits, etc.

- **Fuzzy system programmability**

The implementation of fuzzy systems with this architecture may store the knowledge base into RAMs (thus allowing programmability) or into ROMs or combinatorial blocks. This choice makes it possible to fit the realization for a given application or for a given implementation technique. In this sense, the design of programmable systems is of interest in the development of general-purpose fuzzy systems or systems whose application requires the adjustment of its knowledge base. On the other hand, non-programmable fuzzy designs are more adequate for applications where the knowledge base is completely defined or the implementation technique itself allows programmability, as occurs with FPGAs.

When programmable fuzzy systems are considered, the load of the knowledge base is performed with a specific protocol. The circuits required to process the input information and to write the memories are integrated within the system. If the system is not programmable, this additional circuitry is not needed, thus reducing hardware complexity and, hence, the size of the chip.

- **Antecedent storage and processing**

As noted previously, there are two options, a memory- or an arithmetic-based approach, for providing the pair of data (label, membership degree) for each input. Among the memory-based proposals described in Section 9.3.1, we will focus on the one that stores the two membership degrees and only one fuzzy label (Figure 9.5b) because it optimizes the memory size. And among the arithmetic-based approaches explained in Section 9.3.2, we will focus on the one illustrated in Figure 9.6b, in which the membership functions are triangular and normalized. These options are very different from a hardware point of view. In the memory-based case, one memory is required per input and the identification of the active fuzzy sets and the calculation of the membership degrees is carried out in just one access to this memory. In the arithmetic-based case, an algorithm has to be performed, which means that computing circuits are required, but the antecedent memory size is very small since it stores only the break points and the slopes. Hence, the antecedent memories of all the inputs can be grouped into a single memory, which greatly eases the layout design when considering ASIC implementations.

- **Reduction of the number of consequents of the rules**

Since the consequents are addressed by the fuzzy labels of the antecedents, each rule stores its consequent into a different memory location, so that consequents can be repeated. A solution to reduce the rule memory size is to use as many rows as different consequents required in the system. This can be done by adding a block that associates all the possible antecedent combinations with the adequate locations of the rule memory. This option is only suitable for systems whose rule base is fixed because for general-purpose systems this would force several rules to have the same consequent.

- **Defuzzification methods**

Among the defuzzification methods reported in the literature and described in Chapter 3, the architecture selected herein allows implementing four of them. If the desired method is selected when designing the system, the adequate block among the four depicted in Figure 9.9 is included in the fuzzy IC. Another solution is to include the block proposed in [Jiménez, 1995; Sánchez-Solano 1997] that can be programmed to perform any of the four methods by an external signal. The structure of this block, shown in Figure 9.11, optimizes the circuitry employed by sharing the maximum number of operators.

9.5.2 OPERATION MODES AND TIMING SCHEMES

The circuits capable of implementing programmable fuzzy systems must load the knowledge base before operating. In addition, the knowledge base should admit modifications at any time. This is why programmable implementations have two operation modes, one named "load mode" in which the RAMs are written, and the other one named "normal operation mode" in which the fuzzy system infers the output. Selection between the two modes is done by an external signal.

- **Load mode**

Memories are written during this mode with a specific protocol. Data are synchronously introduced via an 8-bit bus. They begin to enter when a reset

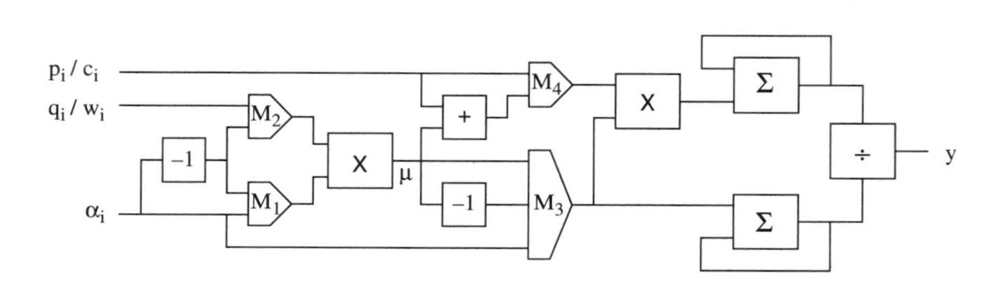

Figure 9.11. Block diagram of a programmable block that implements four defuzzification methods.

signal goes down, and they are stored on the falling edge of the clock. Although the load process can be interrupted whenever the reset signal goes high or the mode selection signal changes the operation, the load cannot begin from an arbitrary position. If the load mode is selected, the storage always starts from the zero row of the first antecedent memory.

The fuzzy IC has to include all the components required to address, activate, and load the antecedent and rule memories. To allow for the case that some memory width may be bigger than 8 bits, the data received through the load bus are stored into intermediate registers so as to write the memory once the word is complete.

- **Normal operation mode**

The circuit performs the fuzzy inferences during this operation mode. The calculation of the antecedent membership functions (for the arithmetic-based approach), the computing of the activation degrees of the active rules, the accumulation of values for the numerator and denominator, and their posterior division are tasks that must be performed sequentially. Pipeline stages are introduced in order to improve the inference speed of the system. Fuzzy systems designed with a memory-based approach to calculate the antecedent membership degrees have two pipeline stages, while those that employ the arithmetic-based approach have three stages. The reason is that the access to a memory is very fast while the arithmetic algorithm is implemented sequentially by several operators working during several clock cycles. The other two pipeline stages are the same in both cases: the first one is dedicated to evaluating the active rules as well as calculating and storing the numerator and denominator needed for calculating the system output, and the second stage performs the division.

The number of clock cycles of the pipeline stages also depends on the antecedent processing. If they are processed by a memory-based approach, this number is:

$$cycles = max(2^{inputs}, N) + 1 \tag{9.8}$$

If an arithmetic approach is considered, another term has to be added:

$$cycles = max(2^{inputs}, N, n\_mf) + 1 \tag{9.9}$$

where N is the number bits of the inputs and *n_mf* is the number of the antecedent membership functions.

9.5.3 PERFORMANCE ANALYSIS

Different implementations of this architecture have been analyzed in terms of the clock cycles required to carry out an inference and in terms of silicon area occupation.

• **Inference speed**

One of the major advantages of this architecture is the capability of providing a high number of FLIPS. Since the architecture description is independent of the implementation technology, the inference speed performance has been analyzed in terms of the clock cycles needed for carrying out an inference. Once this cycle number is known, the inference speed can be deduced from the particular clock frequency at which the fuzzy IC works.

According to (9.8) and (9.9), the number of clock cycles per inference depends on several implementation options and on the approach selected to process the antecedents. This is illustrated in Table 9.1, taking into account different numbers of inputs and resolution bits, and allowing up to 8 membership functions per input. For these examples, the memory- and arithmetic-based approaches only differ when there are two inputs and 4 bits of resolution. In this case, the memory alternative requires 5 instead of the 9 clock cycles required by the arithmetic approach. Regardless, the number of clock cycles is always very small, thus allowing high inference speeds. Inference speed is ultimately determined by the implementation technology, and in any case it is bigger than one MFLIPS.

• **Silicon area occupation**

The tests realized to analyze the area consumption have been divided into two main blocks that consider, respectively, programmable fuzzy systems and systems with a fixed knowledge base. All the data about area occupation refers to a 0.7-μm double metal CMOS technology and complete circuits including the pad ring.

Programmable fuzzy systems

For these circuits, several tests are shown in the following that consider 2, 3, and 4 inputs; either with memory-based realizations that employ 4, 8, and 10 bits of resolution; or with arithmetic-based approaches with 4, 8, and 12 bits. The results obtained for the memory and arithmetic realizations are illustrated in Figure 9.12.

Table 9.1. Clock cycles required to carry out an inference.

Inputs \ bits	4	8	12
2	5 or 9	9	13
3	9	9	13
4	17	17	17

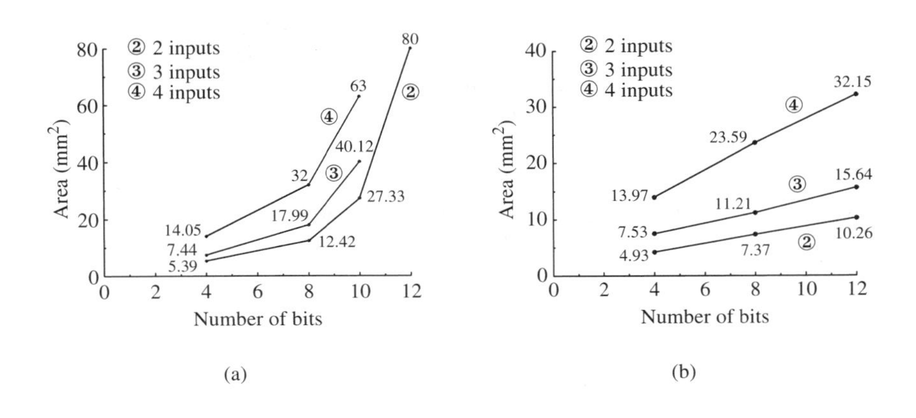

Figure 9.12. Comparison of programmable fuzzy ASICs in terms of area occupation. (a) Memory-based approach; (b) Arithmetic-based approach.

Regarding the memory-based approach, the area does not increase very much as the number of inputs increases (the worst case is an increase of 88%). The most significant variations are caused by an increase in the number of bits. Since these variations are exponential, the number of bits capable of being implemented with this approach is limited. The final reason for that is the exponential growth of the antecedent memories. For example, if a fuzzy system is implemented with 4 bits of resolution, each antecedent memory has 16 rows. If the resolution bits are 8, the memories have 256 rows, and they have 1024 rows for 10 bits of resolution. This shows that antecedent memories occupy most of the silicon area when implementing high resolution systems.

However, in the arithmetic-based approach, the area increase with respect to the number of bits is approximately linear and moderated, allowing implementations with resolutions of up to 12 bits. In addition, the increase with respect to the number of inputs remains similar to that of the memory option.

The size of circuits implemented with both memory- and arithmetic-based alternatives is similar only when the resolution is low, otherwise the second alternative is much more efficient. For systems with two inputs, for example, the difference in size is 9% when considering 4 bits of resolution (5.39 mm^2 for the memory option against 4.93 mm^2 for the arithmetic option) while it is 679% when 12 bits are considered (80 mm^2 against 10.26 mm^2). In summary, the memory approach allows a full freedom to define the antecedents (maintaining the overlapping degree at 2) but it is not efficient in terms of area for high resolution. Conversely, the arithmetic approach restricts the shapes of the antecedents but allows working with more resolution.

Non-programmable fuzzy systems

Implementations of non-programmable fuzzy systems are considered in this second block of tests. In this case, the knowledge base is stored into ROMs or combinatorial logic circuitry, so that the knowledge base should be selected prior to realizing the tests. The examples shown herein consider a knowledge base with two inputs. The resolution is selected as in the former case of programmable circuits, that is, 4, 8, and 10 bits for the memory-based approach to store the antecedents and 4, 8, and 12 bits for the arithmetic alternative. In addition, two defuzzification methods have been analyzed, the Fuzzy-Mean and the Weighted-Fuzzy-Mean methods.

The knowledge base needed for approximating the following function has been selected:

$$F = \frac{1}{2}[1 + sin(2\pi x_1)\cos(2\pi x_2)] \tag{9.10}$$

This knowledge base has been obtained by using the learning tool *xfbpa* of *Xfuzzy*. We refer the reader to Chapter 12 to see some examples of the learning process. Seven triangular membership functions have been used for covering the input universes of discourse while five functions have been required for the consequents. Figure 9.13 shows the resulting knowledge base after the learning process. Although bell-shaped membership functions are illustrated for the consequents, the use of simplified defuzzification methods that calculate the output as a weighted average means that only one or two parameters are employed to represent the consequents. In the case of the Fuzzy-Mean method, this representative parameter is the location of the bell in the universe of discourse. In the Weighted-Fuzzy-Mean method, the location and the width of the bell are the two representative parameters.

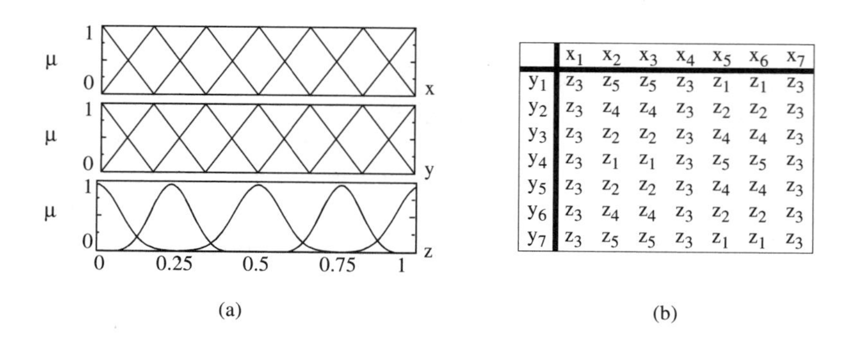

	x_1	x_2	x_3	x_4	x_5	x_6	x_7
y_1	z_3	z_5	z_5	z_3	z_1	z_1	z_3
y_2	z_3	z_4	z_4	z_3	z_2	z_2	z_3
y_3	z_3	z_2	z_2	z_3	z_4	z_4	z_3
y_4	z_3	z_1	z_1	z_3	z_5	z_5	z_3
y_5	z_3	z_2	z_2	z_3	z_4	z_4	z_3
y_6	z_3	z_4	z_4	z_3	z_2	z_2	z_3
y_7	z_3	z_5	z_5	z_3	z_1	z_1	z_3

Figure 9.13. Knowledge base implemented in the non-programmable ASICs. (a) Antecedents and consequents; (b) Rule base.

Figure 9.14 illustrates the sizes obtained for different realizations. The dashed lines correspond, respectively, to implementations of the Weighted-Fuzzy-Mean and Fuzzy-Mean methods. It is worth to point out that, unlike in programmable realizations, the area of ASICs that store the antecedents into memories does not increase exponentially with the resolution. For non-programmable realizations, both memory- and arithmetic-based approaches show a similar increase in area with respect to the number of bits. This increase is somewhat superior when implementing the Weighted-Fuzzy-Mean instead of the Fuzzy-Mean defuzzification method, because the additional circuitry required for the first method increases with the number of bits.

Comparison of programmable and non-programmable realizations

Finally, let us compare the tests carried out for programmable and non-programmable realizations. It can be seen that when the resolution is low, the differences regarding silicon area are not especially significant. For the option of memory storage, the area of a programmable circuit with 4 bits of resolution is 5.39 mm^2, which, compared to the 3.38 mm^2 that a non-programmable circuit occupies, means an increase of 79%. This increase is lower for the arithmetic option, from 4.93 mm^2 in the programmable case to 3.40 mm^2 in the non-programmable one, which means an increase of 42%. Differences are much clearer as resolution grows higher. In this sense, while the area increase from a programmable to a non-programmable realization with 12 bits is 58% when the arithmetic approach is employed, it is almost 400% when resorting to the use of the memory alternative for implementing circuits with 10 bits. In the latter case, the non-programmable circuit has an area of 5.56 mm^2, which increases only slightly with the number of resolution bits, while the programmable circuit occupies 27.33 mm^2.

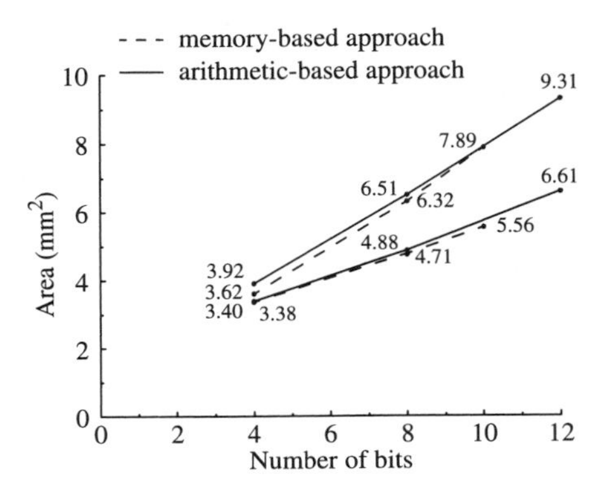

Figure 9.14. Comparison of non-programmable fuzzy ASICs in terms of area occupation.

9.5.4 PROTOTYPES

Two prototypes have been integrated into a 1.0-μm double metal CMOS technology with the architecture previously described. They both store the knowledge base into RAMs to be programmed for different applications. They have 3 inputs and 1 output with 6 bits of resolution and implement the Fuzzy-Mean defuzzification method. The main difference between these digital prototypes is the antecedent processing, which in one case is based on memory and in the other case on arithmetic calculations.

The starting point for the design of both circuits is a functional description with the hardware description language VHDL, which involves the major verification effort. Verification at this level exploits the VHDL power of not only generating adequate test patterns for the circuit but also writing simultaneously those patterns into files whose formats are understood by a logic simulator. In addition, the simulation entities built in VHDL also store the simulation patterns into a file with a format suitable for its posterior transfer to the test equipment. Once the VHDL code has been verified, the design process focuses on two objectives, the synthesis of the VHDL code and the placement and routing of the components to obtain the circuit layout. This has been carried out by using conventional methodologies included within commercial ICs design environments.

• **Memory-based prototype**

The block diagram of this prototype is illustrated in Figure 9.15. This prototype has one memory per input to store the antecedent membership functions. Each of the outputs of these memories enters MFC blocks which provide the

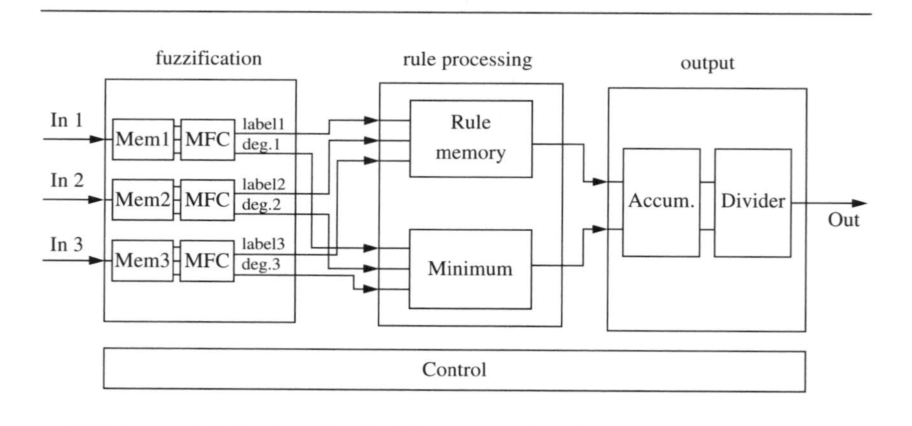

Figure 9.15. Block diagram of memory-based prototype.

membership degrees and the labels of the activated fuzzy sets. The membership degrees are conducted towards a block that calculates the minimum of the degrees corresponding to the same rule. The labels activated by the three inputs address the rule memory. The singleton values provided by the rule memory as well as the activation degrees of the rules coming from the minimum block enter the output block, which performs the accumulation and division. Additionally, a control block is in charge of writing the memories and timing the whole circuit.

Figure 9.16 shows the microphotograph of this prototype depicting the functional blocks. It can be seen from this photograph that the three RAMs of the antecedents (the three blocks on the right) occupy a significant portion of the total area. The block placed at the bottom and on the left corresponds to the rule base RAM. The rest of the circuit is formed by standard cells.

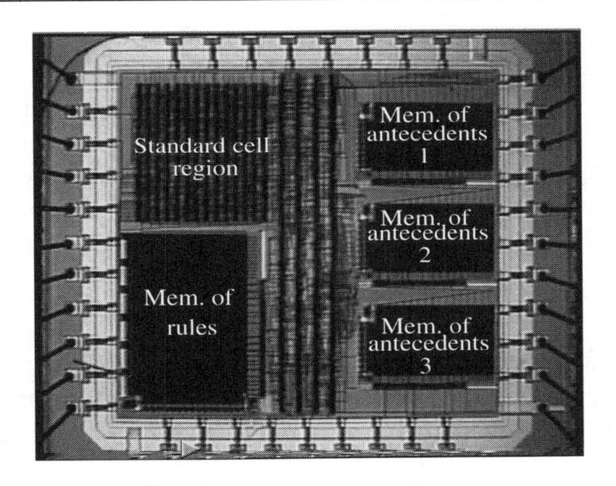

Features	values
Number of inputs	3
Bits of resolution	6
Number of antecedents per input	8
Number of rules	512
Storage approach for antecedents	Memory
Defuzzification method	Fuzzy Mean
Silicon area (including pad ring)	14.36 mm^2
Maximum operation frequency	22.9 MHz
Maximum inference speed	2.54 MFLIPS

Figure 9.16. Microphotograph and features of the memory-based prototype. (©1997 IEEE. Reprinted with permission from [Sánchez-Solano, 1997]).

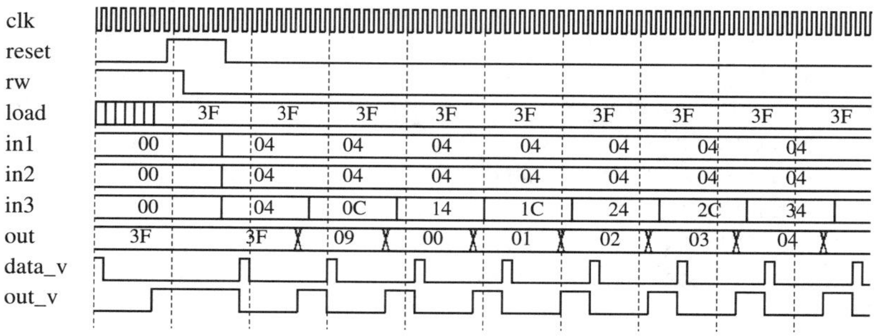

Figure 9.17. Experimental test results of the memory-based prototype.

This prototype has been characterized with the HP82000 test equipment. Figure 9.17 shows the results of one of these tests where a sequence of ascending data is observed at the signal *out*. From the characterization results, the most relevant feature is the maximum operation frequency measured, which is 22.9 MHz. Since the pipeline stages consist of 9 clock cycles, the maximum inference speed is 2.54 MFLIPS.

- **Arithmetic-based prototype**

Figure 9.18 shows the block diagram of the arithmetic-based prototype. Due to the arithmetic option, there is a single antecedent memory shared by the three inputs, although one block per input is added which performs the opera-

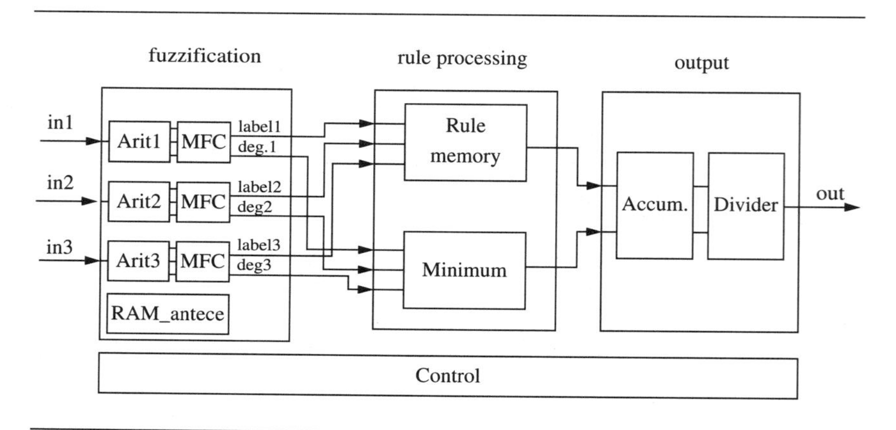

Figure 9.18. Block diagram of the arithmetic-based prototype.

tions required to calculate the linguistic labels of the active fuzzy sets and the corresponding membership degrees. The outputs of these blocks enter the MFC blocks, whose outputs (the membership degree and the label of the selected membership function) are drawn to the rule memory and to the minimum block. Finally, weighted average is carried out through two functional blocks that implement the accumulation and division operations.

Besides the input and output pins associated with the fuzzy inference (*in*1, *in*2, *in*3, and *out*), this prototype (like the memory-based prototype) has input pins for the clock signal, for an asynchronous reset, for a signal that selects the operation mode, and for an input bus to load the memories; and it has output pins indicating that input data are being loaded, and informing of new output data.

Figure 9.19 shows the microphotograph and the main features of this second prototype. It can be seen how a single memory (placed at the left bottom

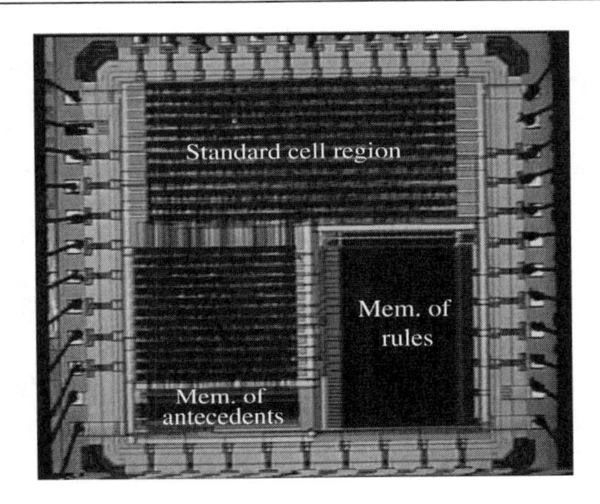

Features	values
Number of inputs	3
Bits of resolution	6
Number of antecedents per input	8
Number of rules	512
Storage approach for antecedents	Arithmetic calculus
Defuzzification method	Fuzzy Mean
Silicon area (including pad ring)	12.3 mm^2
Maximum operation frequency	26 MHz
Maximum inference speed	2.8 MFLIPS

Figure 9.19. Microphotograph and features of the arithmetic-based prototype. (©1997 IEEE. Reprinted with permission from [Sánchez-Solano, 1997]).

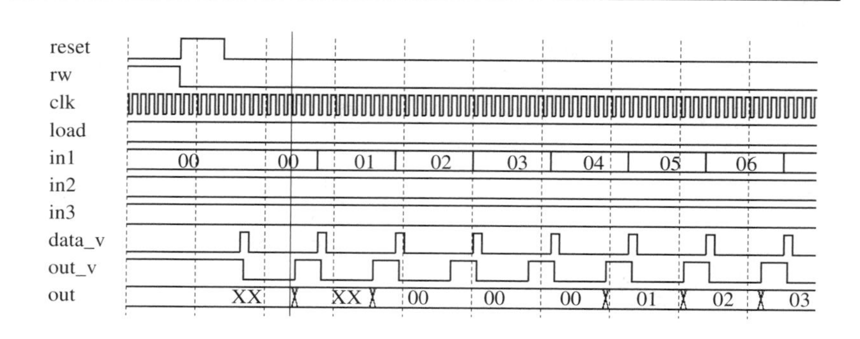

Figure 9.20. Experimental test results of the arithmetic-based prototype.

part) is employed to store the antecedents and how its size is very much smaller than the memories of the first prototype. On the other hand, the rule memory (placed at the right bottom part) has a similar size. The region of standard cells occupies more area than in the memory-based prototype.

This prototype has also been characterized with the HP82000 test equipment. Figure 9.20 illustrates a fragment of one of the experimental results obtained where a sequence of ascending data can be seen in the signal *out*. In this case the maximum frequency of operation measured has been 26 MHz, which means a maximum inference speed of 2.8 MFLIPS because the pipeline stages consist of 9 clock cycles.

REFERENCES

1. **Arikawa, H., Hirota, K.**, Fuzzy inference engine by address-look-up & paging method, in *Proc. 3rd IFSA World Congress*, pp. 45-46, Seattle, 1989.
2. **Ascia, G., Catania, V., Giacalone, B., Russo, M., Vita, L.**, Designing for parallel fuzzy computing, *IEEE Micro*, Vol. 15, N. 6, pp. 62, 1995.
3. **Ascia, G., Catania, V.**, A dedicated parallel processor for fuzzy computation, in *Proc. IEEE Int. Conf. on Fuzzy Systems*, pp. 787-792, Barcelona, 1997.
4. **Chiueh, T.**, Optimization of fuzzy logic inference architecture, *IEEE Computer*, Vol. 25, N. 5, pp. 67-71, 1992.
5. **Eichfeld, H., Löhner, M., Müller, M.**, Architecture of a CMOS fuzzy logic controller with optimized memory organisation and operator design, in *Proc. IEEE Int. Conference on Fuzzy Systems*, pp 1317-1323, San Diego, 1992.

6. **Gabrielli, A., Gandolfi, E., Masetti, M.**, Design of a family of VLSI high speed fuzzy processors, in *Proc. IEEE Int. Conference on Fuzzy Systems*, pp. 1099-1105, New Orleans, 1996.
7. **Gabrielli, A., Gandolfi, E., Masetti, M., Roch, M. R.**, VSLI design and realization of a 4 input high speed fuzzy processor, in *Proc. IEEE Int. Conference on Fuzzy Systems*, pp. 779-785, Barcelona, 1997.
8. **Gandolfi, E., Masetti, M., D'Antone, I., Gabrielli, A., Spotti, M.**, Architecture of a 50 MFLIPs fuzzy processor and the related 1μm VLSI CMOS digital circuits, in *Proc. 4th Int. Conf. on Microelectronics for Neural Network and Fuzzy Systems*, pp. 125-133, Turin, 1994.
9. **Hung, D.L.**, Custom design of a hardware fuzzy logic controller, in *Proc. IEEE Int. Conference on Fuzzy Systems*, pp. 1781-1785, Orlando, 1994.
10. **Ikeda, H., Hiramoto, Y., Kisu, N.**, A fuzzy inference processor with an active-rule-driven architecture, in *Proc. Symposium on VLSI Circuits,* pp. 25-26, 1991a.
11. **Ikeda, H., Hiramoto, Y., Kisu, N., Takahashi, H.**, A fuzzy processor for a sophisticated automatic transmision control, in *Proc. 4th IFSA World Congress,* pp. 53-56, Brussels, 1991b.
12. **Ikeda, H., Kisu, N., Hiramoto, Y., Nakamura, S.**, A fuzzy inference coprocessor using a flexible active-rule-driven architecture, in *Proc. IEEE Int. Conference on Fuzzy Systems*, pp. 537-544, San Diego, 1992.
13. **Jiménez, C. J., Barriga, A., Sánchez-Solano, S.**, Digital implementation of SISC fuzzy controllers, in *Proc. 3rd. Int. Conference on Fuzzy Logic, Neural Networks and Soft Computing,* pp. 651-652, Iizuka, 1994.
14. **Jiménez, C. J., Sánchez-Solano, S., Barriga, A.**, Hardware implementation of a general purpose fuzzy controller, in *Proc. 6th IFSA World Congress*, pp. 185-188, São Paulo, 1995.
15. **Katashiro, T.**, A fuzzy microprocessor for real-time control applications, in *Proc. 5th IFSA World Congress*, pp. 1394-1397, Seoul, 1993.
16. **Ruiz, A., Gutiérrez, J., Felipe-Fernández, J. A.**, A fuzzy controller with an optimized defuzzification algorithm, *IEEE Micro*, Vol. 15, N. 6, pp. 67, 1995.
17. **Sánchez-Solano, S., Barriga, A., Jiménez, C. J., Huertas, J. L.**, Design and application of digital fuzzy controllers, in *Proc. IEEE Int. Conference on Fuzzy Systems*, pp. 869-874, Barcelona, 1997.
18. **Sasaki, M., Ueno, F., Inoue, T.**, An 8-bit resolution 140 KFLIPS fuzzy microprocessor, in *Proc. 5th IFSA World Congress*, pp. 921-924, Seoul, 1993a.
19. **Sasaki, M., Ueno, F., Inoue, T.**, 7.5 MFLIPS fuzzy microprocessor using SIMD and logic-in-memory structure, in *Proc. IEEE Int. Conference on Fuzzy Systems*, pp. 527-534, San Francisco, 1993b.
20. **Togai, M., Watanabe, H.**, A VLSI implementation of fuzzy inference engine: toward an expert system on a chip, in *Proc. 2nd Conference on Artificial Intelligence Applications*, pp. 192-197, Miami, 1985.
21. **Togai, M., Chiu, S.**, A fuzzy logic chip and fuzzy inference accelerator for real-time approximate reasoning, in *Proc. Int. Symposium on Multiple-Valued Logic*, pp. 25-29, 1987.
22. **Ungering, A. P., Thuener, K., Goser, K.**, Architecture of a PDM VLSI fuzzy logic controller with pipelining and optimized chip area, in *Proc. IEEE Int. Conference on Fuzzy Systems*, pp. 447-452, San Francisco, 1993.
23. **Watanabe, H., Dettloff, W.**, Reconfigurable fuzzy logic processor: a full custom digital VLSI, in *Proc. 3rd IFSA World Congress*, pp. 49-50, Seattle, 1989.
24. **Watanabe, H., Dettloff, W., Yount, K. E.**, A VLSI fuzzy logic controller with reconfigurable, cascadable architecture, *IEEE Int. Journal of Solid-State Circuits*, Vol. 25, N. 2, pp. 376-382, 1990.

Chapter 10

FUZZY SYSTEM SYNTHESIS

The application of fuzzy logic–based techniques to problems requiring real-time operations demands the use of structures allowing efficient microelectronic implementations of fuzzy inference mechanisms, as described in the previous chapters. Moreover, as for many other kinds of microelectronic design, the use of computer-aided design (CAD) tools is essential for easing the design tasks and guaranteeing the correct functionality of the final product.

This chapter will complete our overview of the different CAD tools for fuzzy systems, focusing on those tools oriented to easing their hardware synthesis. First, we will analyze the design flow for a fuzzy system and discuss the applicability of hardware description languages like VHDL. Finally, based on the facilities offered by the *Xfuzzy* environment, a design methodology for application-specific fuzzy systems is described, using the different hardware and software concepts introduced along this book.

10.1 HARDWARE SYNTHESIS OF FUZZY SYSTEMS

The design flow used in the microelectronic design of a fuzzy system passes through three different levels: algorithmic, architectural, and circuit. The algorithmic level defines the functional behavior of the system, specifying the membership functions and the rule base, and selecting the different fuzzy operators. In order to obtain a physical implementation, the following stage is the choice of a certain architecture and the identification of the basic building blocks. Finally, using these blocks, the system is implemented as an integrated circuit selecting the most adequate implementation technique and design style for the specific application under development.

The design tasks in each of the above mentioned levels are usually supported by different CAD tools. In the description and simulation stages at the algorithmic level specific tools for fuzzy systems, such as those described in Chapters 4 and 5, are commonly used, while at the architectural and circuit levels the usual practice is to rely on tools included in conventional environments for integrated circuit design. The efficient translation between the different design levels is one of the main potential bottlenecks in the microelectronic implementation of fuzzy inference systems.

The functionality of a fuzzy system synthesis tool fundamentally depends on the strategy selected for implementing the system. As discussed in Chapter 6, the physical implementation of a fuzzy system by means of an off-line strategy usually employs digital memories or programmable logical devices. The synthesis tools applicable in this case are oriented towards obtaining the look-up table and minimizing this table to reduce the required resources. Conversely, when an on-line strategy aimed at producing dedicated hardware is used, the synthesis tools required will be strongly biased by the envisaged implementation technique (namely ASIC or FPGA).

The great diversity of design styles and the high degree of dependence on their particular technologies are characteristics of analog circuits, entangling the development of general purpose synthesis tools. However, some tools particularized for specific application fields (filter design, converters, etc.) are available. These tools are generally called "*silicon compilers*", since they translate a functional description of the system into the layout of an integrated circuit, using a set of already designed blocks and following predetermined placement schemes. Since they are based on already implemented cells, these tools are almost completely linked to a certain circuit fabrication technology, thus complicating their widespread use. Some silicon compilers for fuzzy systems have been recently reported in the literature, based on current mirrors [Lemaitre, 1993] and differential pairs [Franchi, 1998; Manaresi, 1996]. The latter work shows that the time employed in translating the system behavior description into an integrated circuit layout using the silicon compiler is less than 30 minutes.

The circumstances are completely different in the digital case, where the simplification induced by the logic abstraction level has allowed the development of a great number of tools for easing automated circuit synthesis, independent of the target technology. The use of automated synthesis tools provides a series of advantages when facing the implementation of any integrated circuit. First, synthesis automation permits the reduction of the design cycle. This is a very important aspect in the highly competitive market of microelectronics, where it is necessary to react to the demand in a very short timeframe. Related to this point, automated synthesis allows the exploration of design space, making possible the analysis of different alternatives in terms of cost, speed, power consumption, etc. Lastly, a basic aspect in the whole design process is robustness, i.e., the guarantee that the product is free of errors. In this sense, the use of automated synthesis tools ensures the "correction by construction" of the system under design.

Most automated synthesis tools divide the synthesis process into a series of stages, hierarchically ordered, each translating a certain description of the system into another with a higher level of detail. The initial description is usually expressed by means of a high level programming language (Pascal, C, etc.) or a hardware description language (VHDL, Verilog, etc.). The next section focuses on describing the more salient characteristics of VHDL descriptions.

10.2 DESCRIBING FUZZY SYSTEMS WITH A HARDWARE DESCRIPTION LANGUAGE: VHDL

The VHDL language was introduced in the early 1980s as part of a project of the U. S. Department of Defense called VHSIC[1] (*Very High Speed Integrated Circuits*). The project required a language for describing the systems that were developed within it with a double objective: auto-commenting the designs while offering simulation capabilities. In 1985, the IEEE's DTAC (*Design Automation Technical Committee*) considered VHDL an adequate language for describing circuits independent of the design tool and covering the different abstraction levels in the design process. This approach was oriented to solve the problems in design compatibility among different CAD platforms. Since VHDL covered the aspects the DTAC was interested in and it was a non-proprietary language, the VASG (*VHDL Analysis and Standardization Group*) was created, to start the language standardization process. In December 1987 the official VHDL standard was introduced, with the name IEEE 1076-1987 [VHDL, 1988].

The standard is revised approximately every 5 years. These revisions incorporate enhancements to the language according to technological and industrial requirements. This way, although VHDL was conceived for describing digital systems, a new language allowing the description of analog and mixed-signal circuits is available: VHDL-AMS (*VHDL-Analog Mixed Signal*).

VHDL, like most high-level programming languages, is an imperative language. It is based on a declarative syntax, so that the problem to be described is specified as a set of instructions without detailing a solution method, that is, the order of the instructions is irrelevant. However, a procedural syntax, where the desired action is described in terms of a sequence of steps where the order is important, can also be applied in some specific constructions, such as processes and procedures.

A VHDL description consists of a series of design units that allow the specification of the different elements defining the circuit. The basic design unit is called an *entity* and defines the circuit interfaces (like input and output ports). The circuit communicates with its environment through this unit. The entity represents the system as a black box that can only be accessed through its ports. The interior of this black box is described by another design unit called *architecture*. The architecture allows the specification of either the circuit behavior or the circuit structure. Since a system can be described in several ways, a circuit can have several architectures modeling it, while the entity is unique for a certain design.

The specification of an architecture is composed by two areas: a *declarative area*, and another that constitutes the *body of the architecture*. The first area contains the declaration of the elements to be used in the description, such

1. The VHDL acronym corresponds to VHSIC Hardware Description Language.

as the components describing the circuit schematic, inner signals, functions and procedures, data types to be used, etc. The body contains the effective system description. The instructions contained in it are concurrent, that is, all of them are executed simultaneously. These instructions place and interconnect components, execute procedures, assign values to signals, etc.

With these kinds of descriptions it is possible to specify the circuit both in a structural way (schematic) and in a functional way (in terms of system equations). However, it is not possible to produce an algorithmic description, since a procedural syntax is required. For this purpose, VHDL incorporates processes, functions, and procedures. A *process* is a concurrent instruction (since it is used inside the body of an architecture) that contains sequential instructions. These sequential instructions are executed according to the specified control flow, by means of constructions typical in procedural languages, such as loops, "if ... then ... else" blocks, procedure and function calls, etc. Functions and procedures are also executed sequentially and, while functions can only return a single value, procedures can return more than one.

The designer using VHDL has to define the data types, the operators, attributes, and functions. There are specific instructions for creating new data types or for using previously defined types to create others. Operators and functions can be overloaded, that is, it is possible to create different operators and functions with the same name, differentiated by the arguments they accept.

The characteristics of VHDL described thus far make it suitable for system modeling by means of different description techniques (structural, functional, algorithmic). It is, therefore, a powerful language in the application fields it has been conceived for, since it covers different levels of abstraction in digital system designs. In addition, it is a live language, with a revision mechanism to adapt it to new needs. In fact, although VHDL is a language originally conceived for documentation and simulation, the current trend is extending its use towards other application fields such as high level synthesis, electrical circuit modeling, PMS (Processor, Memory, Switch) features modeling, and hardware-software co-specification.

Among the new applications for VHDL, high-level synthesis merits further discussion. When using VHDL for synthesizing a circuit, certain constraints must be imposed on the language. These constraints come, basically, from two factors. First, a key factor is the use that the language makes of time. Since VHDL was conceived for simulation, time is precisely defined in its model. This causes significative differences between simulation-oriented modeling (as foreseen for the language) and synthesis-oriented modeling. In the first case, the designer is able to specify delays in signal assignment and process execution. In synthesis-oriented modeling, the designer cannot impose absolute conditions on the time variable, since it will depend on circuit implementation, the applied technology and the defined goals and constraints. These aspects will determine delays. All these considerations imply constraints on signal assignments, process start and stop operations, etc., and the language is used in a more declarative than procedural way. The second factor influencing the constraints on VHDL when used for high-level synthesis is that certain instructions are only applicable in simulation. For example, the file type and the file object only have meaning in a computer environment and not in a circuit.

10.2.1 MODELING FUZZY SYSTEMS WITH VHDL

Different authors have investigated the possibilities of using VHDL as an appropriate language for supporting the data structures and functions required by fuzzy system simulation [Zamfirescu, 1992; 1993; Galán, 1995]. The advantage of using this hardware description language for describing the algorithmic behavior of a fuzzy system is that it would permit the use of the same simulation environment at both the architectural and circuit levels, so results obtained at the different stages in the design process could be easily compared.

In order to extend the capabilities of VHDL for supporting the description and simulation of fuzzy systems, it is necessary to have a series of data structures holding fuzzy information, together with a set of functions describing the inference algorithms. Incorporating these elements into a VHDL *package* that the designer can reference is extremely useful.

An example of a VHDL library for fuzzy systems is reported in [Galán, 1995]. This package contains the definition of a series of basic types that will be used for storing the values of input and output variables, the description of membership functions, and the activation degrees of rules. Other more complex types, which use the basic ones, are included for grouping data corresponding to the different linguistic variables used inside the rules (*mfc_set*, *input_var* and *output_var*). The VHDL package also includes functions for manipulating these data structures. Initialization functions permit the description of the system database, assigning values to membership functions (*init_mfc*), input variables (*init_inputs*), and output variables (*init_outputs*). Other functions are used for describing the rule base and implementing the inference mechanism: *fzis* performs input fuzzification; *fzand* and *fzor* correspond to the linguistic connectives used inside the rules; *fzthen* performs fuzzy implication and the aggregation of the conclusion of individual rules. Finally, the function *defuzzifier* returns the result of the inference according to the selected defuzzification method.

Figure 10.1 shows an example of a fuzzy system description using the previously mentioned VHDL package. This system corresponds to the truck controller also used as an example in Chapters 4 and 5. The package containing the definitions of types and functions is called "FUZZY4". As discussed in the above section, the VHDL description of the fuzzy system is composed of two parts: the '*entity FLC*' defining the input-output interface of the system, and the '*architecture FLC_arch of FLC*' describing the operational algorithm of the system. In this case the architectural specification is constituted by a process

```
use work.FUZZY4.all;

entity FLC is
    port (FLC_input : in input_array;
          FLC_output: out output_array);
end FLC;

architecture FLC_arch of FLC is
    begin
     FLC:process
       variable mfc_set_PHI :  mfc_set;
       variable mfc_set_X   :  mfc_set;
       variable mfc_set_PSI :  mfc_set;
       variable vars_IN     :  input_var;
       variable vars_OUT    :  output_var;
     begin
-
-- Initialize the membership funcion groups
-
       mfc_set_PHI:=(
       init_mfc("RB",-45.0,56.0,0.0,55.0,0.0,0.0),
       init_mfc("RU",30.0,40.0,0.0,30.0,0.0,0.0),
       init_mfc("RV",70.0,20.0,0.0,20.0,0.0,0.0),
       init_mfc("VE",90.0,10.0,0.0,10.0,0.0,0.0),
       init_mfc("LV",110.0,20.0,0.0,20.0,0.0,0.0),
       init_mfc("LU",150.0,30.0,0.0,40.0,0.0,0.0),
       init_mfc("LB",225.0,55.0,0.0,56.0,0.0,0.0),
       others=>NIL);

       mfc_set_X:=(
       init_mfc("LE",5.0,0.0,10.0,25.0,0.0,0.0),
       init_mfc("LC",40.0,10.0,0.0,10.0,0.0,0.0),
       init_mfc("CE",50.0,5.0,0.0,5.0,0.0,0.0),
       init_mfc("RC",60.0,10.0,0.0,10.0,0.0,0.0),
       init_mfc("RI",95.0,25.0,10.0,0.0,0.0,0.0),
       others=>NIL);

       mfc_set_PSI:=(
       init_mfc("NB",-30.0, 0.0,0.0,13.0,0.0,0.0),
       init_mfc("NM",-15.0,10.0,0.0,9.0,0.0,0.0),
       init_mfc("NS",-6.0,6.0,0.0,6.0,0.0,0.0),
       init_mfc("ZE",0.0,6.0,0.0,6.0,0.0,0.0),
       init_mfc("PS",6.0,6.0,0.0,6.0,0.0,0.0),
       init_mfc("PM",15.0,9.0,0.0,10.0,0.0,0.0),
       init_mfc("PB",30.0,13.0,0.0,0.0,0.0,0.0),
       others=>NIL);

-- Initialize the input and output variables.
-
       vars_IN(1):=init_inputs("M(PHI)",-90.0,
       270.0,-90.0,0.05,0.0,7,mfc_set_PHI);

       vars_IN(2):=init_inputs("M(X)",0.0,100.0,
       25.0,0.05,0.0,5,mfc_set_X);

       vars_OUT(1):=init_outputs(init_inputs
       ("M(PSI)",-30.0,30.0,0.0,0.05,0.0,7,mfc_set_PSI),
       IMP_MIN,AGR_MAX,COA_CRIT,0.0);
-
-- Rule setting and inference mechanism
-
       while (1=1) loop
-
-- Turn to zero the activation degree array
-- place in the output record.
-
       vars_OUT(1).GradeOfActiv:=ini_grade_activ;

       fzthen(fzand(fzis("RB",vars_IN(1)),
                    fzis("LE",vars_IN(2))),
                    "PS",vars_OUT(1));

       fzthen(fzand(fzis("RB",vars_IN(1)),
                    fzis("LC",vars_IN(2))),
                    "PM","vars_OUT(1));

                ...
                ...
                ...

       fzthen(fzand(fzis("LB",vars_IN(1)),
                    fzis("RI",vars_IN(2))),
                    "NS",vars_OUT(1));

       FLC_output(1) <= defuzzifier(vars_OUT(1));

       wait on FLC_input;

       data_assign(2, FLC_input, vars_IN);

       end loop;
     end process FLC;
end FLC_arch;
```

Figure 10.1. Description of a fuzzy system using VHDL.

including the following stages: (1) initialization of membership functions, inputs and outputs; (2) description of the rule base and execution of the inference mechanism; and (3) defuzzification and assignment of output results.

10.2.2 FUZZY SYSTEM SYNTHESIS FROM VHDL

Comparing the VHDL description of the fuzzy system shown in Figure 10.1 with the XFL specification in Figures 4.7 and 4.8 of Chapter 4, we can realize the great similarity between them. This means that it is relatively simple to translate an XFL specification into a VHDL description which, apart from being used for the simulation of system behavior, could be the starting point for different automated synthesis tools. However, current synthesis tools are quite inefficient when selecting an adequate architecture from a high level algorithmical description, since different compromises must be reached at both the algorithmical and physical design spaces for obtaining an optimal cost/performance ratio [Costa, 1995].

This is the reason why most papers in the literature deal with the development of specific tools based on specific fuzzy system architectures [Costa, 1994; Hollstein, 1996; 1997]. This is the strategy used by the *Xfuzzy* tools that will be described in the next section.

10.3 TOOLS FOR HARDWARE SYNTHESIS IN XFUZZY

Perhaps one of the most salient features of the CAD environment described in this text is the inclusion of tools for hardware synthesis capable of obtaining microelectronic implementations of fuzzy systems from the high-level specifications of their behavior. The XFL language (as discussed in Chapter 4) was specifically designed for easing the process of deriving implementations from specifications based on it, either to high-level programming languages or to physical implementations based on integrated circuits.

Hardware synthesis tools based on XFL currently cover two basic strategies for the development of digital implementations. From an XFL specification, these tools are able to generate representations that can be input to environments for integrated circuit design in order to realize the final implementation of the system. These intermediate representations use commonly accepted standards, so hardware synthesis tools based on XFL are compatible with the most widely used environments for integrated circuit design [Synopsys, 1997; Mentor, 1997]. Furthermore, with the objective of easing quick prototyping capabilities, the synthesis tools that will be described in the following sections can produce specific files for the Synopsys logical synthesis tools and the Xilinx FPGA design tools, so the final implementation of a fuzzy system can be directly obtained from its XFL specification [Barriga, 1996].

10.3.1 SYNTHESIS BASED ON OFF-LINE STRATEGIES

xftl is the tool intended for the microelectronic realization of a fuzzy system by means of an off-line strategy. For this purpose, *xftl* uses the XFL to C compiler (*xfc*) to produce a C program that, once compiled, is able to generate a representation of the system behavior in terms of a look-up table associating all the possible input combinations with the outputs that must be obtained according to its XFL specification.

The tool uses as an intermediate format the Berkeley PLA format [Brayton, 1984], so, once the look-up table has been obtained, a wide range of design tools can be used on it for typical operations such as logical minimization and extraction of Boolean equations. This way, the tool can be seamlessly integrated within different development environments. As an example of this, it is worth noting that, if the synthesis tools of Synopsys and Xilinx are available, *xftl* is able to automatically direct them to obtain an implementation of the system on an FPGA.

The design flow for fuzzy inference systems using *xftl* is shown in Figure 10.2. The starting point of the synthesis process is a description of the system behavior in the high-level specification language XFL. The system specification contains the initial knowledge base (membership functions for antecedents and consequents, and the rule base) and information about the inference mechanism and the defuzzification method to be implemented. This description is simulated and refined using the verification tools integrated within *Xfuzzy*.

Once a valid description of the fuzzy system is available, that is, a definition that fulfills the requirements of the problem, the next step consists of using *xftl* to generate a look-up table representing the input/output functional behavior of the inference system. At this point, the designer must establish a set of implementation options defining hardware-related requirements, like width for buses, and the specific options to be applied by the logical minimization and FPGA design tools that will be used in the coming design stages. The output of *xftl* consists of two files. The first file contains a formal description of the look-up table in PLA format. This table can be minimized by any tool able to accept the PLA format and implemented either by a combinational circuit or by a memory. The second of the output files of *xftl* is intended to control the Synopsys logical synthesis tools for minimizing the look-up table and extracting the Boolean equations in order to implement the table using the combinational resources of a Xilinx FPGA. With this purpose, this second output file contains a series of commands to direct the synthesis process performed by Synopsys and Xilinx FPGA design tools.

The output of Synopsys is a file with XNF (*Xilinx Netlist Format*) format containing the description of the fuzzy inference system. Optionally, it is also possible to generate a file containing the connection requirements (number of configurable logic blocks, CLBs, and input/output blocks, IOBs) and the timing constraints. The file(s) are used as input to the tools that perform the mapping and connection of the FPGA. The final result is a file with the physical implementation of the fuzzy system, which can be directly used to program the FPGA.

The algorithm behind *xftl* is rather obvious: it consists of a loop where successive values of the input variables are applied to the system inference engine. These values correspond to the successive steps on the universe of discourse of variables defined by their assigned size in bits. The outputs of the system are coded according to the size in bits assigned to each output variable. The current version of the tool employs the number of bits used for representing each sys-

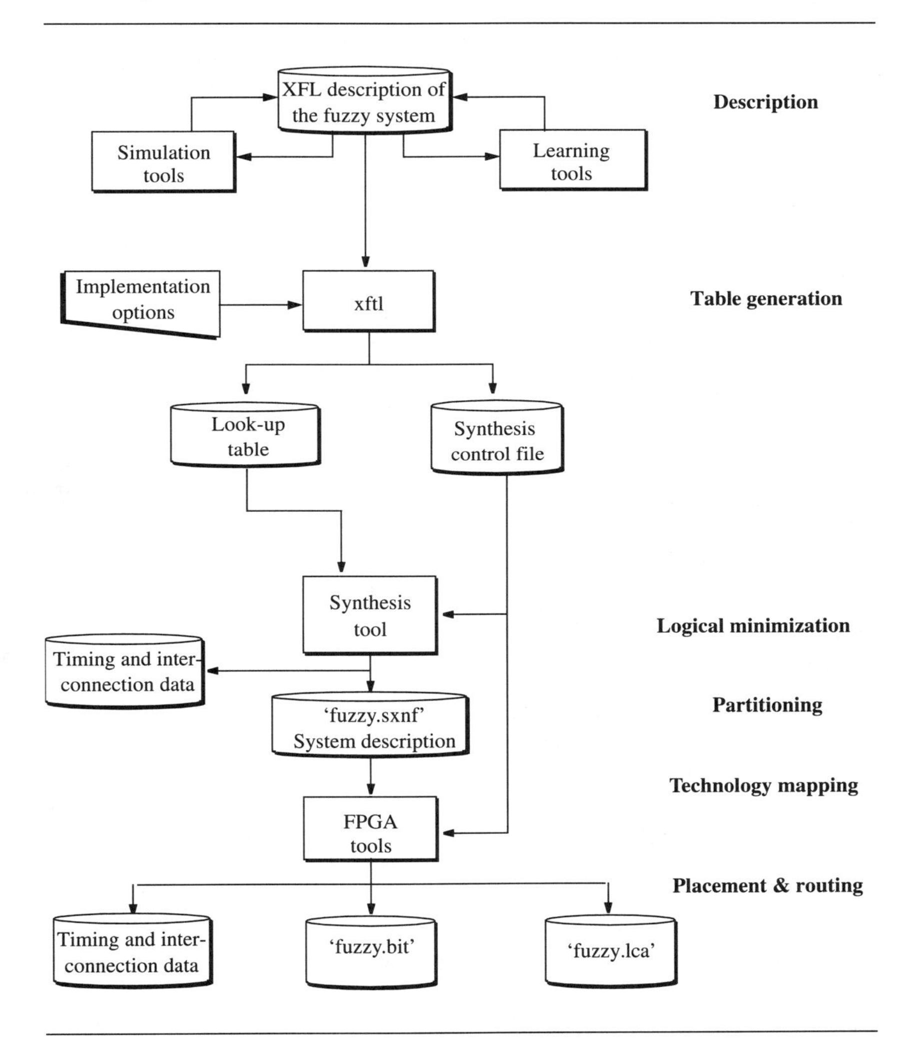

Figure 10.2. Design flow with *xftl*.

tem variable to split the universe of discourse of that variable in equal intervals, coding values in an ascending order.

Therefore, one of the most important parameters when using this tool is the assignment of the number of bits to be employed for coding each one of the system variables. An adequate coding yields a table that is more compact and easier to implement. Moreover, the tool execution time and the hardware resources needed for implementing the look-up table are dramatically influenced by the number of iterations of the main loop required for exploring the input space, so it is convenient to be extremely careful when assigning the bitsize to input variables. If the options that the tool accepts for controlling these sizes are not used, each variable is assigned a bitsize derived from the cardinality of the type it belongs to, using the simple procedure of rounding up to the integer immediately above the base-2 logarithm of the cardinality (Section 4.3.2 in Chapter 4).

As we pointed out when the off-line synthesis strategy was introduced in Chapter 6, one of the main advantages it possesses is that it does not impose any constraint on the system definition at the algorithmic level. Essentially, the tool stores in a look-up table format the coded result of an exhaustive exploration of the system state space obtained from a previous software synthesis. Therefore, *xftl* can generate the microelectronic implementation of any system that can be defined using XFL, independently of the complexity of its internal structure or the type of fuzzy operations in use. In fact, the definition of fuzzy operations applicable for *xftl* is the same that is applied for *xfc*. But this synthesis procedure is limited by the exponential growth of the look-up table as the number of inputs and/or the number of bits used for representing these inputs increases. Therefore, the applicable constraints are due to considerations such as the required area or the availability of components for holding the physical implementation of the system.

10.3.2 USING XFTL

As the rest of tools integrated within *Xfuzzy*, *xftl* can be executed from the Unix shell or through a graphical user interface. When *xftl* is called from the shell the applicable syntax is shown in Figure 10.3. The graphical user interface of *xftl* is shown in Figure 10.4. The upper part of the window contains text areas for setting the general options of the tool: the prefix for output files (*Output file prefix*), the type of fuzzy variables in the program that generates the system representation in PLA format (*Type of fuzzy variables*), and the default bitsize for coding system variables (*Default bitsize*). Below them, a list of the variables of the system (with the label *Bitsize for variables*), provides a text area for each variable to set the particular bitsize for representing it. The lower part of the window, labelled *Synthesis options*, is only visible when the circuit synthesis tools are available. This section provides controls for setting the optimization options applied by the circuit synthesis tools and the FPGA device to be used for the final system implementation. Finally, the control buttons in the lower part of the window are used for building the C source file that can be employed for obtaining an implementation of the system in PLA format (*Build C source*), compiling and executing the C source file (*Build PLA file*), and calling the circuit synthesis tools (*Synthesize to FPGA*).

```
Xftl: Synthesizable look-up table generator for XFL-based fuzzy systems

Synopsis: xftl <-E<c|p|s>> <-O OperationsFile> <-o OutputPrefix> <-b Bitsize>
          <-v Variable:Bitsize> <-t FuzzyType> <-w> <-C<s|x>> <-N<f|s>> <-F<l|m|h>>
          <-M<s|m|n>> <-P> <-T> <-B> <-R> <-H> <-D Device> SourceXFLFile

    -E                      : Mode of execution:
                                  c - Only build C source file.
                                  p - Generate PLA format file.
                                  s - Run circuit synthesis tools.
    -O OperationsFile       : Fuzzy operators definition file.
    -o OutputPrefix         : Prefix to be used for output files.
    -b Bitsize              : Default bitsize for variables.
    -v Variable:Bitsize     : Bitsize for the specified variable.
    -t FuzzyType            : Type for fuzzy variables.
    -w                      : Inhibit warning messages.
    -C                      : Circuit synthesis tool:
                                  s - Synopsys design compiler.
                                  x - Synopsys FPGA compiler.
    -N                      :Deactivate logical optimization processes:
                                  f - Flattening.
                                  s- Structuring.
    -F                      : Select the flattening effort: l low, m medium, h high.
    -M                      : Minimization strategy:
                                  s - single-output.
                                  m - multiple-output.
                                  n - no minimization.
    -P                      : Signal phase assignment.
    -T                      : Omit time-based structuring.
    -B                      : Omit Boolean-based structuring.
    -R                      : Applie place and routing processes.
    -H                      : Block name removal.
    -D Device               : FPGA device the system will be implemented on.
    SourceXFLFile           : XFL input file.
```

Figure 10.3. Command line syntax for *xftl*.

10.3.3 SYNTHESIS BASED ON ON-LINE STRATEGIES

The second tool currently available within the *Xfuzzy* environment follows an on-line implementation strategy. This tool, called *xfvhdl*, starts from the XFL specification of a fuzzy system and produces a VHDL description corresponding to the active rule-driven architecture with simplified defuzzification method that was described in Chapter 9 [Lago, 1998]. Since it is based on a standard hardware description language, the VHDL description can be further synthesized as an integrated circuit by means of any of the logical synthesis tools that accept this language as input (Synopsys, Mentor Graphics, etc.).

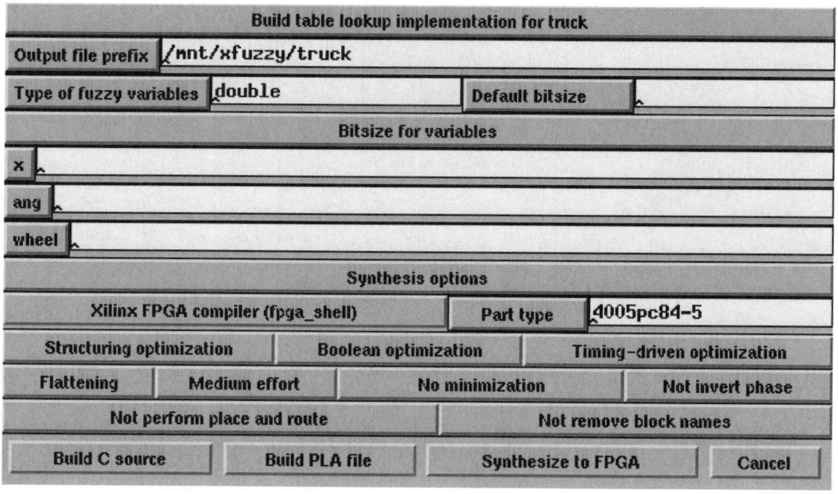

Figure 10.4. Graphical user interface for *xftl*.

In principle, VHDL is independent of the implementation technique selected for integrated circuits. This allows the code generated by *xfvhdl* to be used for implementing the fuzzy system as an ASIC employing any available fabrication process. In addition, *xfvhdl* offers facilities for obtaining a direct synthesis of the fuzzy system on a Xilinx FPGA of the XC4000 family. For this purpose, the Synopsys logical synthesis tool is used, together with the software provided by Xilinx for performing the mapping and connection of the FPGA resources. To control the direct synthesis process, *xfvhdl* generates a command file containing the instructions required by Synopsys according to the options selected by the user when the tool was invoked.

Figure 10.5 shows the design flow for a fuzzy system using *xfvhdl*. The starting point of the synthesis process is, as in the case of the tool described in the previous section, the XFL description of the inference system. Once this description has been refined and validated with the verification tools of *Xfuzzy*, *xfvhdl* takes as input the file with the XFL description and a series of arguments determining the implementation options for the fuzzy system. Upon running *xfvhdl* the following files are produced:

- The VHDL description of the fuzzy system at the structural level, that is, as an interconnection of a set of components
- The VHDL description of the antecedent and rule memories defining the knowledge base of the system

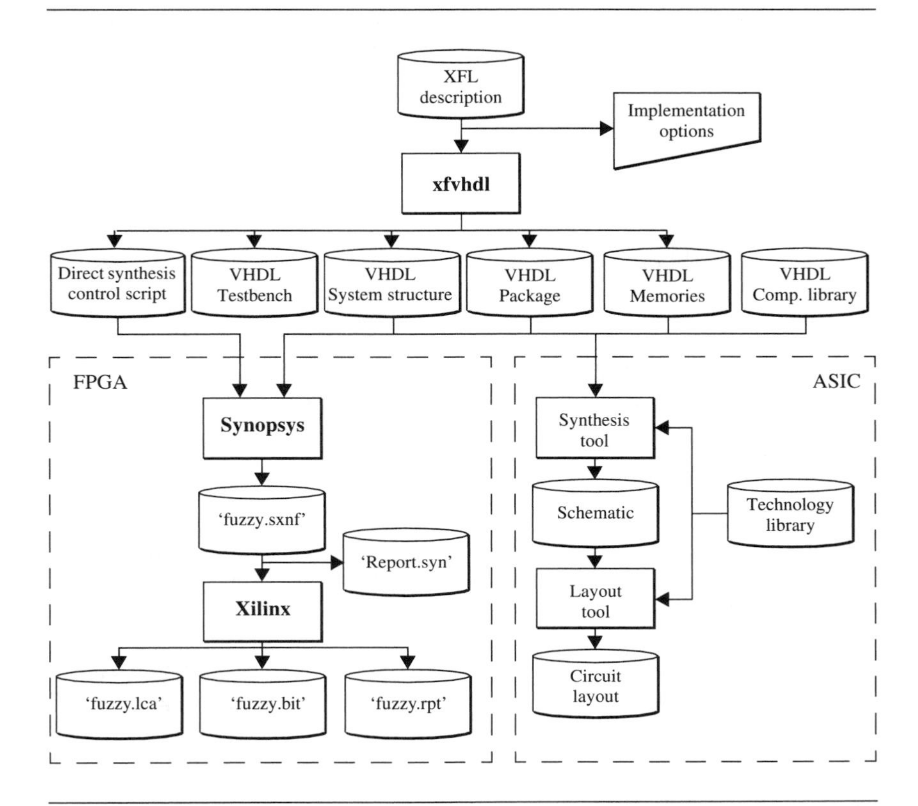

Figure 10.5. Design flow with *xfvhdl*.

- A file with a VHDL package containing the declaration of each of the components that make up the fuzzy system
- A command file with the instructions for directing the Synopsys synthesis tools to build the implementation of the system on the selected Xilinx FPGA
- A VHDL file with the description of a testbench for seamlessly performing circuit simulations of the synthesized fuzzy system

The following stages in the design process depend on the chosen implementation technique. The files produced by *xfvhdl* can be used as input to either ASIC or FPGA synthesis tools. The choice of the implementation technique will depend on specifications and/or cost aspects. The different steps in the design process are completely determined by this decision. ASIC realization is based on technology mapping criteria. In this sense, the implementation of the system means the selection of standard cells and macrocells in the target technology. The files provided by *xfvhdl* are used by the synthesis tool for producing the system schematic according to the library of the selected technology. This schematic will be used by the place and routing tools to obtain the final layout of the circuit. The steps required for implementing the fuzzy system on an FPGA are similar to those described in the previous section.

10.3.4 VHDL COMPONENT LIBRARY

xfvhdl uses a library of parameterized cells containing the VHDL descriptions of the basic blocks in the architecture of the system. This library consists of two kinds of blocks:

- Blocks for building the data path. These blocks store the knowledge base and implement the inference mechanism.
- Control blocks, which direct the read/write operations on memories as well as the signal in charge of scheduling the different tasks.

Figure 10.6 shows a general block diagram corresponding to the implementation architecture. In order to incorporate in this general scheme the different implementation options discussed in Chapter 9, some of these blocks permit more than one possible implementation.

The blocks used in the fuzzification stage depend on the option selected for storing antecedents. In the antecedent memory option, each MFC block is made up of a memory block in charge of providing the label and activation degrees (*AntecedentMem_n*) and an active rule selection block that serializes the flow of these data to the inference stage (*MF_Grade*). In the arithmetic calculation option, an antecedent memory block common to all inputs (*ArithCalc-Mem*) is used. In addition, each MFC block is made up of two blocks: one in charge of generating the label and the activation degrees (*Arithmetic*) and another for selecting the active rule (*MF_Grade*).

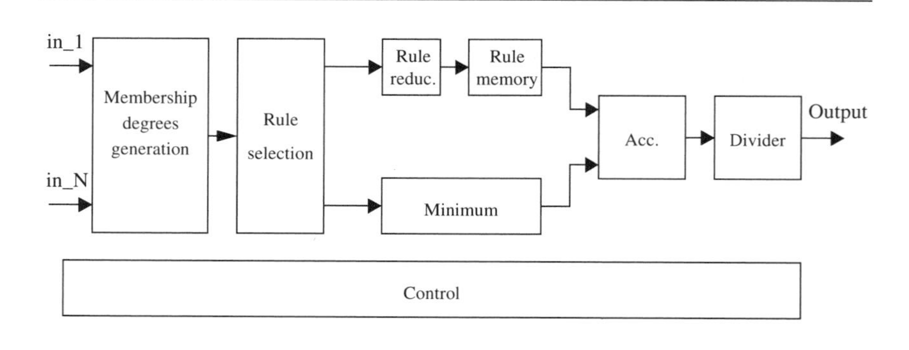

Figure 10.6. Block diagram of the selected architecture.

At the inference stage, the rule memory (*RulesMem*), containing the parameters for the consequents of the fuzzy rules, can be implemented on RAM or ROM depending on the desired programmability of the system. For calculating the activation degree of the different rules a block performing the minimum function on its inputs is provided.

The defuzzification stage is divided into two blocks: the first one performs the accumulation tasks according to the defuzzification method in use, while the second performs the division (*Division*). The component library contains a module for each of the simplified defuzzification methods supported by the active rule-driven architecture: *CenterOfSums*, *FuzzyMean*, *WeightedFuzzyMean*, and *Yager*, plus an additional module (*Defuzzific*) implementing a programmable defuzzifier (See Section 9.5 in Chapter 9).

Apart from these blocks implementing the fuzzy system functionality, other blocks included in the library are in charge of controlling the operation of the rest of the blocks and loading memories. Since the functional behavior depends on the architectural options, different control blocks have been defined to cover all the available possibilities.

Given the XFL specification of an inference system and the VHDL components library, the implementation of a fuzzy system requires the choice, interconnection and dimensioning of those necessary blocks according to the selected architectural options (see Table 10.1).

Table 10.1. Summary of components employed by the different design options.

System block	ARCHITECTURAL OPTION			
	Memory storage		Arithmetic calculation	
	RAM	ROM	RAM	ROM
Membership degree generation	One *ram_antec* per input	One *rom_antec* per input	One *arith* per input One *ram_antec*	One *arith* per input One *rom_antec*
Rule selection	One *fp_grade* per input			
Rule reduction	Only if this option is used			
Defuzzifier	One of *CenterOfSums*, *FuzzyMean*, *WeightedFuzzyMean* or *Yager*			
Rule memory	One *ram_rules*	One *rom_rules*	One *ram_rules*	One *rom_rules*
Minimum	One *minimum*			
Division	One *divider*			
Timing and control	*control1*	*control1_nc*	*control2*	*control2_nc*

The main part of the information that *xfvhdl* uses for synthesizing the inference system comes from the XFL specification. From this specification, the tool determines the inputs and outputs of the system, their universes of discourse and membership functions, its rule base, and the fuzzy operations to be applied. As the first step in the synthesis process, *xfvhdl* tests whether the XFL specification is synthesizable into the target architecture and satisfies the constraints imposed by the tools. The implementation options accepted by *xfvhdl* define the number of bits for representing inputs and outputs, the slope (when arithmetic calculation is used for antecedent assignment), and parameters for some defuzzification methods.

10.3.5 XFVHDL OUTPUT FILES

The execution of *xfvhdl* produces a series of files for the circuit verification and synthesis of the fuzzy system. These files can be grouped into the following categories:

- **VHDL description of the system**

This file contains the structural description of the fuzzy inference system. The system is built by connecting a set of basic blocks according to its XFL description and the selected implementation options. As in the case of software synthesis, this tool produces an element of the target language (a VHDL entity in this case) that can be used independently or further integrated with other elements, building a more complex system that integrates fuzzy logic capabilities. Figure 10.7 shows the VHDL entity resulting from the execution of *xfvhdl* for a fuzzy system with the following XFL definition:

```
system (t1 ? in1, t2 ? in2, to ! output)
```

As this example shows, the implementation of the system includes, together with the input (*in1*, *in2*, etc.) and output (*output*) signals, a set of control sig-

```
entity FLC is
    port( clk: in std_logic;
        reset: in std_logic;
        in1:  in std_logic_vector(N downto 1);
        in2:  in std_logic_vector(N downto 1);
        output: out std_logic_vector(N downto 1);
        valid_out: out std_logic;
        valid_in: out std_logic);
end FLC;
```

Figure 10.7. VHDL entity generated by *xfvhdl*.

nals: a clock signal (*clk*) for synchronizing the rest of the components, a *reset* signal for initializing the whole system, and some protocol signals indicating the availability of valid data at the input and output lines (*valid_in* and *valid_out*, respectively).

Figure 10.8 illustrates the top-level schematic of a fuzzy system with two inputs and one output using antecedent arithmetic calculation and the Weighted-Fuzzy-Mean defuzzification method. The components implementing the membership functions and the rule base are automatically generated by *xfvhdl* from the XFL specification. The rest of the components are taken from the component library and parameterized according to the XFL definition of the system.

- **VHDL description of the knowledge base**

The information for antecedents, rules, and consequents are coded into a set of files. These definitions are based on value tables using VHDL *"case"* sentences. This approach permits the application of logical minimizations when these blocks are implemented using combinational circuits (Figure 10.9).

- **Package file**

The entities file (*entities*) contains the declaration of all the blocks from which the system is constructed. The precise placement of these blocks will depend on the selected architectural options.

- **Test patterns file**

In addition to the files needed to synthesize the circuit, a file with test patterns can be optionally generated for verifying the VHDL description of the fuzzy system. This file connects the system to a VHDL process generating a clock signal and another process that provides the initial reset signal and the inputs for evaluating the system response (Figure 10.10).

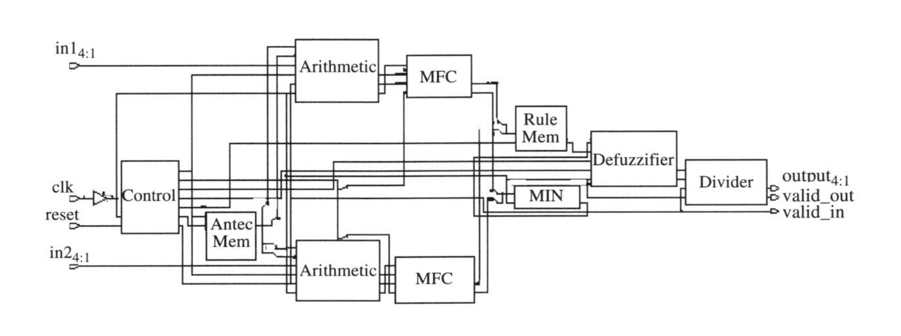

Figure 10.8. Fuzzy system top-level schematic.

```
Case Addr Is
    When "000000" => data <= "0001111111000000";
    When "000001" => data <= "0001110010001100";
    ........................................
    When "111101" => data <= "1010011000110011 0";
    When "111110" => data <= "1010001100111001 0";
    When "111111" => data <= "1010000000111111 1";
    When Others=> data <= "----------------";
End Case;
```

Figure 10.9. An example of VHDL description of a fuzzy system rule base.

10.3.6 USING XFVHDL

When calling *xfvhdl* from the shell, the syntax to be used is summarized in Figure 10.11. The graphical user interface of *xfvhdl* (Figure 10.12) has two differentiated sections. The first one, in the upper half and labelled with the title *VHDL generation options* gives access to the options controlling the characteristics of the VHDL code produced by *xfvhdl*. The second section, in the lower half of the window and labelled as *Synthesis options*, is only shown when the circuit synthesis tools of Synopsys are available and contains the elements that define how the circuit synthesis tool is called. When the window appears, the text fields and buttons show the values *xfvhdl* uses by default.

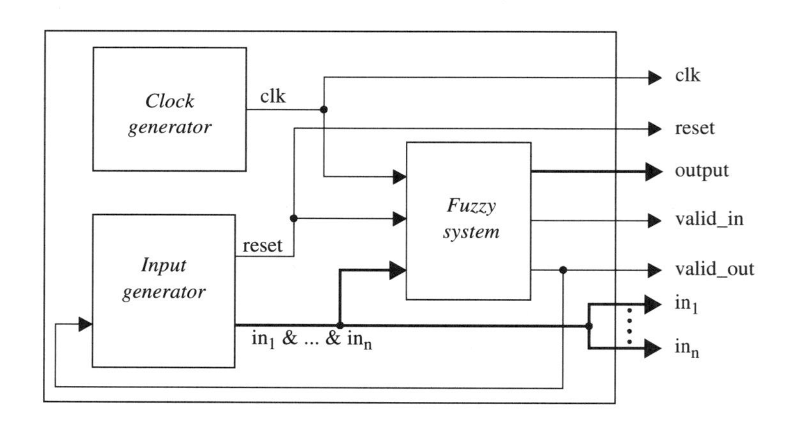

Figure 10.10. Testbench structure.

```
Xfvhdl: Synthesizable VHDL code generator for XFL-based systems

Synopsis: xfvhdl <-o OutputPrefix> <-s>
          <-n Bitsize> <-d Bitsize> <-p Bitsize> <-k Bitsize>
          <-S> <-L LibraryDirectory> <-C<s|x>>
          <-E<l|m|h>> <-V> <-B> <-O<a|s>> <-R> <-D Device>
          SourceXFLFile

  -o OutputPrefix      : Prefix to be used for output files.
  -s                   : Generate code for simulation.
  -n Bitsize           : Bitsize for inputs and output.
  -d Bitsize           : Bitsize for membership degrees.
  -p Bitsize           : Bitsize for slopes (for arithmetic calculation of antecedents).
  -k Bitsize           : Bitsize for the second defuzzification parameter (if needed).
  -S                   : Run circuit synthesis tools.
  -L LibraryDirectory  : Components library directory.
  -C                   : Circuit synthesis tool:
                                 s - Synopsys design compiler.
                                 x - Synopsys FPGA compiler.
  -E                   : Select the mapping effort: l low, m medium, h high.
  -V                   : Verify design.
  -B                   : Perform boundary optimization.
  -O                   : Select the optimization type: a area, s speed.
  -R                   : Generate Synopsys report.
  -D Device            : FPGA device the system will be implemented on.
  SourceXFLFile        : XFL input file.
```

Figure 10.11. Command line syntax for *xfvhdl*.

In the section for the VHDL generation options it is possible to set the output file prefix, the number of bits for the inputs and outputs of the circuit (*Bits for I/O*), for storing the membership degrees, and for the specific parameter of certain defuzzification methods (*Bits for defuzzification weight*). There is a button for setting the method for calculating the antecedents. In the case of arithmetic calculation, the value of the text area labelled *Bits for membership function slope* is used. Finally, a button for requesting the verification of the VHDL is also available.

The section for the direct circuit synthesis options permits the configuration of the optimization to be applied by the circuit synthesis tool, the FPGA device to be used and the directory where the VHDL module library resides.

The usual control button area permits calling *xfvhdl* in its two options, either to synthesize the VHDL description of the system or to directly produce an FPGA implementation from VHDL once the description has been generated.

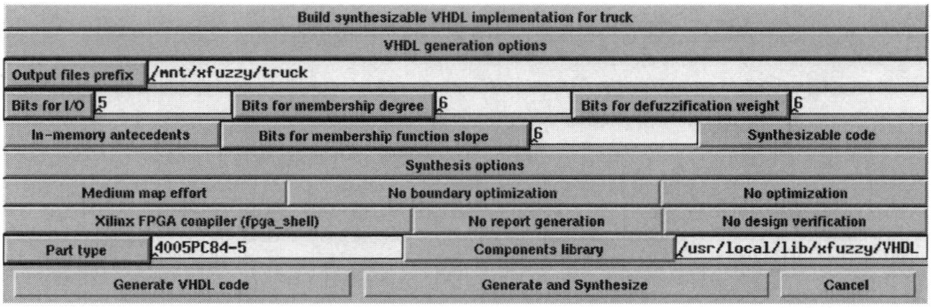

Figure 10.12. Graphical user interface for *xfvhdl.*

10.3.7 LIMITATIONS TO THE XFL MODEL

In contrast to the synthesis tools based on off-line strategies, when a dedicated hardware on-line implementation is intended, the target architecture for the fuzzy system imposes a set of significative constraints on the input XFL model. These constraints affect both the structure and definition of the knowledge base and the inference mechanisms and fuzzy operations that can be used. More precisely, the constraints that *xfvhdl* imposes on the XFL definition of the system to be synthesized are:

- Module composition is not currently supported by *xfvhdl.* Only those systems with a module *'system'* that directly contains a rule base can be synthesized by the tool.
- The tool assumes that there is only one output variable. Although this can be seen as a strong limitation, it is not the case. Systems with more than one output can be split into several XFL definitions for synthesizing them with *xfvhdl* (one for each output variable). The import capabilities and modularity of XFL make this a very simple task.
- All the types of the input variables must have the same number of membership functions. Functions of class *'delta'* are not supported.
- For any type applicable to an input variable, there must be no point in the universe of discourse where more than two membership functions take a non-zero value. In other words, the overlapping degree of the membership functions cannot be higher than two.
- If arithmetic calculation of antecedents is used, the only acceptable class of membership functions are triangular ones, with an overlapping degree of two. These are the constraints we refer to when discussing the condition *'LINKED'* in the specific learning file used by *xfbpa* (Section 5.4 in Chapter 5).

When *xfvhdl* starts, it first analyzes the characteristics of the abstract syntax tree representing the XFL source definition and, according to them and to the parameters used in its invocation, it decides whether these constraints are satisfied and the synthesis into VHDL is possible.

In the same way that it imposes constraints on the characteristics of synthesizable XFL definitions, the model used by *xfvhdl* only supports a reduced set of fuzzy operations. In fact, apart from the case of defuzzification methods, the tool employs predetermined fuzzy connectives: minimum as t-norm, maximum as s-norm, and 1's complement (the default negation for all XFL-based tools) as negation. We have not mentioned implication function, since the tool only allows the use of simplified defuzzification methods, where the implication function is implicitly included by the defuzzification process.

With respect to defuzzification methods, the current version of *xfvhdl* supports four different types (Those shown in Table 10.1). The specifications of these methods as VHDL entities are contained in files residing in the VHDL module library used by the tool and described above. The identifier of each method corresponds to the name of the file (with the extension *.xfl*). To illustrate this point, Figure 10.13 shows part of the contents of the file *FuzzyMean.vhdl* in the VHDL used by the tool.

To make *xfvhdl* use this module for the implementation of the defuzzification method, it is sufficient to include in the source XFL file the line:

```
#defuzzification FuzzyMean
```

10.4 A DESIGN METHODOLOGY FOR FUZZY SYSTEMS

The choice of an efficient architecture for translating into hardware the inference mechanisms based on fuzzy logic and the availability of a set of tools easing the tasks of description, verification, and synthesis of a fuzzy system, permit us to propose a design methodology for this kind of system. The main stages in the design of a fuzzy system with this methodology are shown in Figure 10.14.

The first stage in the design process is the specification of the system functionality. The high-level specification language used by the *Xfuzzy* environment maintains a unified description of the fuzzy system along the whole design process. As we saw in Chapter 4, XFL provides a great flexibility for defining both membership functions (using a wide range of predefined functions and the possibility of defining piecewise linear approximations to arbitrary functions) and the rule base (using compositional operators that allow the use of hierarchical rules). Furthermore, the graphical user interface provided by the environment eases the tasks of editing the different components of the system knowledge base.

As explained in Chapter 5, the XFL specification describing a fuzzy system can be refined using the verification tools provided by the environment.

```
entity FuzzyMean is
port  (point: in  std_logic_vector(N downto 1);
        degree: in  std_logic_vector(DG downto 1);
        acc: in  std_logic;
        clear_st: in  std_logic;
        clk: in  std_logic;
        div_n: out std_logic_vector(D  downto 1);
        div_d: out std_logic_vector(DR downto 1)
        );
end FuzzyMean;
architecture FPGA of FuzzyMean is
        signal S1_1: std_logic_vector(N + DG downto 1);
        signal S2_1: std_logic_vector(DG downto 1);
        signal S1: std_logic_vector(D  downto 1);
        signal S2: std_logic_vector(DR downto 1);
begin
        S1_1 <= point * degree;
        S2_1 <= degree;
        Algorithm:process
        begin
                wait on clk until clk = '0';
                if (clear_st = '1') then
                        S1 <= (others => '0');
                        S2 <= (others => '0');
                else
                        case acc is
                        when '1' =>
                                S1 <= S1_1 + S1;
                                S2 <= S2_1 + S2;
                        when others =>
                                S1 <= S1;
                                S2 <= S2;
                        end case;
                end if;
        end process Algorithm;
        div_n <= S1;
        div_d <= S2;
end FPGA;
```

Figure 10.13. VHDL definition for the defuzzification method *FuzzyMean*.

When the expected behavior of the fuzzy system is described in terms of a set of input/output data, the *Xfuzzy* learning tool, *xfbpa*, can use these data as training samples and apply different supervised learning algorithms. This tool is able to adjust the parameters of the membership functions and to identify the rules whose activation degrees are below a specific threshold.

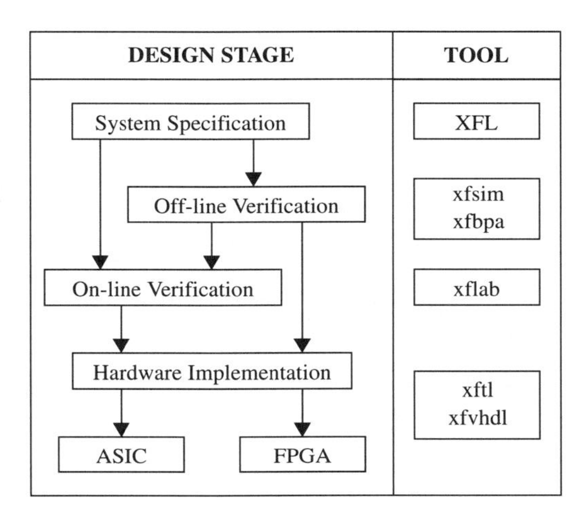

Figure 10.14. Design methodology and related tools.

If the fuzzy system should be combined with other systems (a given environment), like a plant to control, the *Xfuzzy* off-line simulation tool, *xfsim*, permits us to verify the whole system behavior (fuzzy system and environment).

The environment models used in the above simulation stage are usually first-order models that do not consider some effects difficult to be coded into a computer algorithm. When these second-order effects are relevant, simulation results may be very far from real behavior. To solve this problem, *Xfuzzy* incorporates the tool *xflab* able to connect a software implementation of the fuzzy system to its real operational environment through a data acquisition board.

The hardware implementation stage can be started once the fuzzy system specification has been validated. To perform this task the XFL specification has to meet the requirements imposed by synthesizable architectures. The designer can choose at this point the implementation technique that better fits the problem to be solved. It is worth remembering that the use of FPGAs provides quick prototyping capabilities and that these devices are intrinsically programmable, so they offer mechanisms for changing or adjusting system functionality. On the other hand, ASIC implementations are more efficient in terms of silicon area and inference speed, becoming more appropriate when the number of integrated circuits to be produced is high.

This design methodology will be used in the application examples shown in Chapters 11 and 12.

REFERENCES

1. **Barriga, A., Sánchez-Solano, S., Jiménez, C. J., Gálan, D., López, D. R.**, Automatic synthesis of fuzzy logic controllers, *Mathware & Soft Computing*, Vol. III, N. 3, pp. 425-434, 1996.
2. **Brayton, R. K., Hachtel, G. D., McMullen, C. T., Sangiovanni-Vincentelli, A. L.**, *Logic Minimization Algorithms for VLSI Synthesis*, Kluwer Academic Pub., 1984.
3. **Costa, A., De Gloria, A., Faraboschi, P., Pagni, A.**, A tool for automatic synthesis of fuzzy controllers, in *Proc. IEEE Int. Conference on Fuzzy Systems,* pp. 1771-1775, Orlando, 1994.
4. **Costa, A., De Gloria, A., Faraboschi, P., Pagni, A., Rizzotto, G.**, Hardware solutions for fuzzy control, *Proceedings of the IEEE*, Vol. 83, N. 3, pp. 422-434, 1995.
5. **Franchi, E., Manaresi, N., Rovatti, R., Bellini, A., Baccarani, G.**, Analog synthesis of nonlinear functions based on fuzzy logic, *IEEE Int. Journal of Solid-State Circuits*, Vol. 33, N. 6, pp. 885-895, 1998.
6. **Galán, D., Jiménez, C. J., Barriga, A., Sánchez-Solano, S.**, VHDL package for description of fuzzy logic controllers, in *Proc. European Design Automation Conference*, pp. 528-533, Brighton,1995.
7. **Hollstein, T., Halgamuge, S., Glesner, M.**, Computer aided design of fuzzy systems based on generic VHDL specifications, *IEEE Transactions on Fuzzy Systems*, Vol. 4, N. 4, pp. 403-417, 1996.
8. **Hollstein, T., Kirschbaum, A., Halgamuge, S., Glesner, M.**, Application specific fuzzy processors, in *Proc. Int. Symposium on Nonlinear Theory and its Applications*, pp. 133-136, Honolulu, 1997.
9. **Lago, E., Jiménez, C. J., López, D. R., Sánchez-Solano, S., Barriga, A.**, XFVHDL: A tool for the synthesis of fuzzy logic controllers, in *Proc. Design, Automation and Test in Europe*, pp. 102-107, Paris, 1998.
10. **Lemaitre, L., Patyra, M. J., Mlynek, D.**, Synthesis and design automation of analog fuzzy logic VLSI circuits, in *Proc. Int. Symposium Multiple_Valued Logic*, pp. 76-79, Sacramento, 1993.
11. **Manaresi, N., Rovatti, R., Franchi, E., Guerrieri, R., Baccarani, G.,** A silicon compiler of analog fuzzy controllers: from behavioral specifications to layout, *IEEE Transactions on Fuzzy Systems,* Vol. 4, N. 4, pp. 418-428, 1996.
12. **Mentor Graphics** documentation set, *VHDL Style Guide for AutoLogic II*, Mentor Graphics software version B.2, 1997.
13. **Synopsys** documentation set, *VHDL Compiler Reference Manual*, Synopsys version 1997.08, 1997.
14. **VHDL, IEEE Std 1076-1987**, *IEEE Standard VHDL Language Reference Manual*, The Institute of Electrical and Electronic Engineers, Inc., 1988.
15. **Zamfirescu, A., Ussery, C.**, VHDL and fuzzy logic if-then rules, in *Proc. Euro-VHDL'92*, pp. 636-641, Hamburg, 1992.
16. **Zamfirescu, A.**, Logic and arithmetic in hardware description languages, in *Fundamentals and Standards in Hardware Description Languages,* Mermet, J. P., Ed., pp. 109-151, NATO ASI Series, 1993.

Chapter 11

FUZZY SYSTEMS AS CONTROLLERS

The use of fuzzy logic–based inference techniques in control problems has motivated one of the main application fields for fuzzy systems. As discussed in Chapter 1, this is mainly due to two reasons: on the one hand, the capability of fuzzy logic to express the knowledge of an expert operator using simple natural language rules; on the other hand, the fact that using fuzzy logic it is not necessary to have an analytical model of the system to be controlled.

In many industrial applications of fuzzy control, the controller acts independently, replacing other conventional control elements. Some examples of this kind of application are the control of the movement of crane loads [Yasunobu, 1986; Itoh, 1993], welding control [Yamane, 1993], and the control of the components in chemical baths [Dohmann, 1994]. In some other cases, the fuzzy controller is a subsystem within a hybrid control architecture, as in the temperature control for industrial systems described in [Isaka, 1993], or it is in charge of monitoring and supervising the whole control environment, as in the case of the control of the processes in an iron mine introduced in [Hall, 1993]. Apart from industrial control, other application fields of fuzzy control have become of great importance in recent years. We can include here the application of fuzzy logic to household appliances [Quail, 1992] or to electronic components for image processing equipment [Takagi, 1992]. Finally, since fuzzy systems are not conditioned by the availability of a well-defined mathematical model, they are especially suited for biomedical applications [Stimann, 1998]. This characteristic is very useful when controlling human physiological systems, which are very complex and suffer from a great number of perturbations and, in many cases, include subjective components that complicate their modeling. Among other applications, fuzzy controllers have been used for controlling the plasmatic glucose level for diabetic patients, the cardiac rhythm [Sugima, 1991], the level of muscular relaxation [Linkens, 1988], blood pressure [Isaka, 1989], or post-operative pain.

This chapter is divided into three parts. The first part introduces some basic concepts of control systems and describes the control strategies more often used in practical applications, like PID (proportional-integral-derivative) control and the strategies derived from it (P, PD, and PI). Afterwards, fuzzy control will be introduced as an alternative strategy, describing the general characteristics of a fuzzy controller and discussing the more widely used techniques for its realization. The last part of the chapter describes in detail the development of an application example. This example corresponds to the control of the height of a ball sustained by an airflow inside a tube. The development of this controller is performed according to the methodology introduced in Chapter 10. Furthermore, the final implementation of the system will allow the application of some of the circuits described in the preceding chapters.

11.1 CONVENTIONAL CONTROL SYSTEMS

The objective of any control system is to apply a *control action* on another system (known as "the plant"), so as to make it comply with a predetermined behavior. Control systems are usually divided into two categories: *open loop systems* (Figure 11.1a), and *closed loop systems* or systems with feedback (Figure 11.1b). In feedback control systems, the output of the plant is measured by some sensing devices and is compared with the reference signal. The error between the reference signal and the measured output is used for calculating the correcting action to be applied on some of the plant inputs. Closed loop control systems have many advantages over open loop systems, since they are less sensitive to variations in the plant parameters, can control transient plant response, provide lower errors in stationary states, and allow the control and partial suppression of the effects of perturbation signals (as in the case of noise produced by sensors) [Passino, 1998]. In fact, a closed loop system is the only feasible solution in many cases. This is why most of the conventional control systems (and the proposed fuzzy controllers) use feedback.

The use of feedback, however, implies an associated cost. As the number of required components increases, the complexity of the system also increases. In particular, sensors become key components since they are the most expensive elements (defining the cost of the solution) and limit the precision (inducing noise and non-linearities). Furthermore, feedback may cause the system to become unstable. The main problem when designing these closed loop systems is related to the response times of the elements composing the feedback loop: plant, sensors, controller, etc. These delays imply that the controller is not correcting the current plant state but the state it had some time ago. Therefore it is

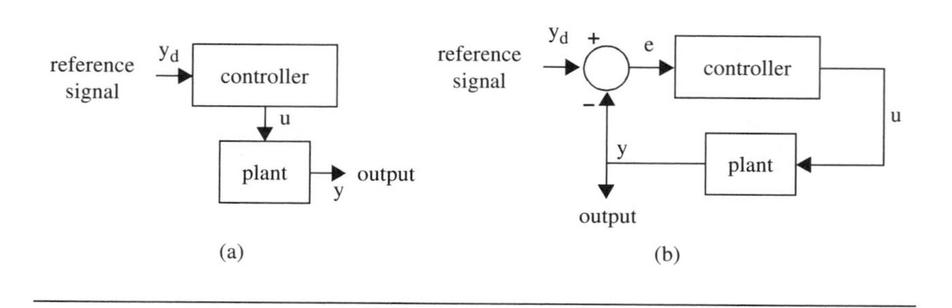

Figure 11.1. (a) Open loop and (b) closed loop control systems.

difficult for a simple negative feedback scheme to cope with both dynamic errors (like plant oscillations inside a certain range) and static errors (like differences between the output and the reference signal below certain limits).

Conventional control theory requires a mathematical model of the plant in order to design and adjust the controller. This model can be derived either from the physical laws applicable to the processes governing plant behavior, or from data obtained by actually operating the plant, or even using a combination of both procedures. Regardless, since plants are of dynamic nature, their descriptive equations are often differential equations. The complexity of the system and/or the uncertainty about significative factors usually requires making assumptions about the plant operation. In many cases, a set of physical laws describing an equivalent linear system is used, thus yielding a set of linear differential equations. This approach is based on the fact that most physical systems are linear inside some range of the variables defining them (all real systems become non-linear if their variables are not bounded). The linear approximation of the plant permits the use of the Laplace transform and makes the substitution of the differential equations by algebraic ones easier to solve. The introduction of frequency diagrams and characteristics (by Nyquist, Bode and Nichols) in the 1940s, and the root loci diagrams (by Evans) in the 1950s, provide a series of mathematical tools for studying system stability.

The types of conventional control most widely employed correspond to the so-called *proportional-integral-derivative* (PID) control and those derived from it (*proportional*, P, *proportional-derivative*, PD, and *proportional-integral*, PI). Undoubtedly, PID is the type most used, corresponding to about 90% of the existing controllers [Yamamoto, 1991].

In the case of proportional control (P), the control action is directly proportional to the current error and is given by the following expression:

$$u(t) = k_P \cdot e(t) \tag{11.1}$$

where u is the output of the controller, e is the error defined by the difference between the reference signal, y_r, and the plant output, y, ($e = y_r - y$) and k_P is a constant.

In a PD (proportional-derivative) control system, the control action incorporates a term proportional to the derivative of the error:

$$u(t) = k_P \cdot e(t) + k_D \cdot \frac{de(t)}{dt} \tag{11.2}$$

In the case of discrete time systems, with a sampling interval T, (11.2) can be expressed as:

$$u(nT) = k_P \cdot e(nT) + k_D \cdot \Delta e(nT) \tag{11.3}$$

Since the derivative indicates the direction of error change (ascent or descent), this type of control has a short-term predictive effect, allowing a more stable system. A correct value of the constant k_D permits the correction of overshooting effects in the transient response of the system under control.

In a PI (proportional-integral) controller, the control action incorporates a term accounting for the integral of the signal error over an interval of time. In the case of systems operating in continuous time, the control action is given by:

$$u(t) = k_P \cdot e(t) + k_I \cdot \int e(t)dt \qquad (11.4)$$

For discrete time systems, (11.4) becomes:

$$u(nT) = k_P \cdot e(nT) + k_I \cdot \sum_{i=0}^{n} e(i \cdot T) \qquad (11.5)$$

An alternative solution, which also takes into account the cumulative error effect, consists in configuring the controller as an incremental controller, where the output is the change in the control action with respect to the former instant, Δu, [Bollinger, 1989]:

$$\Delta u(nT) = k_I \cdot e(nT) + k_P \cdot \Delta e(nT) \qquad (11.6)$$

So the control action to be applied to the plant at a certain moment is obtained by combining the previous value with the output of the controller:

$$u(nT) = u((n-1)T) + \Delta u(nT) \qquad (11.7)$$

By considering the integral of the error or using an incremental controller it is possible to compensate for problems derived from persistent errors, increasing the precision in the stationary state without introducing additional instability. However, with respect to the transient response, the rising time of the system is slow. This characteristic can be easily understood taking into account that a PI controller is essentially a low-pass filter that attenuates high-frequency signals [Kuo, 1982].

Finally, in a PID (proportional-integral-derivative) controller, the control action incorporates the three terms introduced above. Depending on the system operating in continuous or discrete time, the control action is given by the following expressions:

$$u(t) = k_P \cdot e(t) + k_I \cdot \int e(t)dt + k_D \cdot \frac{de(t)}{dt} \qquad (11.8)$$

$$u(nT) = k_P \cdot e(nT) + k_I \cdot \sum_{i=0}^{n} e(i \cdot T) + k_D \cdot \Delta e(nT) \qquad (11.9)$$

PID controllers combine the characteristics of PD and PI controllers, permitting the simultaneous tuning of the transient and stationary system responses by the selection of the adequate values for the constants k_P, k_D, and k_I.

This type of controller is easily designed when a linear model of the plant, incorporating a few variables, is available. In fact, the control surfaces provided by PID controllers are hyperplanes intersecting the origin, with a slope that depends on the constants k_P, k_D, and k_I. The problem is that most real processes are linear only inside very narrow ranges of their variables. Furthermore, the parameters defining the range of linear operation of the plant depend on its point of operation, and even change with time. This means that many times a conventional controller is only well fitted for a particular point of operation and a limited time. Finally, in the design of a good conventional PID controller it is unavoidable to consider aspects beyond its analytical formulation such as, for example, the operator interfaces, switching from automatic to manual control and back, etc. Therefore an industrial PID controller always incorporates a certain heuristics.

There are three general procedures for evaluating the performance of a controller [Passino, 1998]. The first procedure consists of performing a mathematical analysis based on formal models, which permits checking the closed loop system specifications (stability, observability, response times, etc.). The second procedure is based on the use of simulation models of the physical system, which accounts for the implementation details of the system under development. The third alternative, which can be impractical in some applications, consists of performing experiments on the actual system.

11.2 FUZZY CONTROL SYSTEMS

Fuzzy logic and fuzzy inference systems provide mechanisms that allow the application of human knowledge heuristics to the control of a system. The design of a fuzzy controller starts from the information defining how the controller must behave when connected to the plant. Sometimes this information comes from the experience of a person in charge of this control task (an expert), while in other cases this information can be extracted from the dynamic behavior of the plant. Fuzzy controllers inherit from conventional control theory the idea of including proportional, derivative, and integral elements. This way, the terms Fuzzy-P, Fuzzy-PD, Fuzzy-PI, or Fuzzy-PID controllers are used [Yager, 1994]. However, these elements are not incorporated by means of analytical expressions such as (11.1)-(11.9), but by heuristic control rules of the type:

"*If* e *is {ling. val.} and* Δe *is {ling. val.}* $\rightarrow$ u *is {ling. val.}*", or

"*If* e *is {ling. val.} and* Δe *is {ling. val.}* $\rightarrow$ Δu *is {ling. val.}*"

for the cases of Fuzzy-PD and incremental Fuzzy-PI controllers, respectively.

The rule base for a typical incremental controller, like the one originally introduced by Mamdani and Assilian in [Mamdani, 1975], describes the change in the control action, Δu, depending on the error, e, and the error variation, Δe, of the system with respect to its intended behavior. This control law can be formalized as:

$$\Delta u(nT) = F(e(nT), \Delta e(nT)) \tag{11.10}$$

Figure 11.2a shows a very common rule base for this kind of controller. The universe of discourse of each of the inputs is covered by the three membership functions shown in Figure 11.2b. This knowledge base can be applied to any system whose behavior can be approximated by a first-order system [Yager, 1994].

A certain similarity can be observed when comparing the expression that defines the fuzzy controller proposed by Mamdani (11.10) with one that corresponds to an incremental PI controller (11.6). A fuzzy controller of the type proposed by Mamdani can be considered a Fuzzy-PI controller. The difference

1. **IF** error $e(k)$ is *positive* AND error variation $\Delta e(k)$ is *near zero*
   THEN control variation $\Delta u(k)$ is *positive*.

2. **IF** error $e(k)$ is *negative* AND error variation $\Delta e(k)$ is *near zero*
   THEN control variation $\Delta u(k)$ is *negative*.

3. **IF** error $e(k)$ is *near zero* AND error variation $\Delta e(k)$ is *near zero*
   THEN control variation $\Delta u(k)$ is *near zero*.

4. **IF** error $e(k)$ is *near zero* AND error variation $\Delta e(k)$ is *positive*
   THEN control variation $\Delta u(k)$ is *positive*.

5. **IF** error $e(k)$ is *near zero* AND error variation $\Delta e(k)$ is *negative*
   THEN control variation $\Delta u(k)$ is *negative*.

(a)

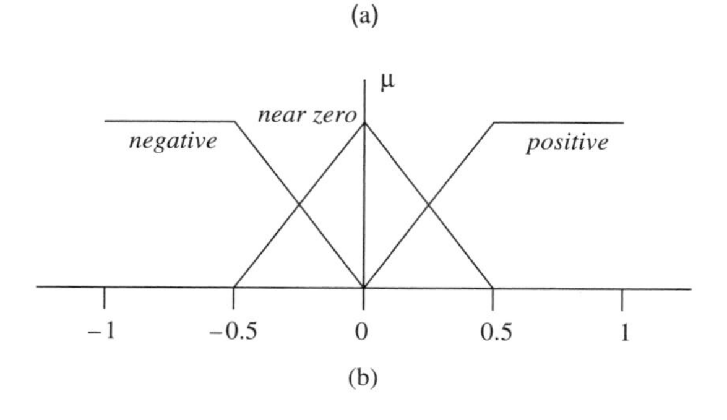

(b)

Figure 11.2. Knowledge base for a fuzzy controller.

of this fuzzy controller with respect to a conventional PI controller resides in the type of input/output relationship, linear for a conventional PI, while in the case of the Fuzzy-PI it is generally non-linear.

When the output of the fuzzy controller does not represent the variation in the control action, $\Delta u(n)$, but rather the control action itself, $u(n)$, an alternative type of fuzzy controller is obtained. In this case, the control function resembles that of a conventional PD controller given by (11.3). Therefore, a fuzzy controller with the following control function is called a Fuzzy-PD controller:

$$u(nT) = F(e(nT), \Delta e(nT)) \tag{11.11}$$

Lastly, the Fuzzy-PD controller can be extended considering the sum of errors as a third input, giving the so-called Fuzzy-PID controller, which can be considered a fuzzy version of the conventional PID controller defined by (11.9):

$$u(nT) = F(e(nT), \Sigma e(nT), \Delta e(nT)) \tag{11.12}$$

Hence, conventional and fuzzy control relate the same input and output variables. However, while conventional control usually relates them by using differential equations typical of mathematical language, fuzzy control employs rules expressed in terms of natural language. The control surface provided by a fuzzy controller is not necessarily linear (as in the case of a conventional controller) and can be locally modified by changing the rules applicable in the corresponding area of the universes of discourse, as is explained in Chapter 12.

11.2.1 COMPONENTS OF A FUZZY CONTROL SYSTEM

The design of any control system (in particular, of a fuzzy control system) cannot be based only on the design of the controller itself, but must contemplate the whole controller-plant system. This is especially important when the controller is going to be implemented in hardware and the circuitry has to be optimized. A control system normally operates in real time: input acquisition and output provision occur at determined moments, and the dynamic behavior of the controller (response time, data acquisition period, etc.) is of great importance.

Figure 11.3 shows the general structure of a fuzzy control system. It can be observed that, apart from the plant and the fuzzy system that constitutes the kernel of the controller, the set includes a series of sensors and actuators, together with signal conditioning modules for the input and output signals. Sensors are devices that translate a physical signal into an electrical one. Actuators provide the inverse functionality. Perhaps, from the designer's point of view, the most important aspect is the type of signal provided by sensors and accepted by actuators, since this directly influences the hardware characteristics of the controller.

Depending on the signals they provide, sensors can be classified into two groups:

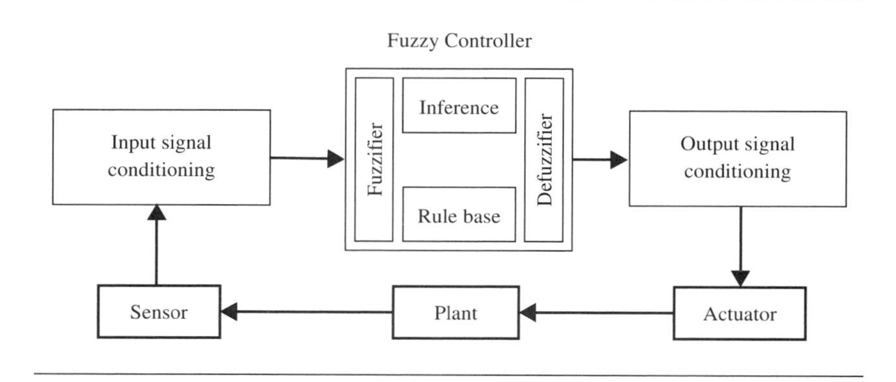

Figure 11.3. Components of a real-time control system.

- Sensors producing binary signals, normally an output voltage with a high or low level, depending on the detection of a certain physical event.
- Sensors producing continuous signals. The output signal is a monotonic and ideally linear function of the variable under measurement. Many sensors belong to this group, such as pressure and temperature sensors. Most of them must be calibrated and, even then, their precision may not be very high [Lawrence, 1987].

With respect to actuators, they can also be classified into two groups:

- Two-state actuators, one with applied energy (ON state) and the other without it (OFF state). The connection between the controller and the actuator can be implemented by means of a switch that applies the necessary power to the actuator.
- Continuous actuators. In these cases, when the controller provides a digital output, a D/A converter may be necessary, producing a voltage proportional to the digital data. This signal is applied to the control input of a linear amplifier controlling the power applied to the actuator.

The signal conditioning blocks perform functions such as gain and offset conversion, isolation (for example, controller outputs are often isolated from the actuators since they work with a much higher power) or filtering (for example, an analog signal converted into a digital format to provide a controller input is previously filtered in order to avoid information loss). Other functions of these blocks are current-voltage conversions, and format changes from analog to digital and back, in the cases they are necessary.

11.2.2 FUZZY CONTROLLER DEVELOPMENT

A fuzzy controller is no more than a particular kind of fuzzy system and, therefore, the necessary stages for developing it are basically the same that were discussed in Chapter 4, that is: (1) selection of system variables; (2) definition of the rule base; and (3) selection of fuzzy operators. Nevertheless, there is a series of peculiar characteristics worth discussing.

The input variables of a fuzzy control system are determined by the type of controller to be used (Fuzzy-PD, Fuzzy-PI, Fuzzy-PID, etc.). In all cases, these inputs are concrete values, provided by the sensors and processed by the signal conditioning modules. The range of values that those variables can take is often constrained to an interval that determines the corresponding universe of discourse. Nevertheless, looking for simplicity, in most cases those universes of discourse are normalized by the signal conditioning stage. As discussed elsewhere, the number of linguistic labels is usually bounded between 3 and 7, and the shape of the fuzzy sets associated to each label is strongly conditioned by the implementation technique to be applied (see Section 4.1.1 in Chapter 4).

The outputs of the fuzzy controller correspond to control actions or variations of the control actions that will be applied to the plant. Therefore, they are also concrete values, so a defuzzification stage must be included in the controller. Depending on the defuzzification method in use, the linguistic labels assigned to the different outputs correspond to fuzzy sets or singletons. This latter option is the most widely employed, since practical reasons justify the common use of simplified defuzzification methods in control applications.

With respect to the definition of the rule base, apart from the two general methods described in Section 4.2.2 of Chapter 4 (based on intuition and experience, or based on a set of input/output data or a simulated model), in the case of fuzzy controllers a third method can be applied. This method consists of building the rule base from a standard rule base [Yager, 1994]. For this kind of strategy, in fuzzy controllers using the error and the variation of the error as inputs, the starting rule base is often the one proposed by MacVicar-Whelan [MacVicar-Whelan, 1976], developed by extending the rule bases used in the first fuzzy controllers [Mamdani, 1975; King, 1977]. The MacVicar-Whelan rule base comprehends the rule bases of these controllers and adds some rules for antecedent combinations they do not contemplate. The expansion of the original rules is performed according to the three following metarules [Tang, 1987]:

MR1: "*If error $e(n)$ and its variation $\Delta e(n)$ are zero,*
keep current control output"

MR2: "*If error $e(n)$ satisfactorily tends towards zero,*
keep current control output"

MR3: *"If error e(n) is not being corrected,*
control action Δu(n) is not zero and depends
on the sign and magnitude of e(n) and Δe(n)"

The MacVicar-Whelan rule base divides the universes of discourse of the two inputs, *e(n)* and *Δe(n)*, and the universe of discourse of the output, *Δu(n)*, into eight linguistic terms (+L, +M, +S, +Z, –Z, –S, –M, –L). The input/output relationship is shown in Table 11.1.

For certain applications, the rule base can be derived by making modifications to this scheme. Thus, a modification could be a change in the membership functions, combining the linguistic labels –Z and +Z into one and/or reducing or increasing the number of the rest of labels, while some other applications could permit the removal of unnecessary rules.

A fuzzy controller using a rule base similar to that described in Table 11.1 provides a (usually non-linear) control surface, which is defined by a great number of parameters. This makes the process of adjusting this controller more complicated than adjusting a conventional controller. Due to this increased complexity, different authors have analyzed the relationships between the design parameters of both kinds of controllers [Filev, 1994; Santos, 1996], and proposed different strategies for adjusting the parameters of Fuzzy-PID controller from a PID controller [Li, 1997; Jantzen, 1998]. The usual idea behind these strategies is to «*start from a well-adjusted PID controller, replace it with an equivalent fuzzy linear controller, make the fuzzy controller non linear, and eventually perform a fine tuning of the non-linear fuzzy controller*» [Jantzen, 1998].

Table 11.1. Rule base proposed by MacVicar-Whelan.

		e(n)							
		–L	**–M**	**–S**	**–Z**	**+Z**	**+S**	**+M**	**+L**
Δe(n)	**–L**	–L	–L	–L	–L	–L	–M	–S	–Z
	–M	–L	–L	–M	–M	–M	–S	–Z	+S
	–S	–L	–M	–S	–S	–S	–Z	+S	+M
	–Z	–M	–M	–S	–Z	+Z	+S	+M	+M
	+Z	–M	–M	–S	–Z	+Z	+S	+M	+M
	+S	–M	–S	+Z	+S	+S	+S	+M	+L
	+M	–S	+Z	+S	+M	+M	+M	+L	+L
	+L	+Z	+S	+M	+L	+L	+L	+L	+L

11.3 APPLICATION EXAMPLE: BALL SUSPENDED BY AN AIRFLOW

To illustrate the process of developing a fuzzy controller, the application example considered has been the problem of regulating the speed of a fan situated in the lower part of a glass tube, so the airflow provided by the fan keeps a ball suspended inside the tube at a certain objective position. Although it is a conceptually simple system, which can be practically implemented with an easy experimental assembly, the complexity of the physical phenomena governing the behavior of the ball and the influence of non-linear effects make it a difficult control problem [Pereira, 1996].

Figure 11.4 shows the photograph of the experimental set up built for the system. It consists of a transparent tube open at both ends. At the lower end, a fan provides the airflow needed for sustaining the ball inside the tube. The speed of the fan is controlled by a pulse-width-modulation signal. The position of the ball is detected by a sensor device positioned at the upper end of the tube. This device is made of a series of light emitting diodes and a light detector diode that receives the light reflected by the ball. Depending on the intensity of the light signal, the detector produces a current proportional to the ball position. Since the response of the sensor is non-linear, some additional elements must be included to linearize it.

11.3.1 CONTROLLER DEFINITION

According to the design methodology proposed in Chapter 10, the first stage in the process of the fuzzy controller development is defining its behavior

Figure 11.4. Photograph of the experimental set up.

by means of a fuzzy system specification language (XFL). When performing this task, we have to bear in mind that the final objective of this application example is obtaining a hardware implementation of the control system, and this, as has been pointed out in many other places, will influence both the definition of the knowledge base (variables, membership functions, and rule base) and the selection of the inference mechanisms (connectives, implication function, and defuzzification method).

The selected controller is of the Fuzzy-PID type. This will enable us to compare this control strategy with those obtained from Fuzzy-PI or Fuzzy-PD controllers, by just simplifying the original rule base to remove the influence of the derivative or integral terms, respectively. The input variables for the controller (e, De, and Ie, in the most general case) are generated by the input signal conditioning stage from the value provided by the position sensor and the reference position established by the user. In order to normalize the inputs to interval [–1, 1], three proportionality constants (K_e, K_{De}, and K_{Ie}) are used. These constants are to be adjusted during the simulation and on-line verification stages in the development of the system.

Three membership functions of trapezoidal and triangular shape (associated with the labels N, Z, and P) have been chosen for representing the antecedent fuzzy sets. Seven singleton values (associated with the labels F0, F1, ..., F6) are used for the consequent. The normalized values inside interval [0, 1] will be converted by the output signal conditioning stage into a percentage of the control signal cycle during which the signal is high.

The rule base for the system is illustrated in Figure 11.5. It includes 27 rules defining system behavior. The figure represents this rule base by means of three tables corresponding to different values of Ie. The table at the center of the figure corresponds to the case (Ie=Z), while the other two mappings contain corrections or offsets from the central mapping.

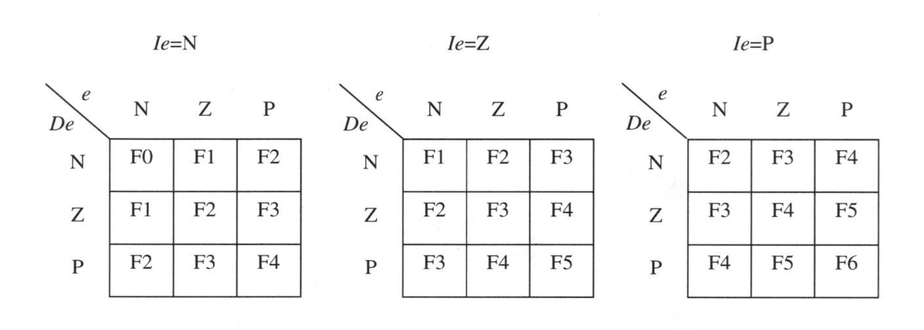

Ie=N

De \ *e*	N	Z	P
N	F0	F1	F2
Z	F1	F2	F3
P	F2	F3	F4

Ie=Z

De \ *e*	N	Z	P
N	F1	F2	F3
Z	F2	F3	F4
P	F3	F4	F5

Ie=P

De \ *e*	N	Z	P
N	F2	F3	F4
Z	F3	F4	F5
P	F4	F5	F6

Figure 11.5. Rule base.

The rule base and membership functions have been selected according to the heuristics of the problem and to the tuning strategy described in Section 11.2.2. Figure 11.6a shows the membership functions for antecedents and consequents for the case of a linear Fuzzy-PID controller. Figure 11.6b shows the membership functions for the non-linear Fuzzy-PID controller obtained after the simulation and on-line verification stages. Figure 11.7 shows the control

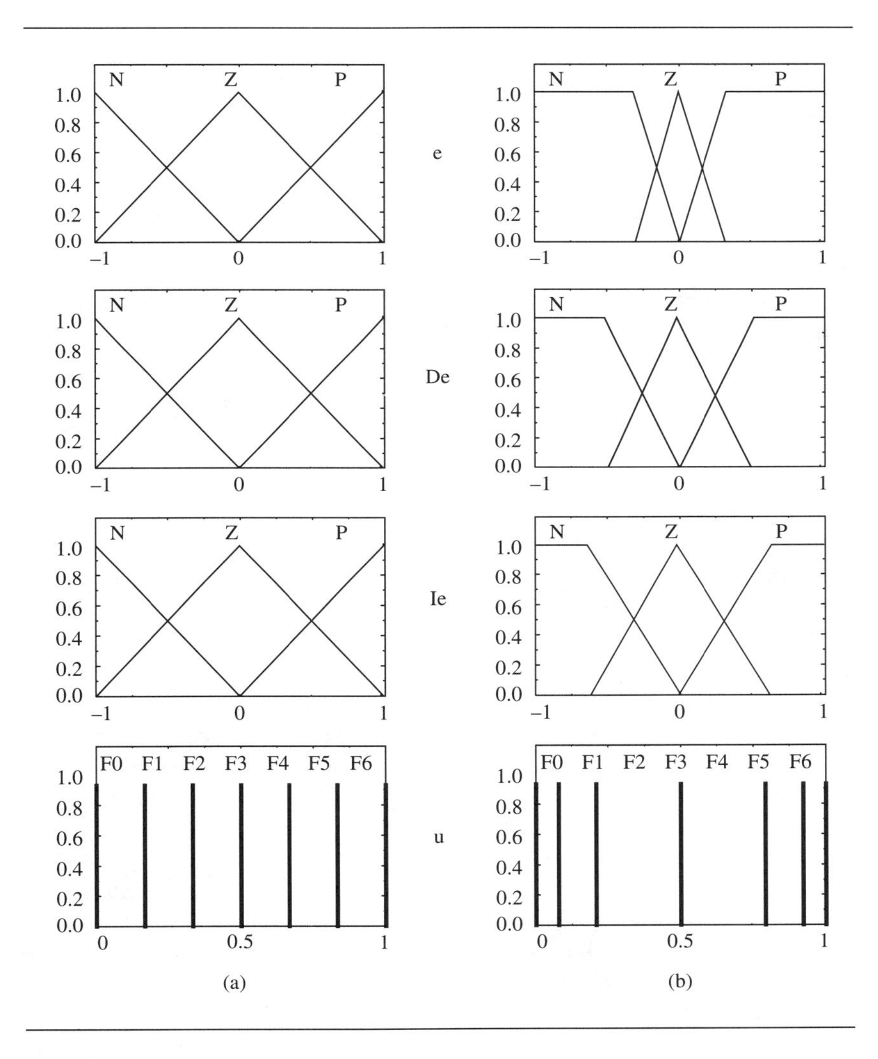

Figure 11.6. Membership functions for antecedents and consequents. (a) Linear Fuzzy-PID; (b) Non-linear Fuzzy-PID.

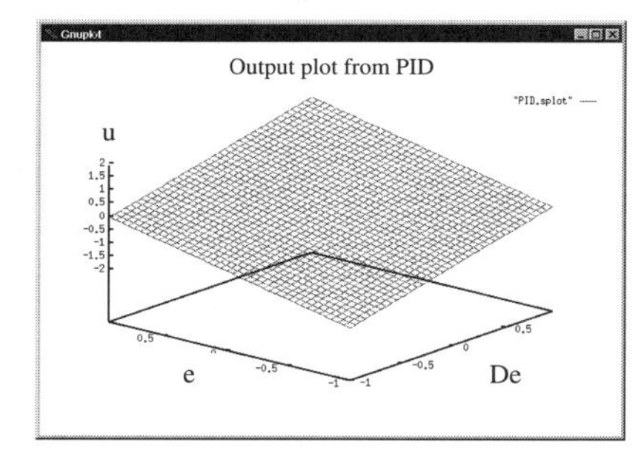

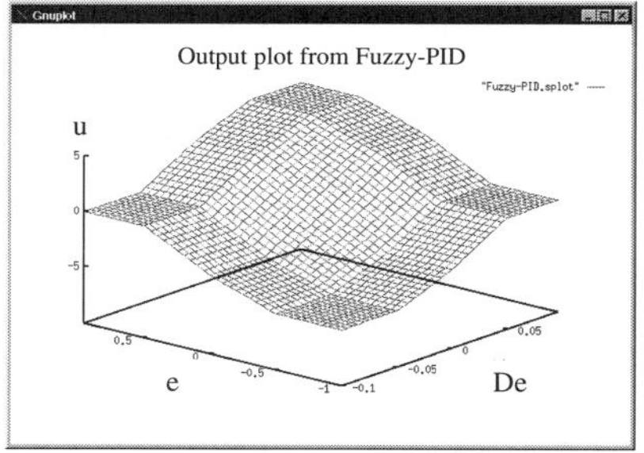

Figure 11.7. Control surfaces.

surfaces corresponding to both knowledge bases. The product operator has been used in both cases as the antecedent connective.

These definitions fulfill the synthesis constraints that permit the microelectronic implementation of the fuzzy system with active rule processing and simplified defuzzification method, using the architecture described in Section 9.5 of Chapter 9.

11.3.2 SYSTEM SIMULATION

Obtaining a simulation model for the controller from its XFL definition is an immediate task with the XFL to C compiler integrated within *Xfuzzy* (*xfc*). However, a model describing the plant behavior is required in order to simulate the whole system. It is important to recall that the objective of this stage is just to obtain a first evaluation of the “fitness” of the controller under development, so the model of the plant does not have to be excessively complex. In fact, the model used for this application example is a C program implementing a first-order model (an “ideal” plant) where the ball is only subject to a couple of forces: gravity and the steam of the airflow provided by the fan. This model does not consider second-order effects, such as friction, turbulences, collisions of the ball against the inner surface of the tube, etc.

The simulation tool of *Xfuzzy* (*xfsim*) eases the construction of an executable file combining the models of the plant and the controller. The environment also permits the integration of the simulation program thus produced with application-specific graphical interfaces built with scripting languages like *Tcl-Tk* [Welch, 1997]. Figure 11.8 shows the aspect of a graphical interface built with the free development tool *visual Tcl* in order to simulate the behavior of the system when using PID and Fuzzy-PID controllers. This interface contains several work areas for visualizing system behavior, modifying its parameters, and performing actions such as stopping the simulation, starting it, or invoking

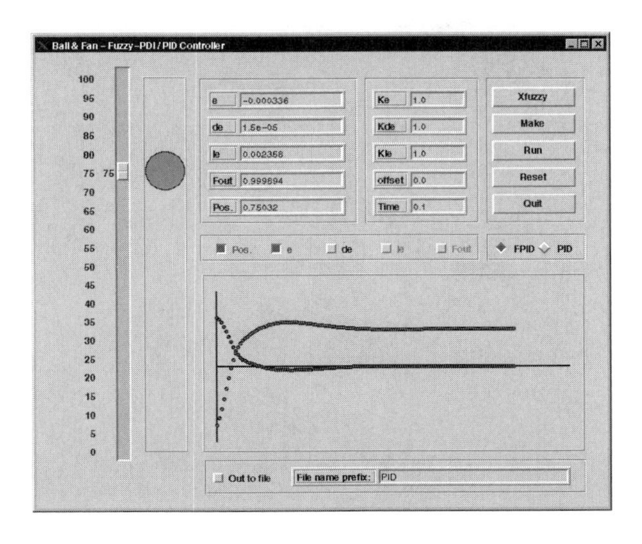

Figure 11.8. Graphical interface using in the simulation stage.

Xfuzzy. The drawing area at the left side of the window shows a simplified representation of the system. A scale permits the selection of the reference height. When the simulation is started, the ball moves according to its physical model and the output of the fuzzy controller. The upper center part of the window shows two groups of values. The group at the left contains the current values of the system variables: the error, *e*, the error variation, *De*, the error integral, *Ie*, the control action, *Fout*, and the ball position, *Pos*. The group at the right is intended for changing the values of the adjustable parameters: the normalization constants for the input variables, K_e, K_{De}, and K_{Ie}, the possible offset of the control action, *offset*, and the duration of the control loop iteration, *Time*. The evolution with time of the different system variables is graphically represented in the lower right part of the window. This data can be also stored into a file for further analysis. Finally, the button area at the upper right part of the window permits invoking the *Xfuzzy* environment, rebuilding the simulation program with the *make* utility, starting the simulation, resetting the system to its initial conditions, and closing the application.

Figure 11.9a illustrates some simulation results for different types of fuzzy controllers. In all cases the starting point is a situation where the ball is in a low position in the tube and the established objective is a higher position. The system evolves until the ball reaches the intended position, and then the reference position is changed to the lower part of the tube. It can be noted that the system follows the defined reference and that, after a certain response delay, it reaches

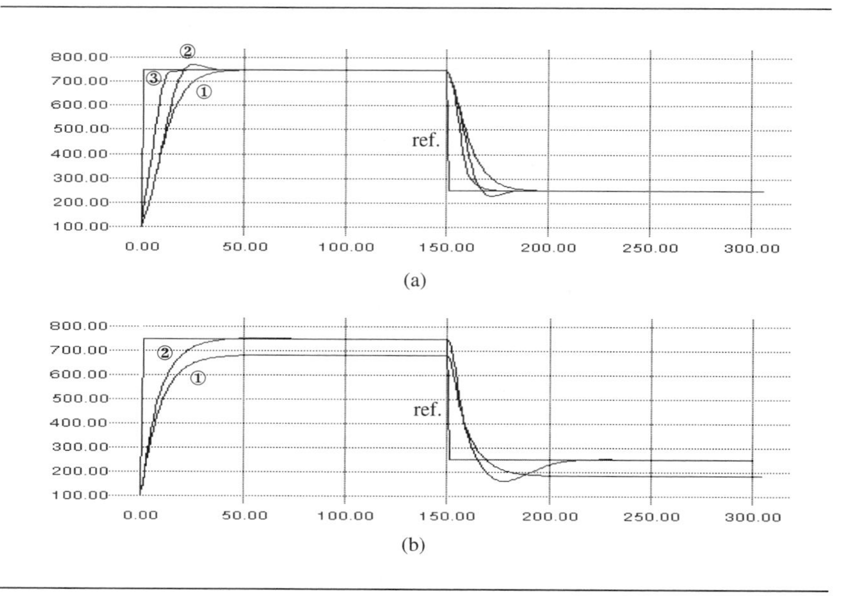

Figure 11.9. Results of the simulation stage. (a) Linear and non-linear Fuzzy-PD controllers; (b) Fuzzy-PD vs. Fuzzy-PID.

the selected objective. Curves ① and ② in Figure 11.9a correspond to linear Fuzzy-PD controllers. It can be observed that by tuning the normalization constants (K_e and K_{De}), it is possible to obtain a compromise between rising time and overshooting. The use of a non-linear Fuzzy-PD controller, like the one corresponding to curve ③, permits the simultaneous enhancement of both characteristics of the transient system response. Figure 11.9b compares the trajectories of the ball, with a negative offset in the output signal, when using Fuzzy-PD and Fuzzy-PID controllers. In case ①, the ball never reaches the reference position, while in case ②, the correction introduced by considering the cumulative error permits reaching the reference, enhancing in this way the stationary system response.

11.3.3 ON-LINE VERIFICATION OF THE CONTROLLER

The simulation results obtained in the above section verify the behavior of the controller for an "ideal" plant. However, the first-order model used thus far does not take into account some other effects that, in this case, have a great influence on the real system. A possible solution could consist of refining the model of the system under control. But, in this case, the refinement would introduce a great mathematical complexity, yielding a problem with a non-trivial

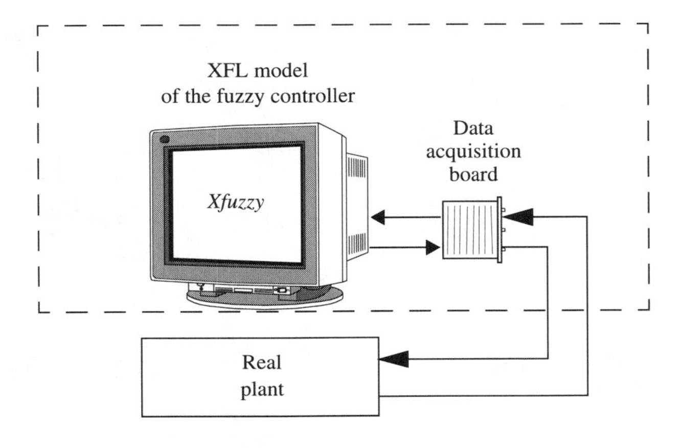

Figure 11.10. Conceptual representation of the on-line verification strategy.

solution. Another option, according to the design methodology described in Chapter 10, consists of introducing the real plant into the feedback loop. This second alternative can be accomplished employing *xflab*, the on-line verification tool provided by *Xfuzzy* (Figure 11.10).

When this verification strategy is used, the plant model is substituted by a program performing the read and write functions on a data acquisition board connected to the plant. This program initializes the input and output ports of the board, provides initial values for the system variables, and executes the control loop. In each iteration of the control loop, the current position of the ball and the reference position are read. From these values, the error, its derivative, and its integral are calculated. Once these values have been normalized, they are passed to the C function implementing the fuzzy controller. The controller output is post-processed and sent to the fan control circuit through the output ports of the data acquisition board. In this example, one of the analog input ports of the board has been used for reading the signal provided by the position sensor, and the three digital counters provided by the board have been used for generating the pulse-width-modulated control action, which is transmitted to the fan.

The graphical interface of *xflab*, shown in Figure 11.11, eases the on-line verification of fuzzy controllers described using XFL, combining all the necessary information in a single window. This information basically consists of the inputs and outputs of the controller, the inputs and outputs of the plant, the intermediate variables in use, and the three code sections corresponding to the processes for initialization, controller input pre-processing, and controller output post-processing. With the graphical user interface it is also possible to

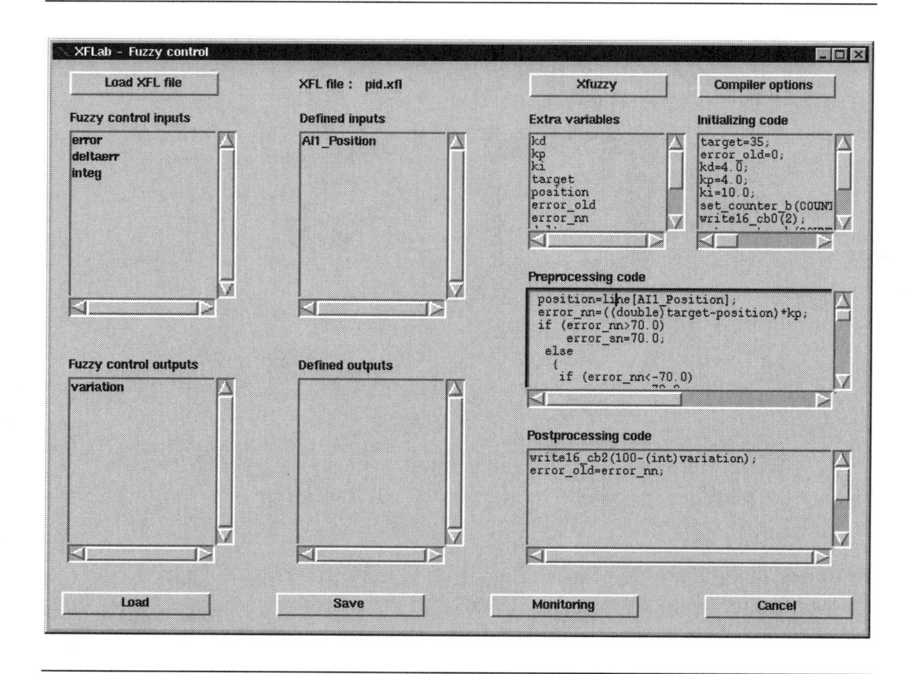

Figure 11.11. Graphical interface of *xflab* for the application example.

select the options applied when compiling programs and to start the monitoring facilities that provide mechanisms for capturing and representing execution data.

For starting the on-line verification process, some results from the high-level simulations are used. One of them is the rule base, although the range and values of the consequents obtained from the simulation must be changed, since the simulated controller output corresponds to the time that the power input of the fan is high, while the actual consequent values represent the percentage of time the power input of the fan is high. Apart from this, the plant suffers a series of second-order effects requiring a fine tuning process of the rule antecedents for obtaining the best possible control. It is also possible to derive the required sampling frequency from the simulation. This value must be as high as possible, while preserving an adequate rate in the changes of the error variation signal.

Figure 11.12 shows experimental results obtained during the on-line verification stage. They illustrate the trajectory of the ball for different reference positions. The y-axis represents the height of the ball in centimeters and the x-axis the number of the data acquisitions and the control actions performed.

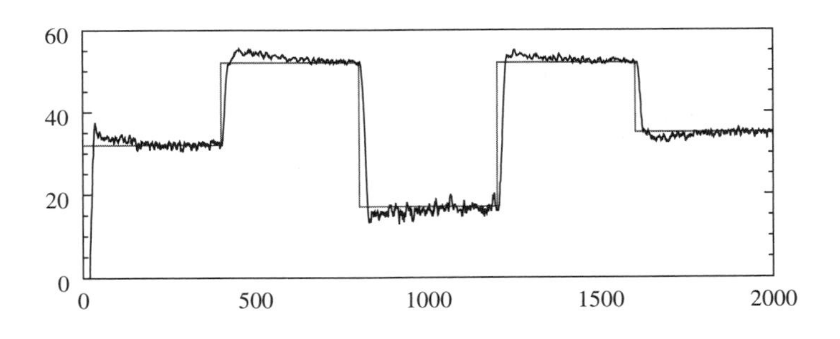

Figure 11.12. Experimental results obtained during the on-line verification stage.

11.3.4 HARDWARE DESIGN AND IMPLEMENTATION OF THE CONTROLLER

The last stage in the development process is the hardware design and implementation of the control system. Apart from the physical implementation of the controller, we must include in the experimental setup all the signal conditioning circuitry that was emulated by software during the on-line verification stage. These elements are placed on a printed circuit board that we will call the control board. This board receives as input the signal provided by the photosensitive diode and generates the pulse-width-modulation signal, which is sent to the fan.

The block diagram of the system is shown in Figure 11.13. The fuzzy controller circuit corresponds to one of the digital prototypes described in Chapter 9. It is a controller with three inputs and one output (with six bits of precision for each of them). It allows up to eight membership functions per input. Antecedents are stored in a memory, so they can have any shape with the only constraints of a maximum overlapping degree of two. The rule memory has a capacity of up to 512 rules. This prototype stores the knowledge base into static RAM. Therefore, during the system initialization stage, the knowledge base must be read from an external source. This source is an EEPROM, whose contents are loaded into the fuzzy circuit through an 8-bit bus. The reference height is selected on the control board by means of a set of switches.

The first function of the signal conditioning circuitry is to transform the (analog) signal generated by the photosensor into a digital signal, by means of an A/D converter. Since the relation current/distance to the ball is not linear, an EEPROM is used to linearize this signal. Next, the calculation of the inputs to the controller (error, error variation, and error integral) is performed using the reference position and the current ball position. The signal provided by the photosensor presents an important noise component, due to both internal (the

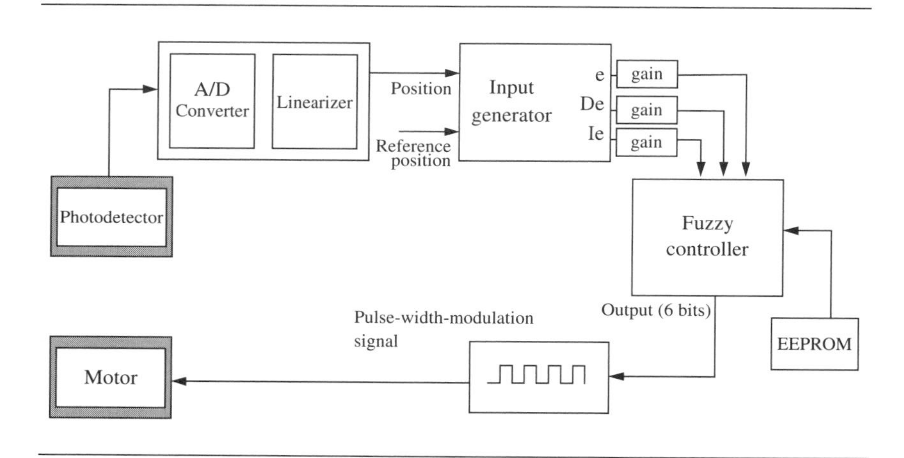

Figure 11.13. Block diagram of the experimental system.

photodiodes and photodetector themselves) and external (the effect of room lights) sources. The output signal conditioning circuitry converts the digital output of the controller into a pulse-width-modulation signal of 5 V which is finally transformed into the voltage applied to the fan motor by means of a power transistor.

Figure 11.14 shows a photograph of the control board. Apart from the integrated circuits corresponding to the fuzzy controller and the EEPROM used for storing the knowledge base, the board includes two FPGAs, one of them in charge of generating the inputs to the controller, and the other for producing

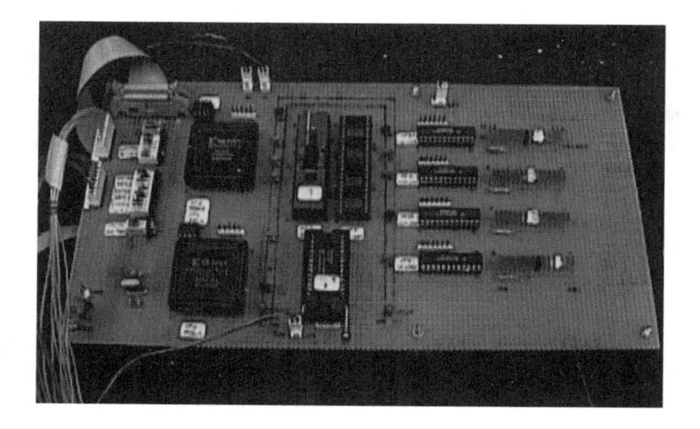

Figure 11.14. Photograph of the control board.

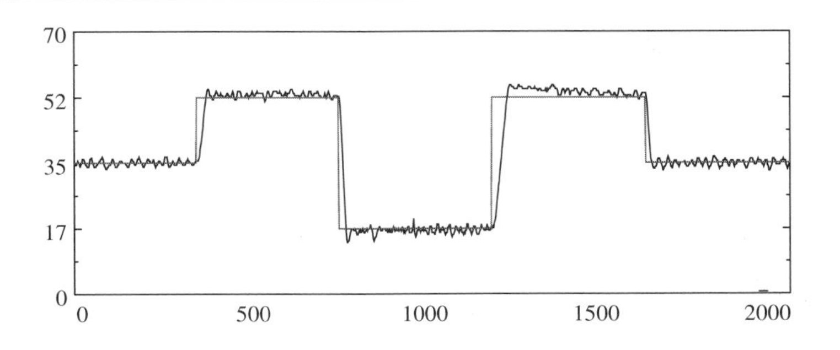

Figure 11.15. Experimental results obtained with the hardware controller.

the clock signal, controlling the load of the knowledge base from the EEPROM into the controller, and translating the controller output into the pulse-width-modulation signal which, once processed by the power stage, will be applied to the fan motor. The board is completed by a series of switches for coding the reference position and modifying the normalization constants, and by a set of LEDs (light emitting diodes) that permit the monitorization of the inputs and outputs of the controller, providing a visual reference of their evolution.

The knowledge base obtained from the on-line verification has been applied to this final implementation. Figure 11.15 shows the experimental results obtained when the ball follows the same trajectory used in the previous test. It can be observed that the hardware controller provides slightly better results. This can be due to the finer control of the sampling times and, mainly, to the increase in the inference speed of the hardware realization.

REFERENCES

1. **Bollinger, J. G., Duffie, N. A.**, *Computer Control of Machines and Processes*, Addison-Wesley, 1989.

2. **Dohmann, E.**, Fuzzy logic control of a chemical bath, in *Proc. IEEE Int. Conference on Fuzzy Systems*, pp. 958-961, Orlando, 1994.

3. **Filev, D. P., Yager, R. R.**, On the analysis of fuzzy logic controllers, *Fuzzy Sets and Systems*, Vol. 68, pp. 39-66, 1994.

4. **Hall, M., Harris, A.**, Fuzzy logic expert system for iron ore processing, in *Proc. IEEE Industry Applications Conference*, pp. 2190-2199, Toronto, 1993.

5. **Isaka, S., Sebald, A. V.**, An adaptative fuzzy controller for blood pressure regulation, in *Proc. Int. Conf. of IEEE Engineering in Medicine & Biology Society*, pp. 1763-1764, Seattle, 1989.

6. **Isaka, S.**, Fuzzy temperature controller and its applications, in *Proc. Applications of Fuzzy Logic Technology*, pp. 59-65, Boston, 1993.

7. **Itoh, O., Migia, H., Itoh, H., Ireie, Y.**, Application of fuzzy control to automatic crane operation, in *Proc. Int. Conference on Industrial Electronics, Control and Instrumentation*, pp. 161-164, Maui, 1993.

8. **Jantzen, J.**, Tuning of fuzzy PID controllers, Tech. report N° 98-H 871, Technical University of Denmark, Sept. 1998.

9. **King, P. J., Mamdani, E. H.**, The application of fuzzy control systems to industrial processes, *Automatica*, Vol. 13, pp. 235-242, 1977.

10. **Kuo, B. C.**, *Automatic Control Systems*, Prentice-Hall, Inc., 1982.

11. **Lawrence, P. D., Mauch, K.**, *Real-time Microcomputer System Design: An Introduction*, McGraw-Hill, 1987.

12. **Li, H. X.**, A comparative design and tuning for conventional fuzzy control, *IEEE Transactions on Systems, Man, and Cybernetics*, Vol. 27, N. 5, pp. 884-889, 1997.

13. **Linkens, D. A., Mahkouf, M.**, Fuzzy log knowledge-based control for muscle relaxant anaesthesia, in *Proc. IFAC Symposium*, pp. 185-190, Fulda, 1988.

14. **MacVicar-Whelan, P. J.**, Fuzzy sets for man-machine interactions, *Int. Journal of Man-Machine Studies*, Vol. 8, pp. 687-697, 1976.

15. **Mamdani, E. H., Assilian, S.**, An experiment in liguistic synthesis with a fuzzy logic controller, *Int. Journal of Man-Machine Studies*, Vol. 7, pp. 1-13, 1975.

16. **Passino, K. M., Yurkovich, S.**, *Fuzzy Control*, Addison-Wesley, 1998.

17. **Pereira, J. S., Bowles, J. B.**, Comparing controllers with the ball in a tube experiment, in *Proc. IEEE Int. Conference on Fuzzy Systems*, pp. 504-510, New Orleans, 1996.

18. **Quail, S., Adnan, S.**, State of the art in household appliances using fuzzy logic, in Yen, J., Langari, R., Zadeh, L., Eds., *Proc. 2nd Int. Workshop on Industrial Fuzzy Systems and Intelligent Systems*, pp. 204-213, IEEE Press, 1992.

19. **Santos, M., Dormido, M., De la Cruz, J. M.**, Fuzzy-PID controllers vs. Fuzzy-PI controllers, in *Proc. IEEE Int. Conference on Fuzzy Systems*, pp. 1598-1604, New Orleans, 1996.

20. **Steimann, F., Adlassnig, K. P.**, Fuzzy medical diagnosis, in *Handbook of Fuzzy Computation*, Ruspini, E. H., Bonissone, P. P., Pedrycz, W., Eds., pp. G13.1:1-14, Institute of Physics Pub., 1998.

21. **Sugima, F., Mizushima, S., Kimura, M., Fukuri, Y., Harada, Y.**, A fuzzy approach to the rate control in an artificial cardiac pacemaker regulated by respiratory rate and temperature: a preliminary report, *Journal of Medical Engineering & Technology*, Vol. 3, N. 15, pp. 107-110, 1991.

22. **Takagi, H.**, Survey: fuzzy logic applications to image processing equipment, in Yen, J., Langari, R., Zadeh, L., Eds., *Proc. 2nd Int. Workshop on Industrial Fuzzy Systems and Intelligent Systems*, pp. 1-9, IEEE Press, 1992.

23. **Tang, K., Mulholland, R. J.**, Comparing fuzzy logic with classical controller design, *IEEE Transactions on Systems, Man, and Cybernetics*, Vol. 17, pp. 1085-1087, 1987.

24. **Welch, B.**, *Practical Programming in Tcl and Tk*, Prentice-Hall, 1997.

25. **Yager, R. R., Filev, D. P.**, *Essentials of Fuzzy Modeling and Control*, John Wiley & Sons, Inc., 1994.

26. **Yamamoto, S., Hashimoto, I.**, Present status and future needs: The view from Japanese industry, *Chemical Process Control, CPCIV*, pp. 1-28, Padre Island, 1991.

27. **Yamane, S., Kitahara, N., Ohshima, K., Yamamoto, M.**, Neural network and fuzzy control of weld pool with welding robot, in *Proc. IEEE Industrial Applications Conference*, pp. 2175-2180, Toronto, 1993.

28. **Yasunobu, S., Hasegawa, T.**, Evaluation of an automatic container crane operation system based on predictive fuzzy control, *Control Theory and Advanced Technology*, Vol. 2, N. 3, pp. 419-432, 1986.

Chapter 12

FUZZY SYSTEMS AS APPROXIMATORS

Fuzzy systems provide in general a non-linear mapping between the input and the output space. The parameters allowing the adjustment of this input/output mapping are related to the fuzzy partition of the input domain (including the shape of antecedent membership functions and their overlapping degrees), the partition of the output domain (the representation of the consequents), the antecedent connectives, the implication function, and the ways of implementing rule aggregation and defuzzification (if needed). What is very remarkable is that any mapping between the input and the output space can be approximated to any degree of accuracy by properly choosing the above parameters. This means that fuzzy systems are universal approximators.

From many points of view, in particular from a microelectronic point of view, the objective is to work with fuzzy systems as simple as possible that feature universal approximation capability. In this sense, we have already seen in Chapters 7, 8, and 9 that the systems most suitable for microelectronic implementation are those that summarize the fuzzy nature of the consequents into several parameters. In addition, we have also seen that a grid or lattice partition of the input spaces allows a relevant circuitry reduction when implementing systems with many rules since a rule active-driven architecture can be employed. The first section of this chapter describes the approximation capability of these systems, in particular of singleton fuzzy systems that represent the consequents by only one parameter.

The universal approximation capability of fuzzy systems explains their successful application to many fields. The applications illustrated in the current chapter will employ fuzzy systems to reproduce a desired input-output mapping, such as identification and modeling of highly non-linear static and dynamic systems. In these cases, the required fuzzy systems are usually designed by using supervised learning algorithms taken from the neural network practice, as was explained in Chapter 5. This is why these systems are often called *neuro-fuzzy systems*.

Concerning the microelectronic implementation of these systems, two situations can be distinguished. One of them is that the learning mechanism is implemented off chip. This solution is discussed in the second section of this chapter and illustrated with several examples. The other situation is that the learning mechanism is implemented on chip together with the inference mechanism. This solution is more ambitious since the resulting chip can work autonomously within embedded systems offering real-time adaptability. This will be addressed in the third section of this chapter.

12.1 APPROXIMATION CAPABILITY OF FUZZY SYSTEMS

Many authors have demonstrated that fuzzy systems are universal approximators, as can be seen, for instance, in [Wang, 1992a; Buckley, 1992; Ying, 1994; Zeng, 1994; 1995; 1996; Kosko, 1994; Castro, 1995]. In particular, we will focus on fuzzy systems that employ a grid partition of the input spaces and that represent the consequents by several parameters (such as singleton fuzzy systems). In the case of singleton fuzzy systems, universal approximation capabilities can be demonstrated by relating them with neural networks such as radial basis function (RBF) networks [Park, 1991; Jang, 1993] or with approximator systems known in the context of mathematical approximation theory, such as Lagrange and Spline interpolators [Zeng, 1994; 1995; 1996; Bossley, 1997; Rovatti, 1998; Baturone, 1999]. In the following, these analogies are briefly described.

12.1.1 LOCAL PIECEWISE INTERPOLATION

The use of a grid partition of the input spaces has advantages other than its simple microelectronic implementation. Considering knowledge representation, a grid partition has semantic meaning. Considering approximation theory, the problem is simplified to local piecewise interpolation.

A fuzzy system with *u* inputs, $\bar{x}=(x_1, ..., x_u)$ and one output (y) establishes in general a non-linear relation between the input $(I_1, ..., I_u)$ and output (R) universes of discourse. Formally, this means that:

$$y = y(\bar{x}): I_1 \times I_2 \times \ldots \times I_u \subset R^u \rightarrow R \tag{12.1}$$

A grid partition distinguishes L_i linguistic labels with a maximum overlapping degree of α within each input universe of discourse, I_i. Hence, $L_i+1-\alpha$ intervals can be distinguished per input. They are separated by the points $\{a_{i1}, a_{i2}, ..., a_{iNi}\}$, where N_i is $L_i-\alpha$. This is illustrated on the left of Figure 12.1 for the case of one input. The number of grid cells that can be distinguished in the output surface is $(L_i+1-\alpha)^u$. Given an input vector $\bar{x}_o$, the fuzzy system identifies the *u* intervals, that is, the particular grid cell, GC_p, to which the input belongs, and provides the corresponding output $y(\bar{x}_o)$ by evaluating the α^u active rules. If the system represents the consequents by several parameters, it was seen in Chapter 7 (Section 7.5.2) that the output is given by:

$$y(\bar{x}_o) = \left. \frac{\sum_{k=1}^{\alpha^u} w_{pk}(\bar{x}) \cdot f_{pk}}{\sum_{k=1}^{\alpha^u} w_{pk}(\bar{x})} \right|_{\bar{x}=\bar{x}_o} = y_p(\bar{x})\Big|_{\bar{x}=\bar{x}_o} \tag{12.2}$$

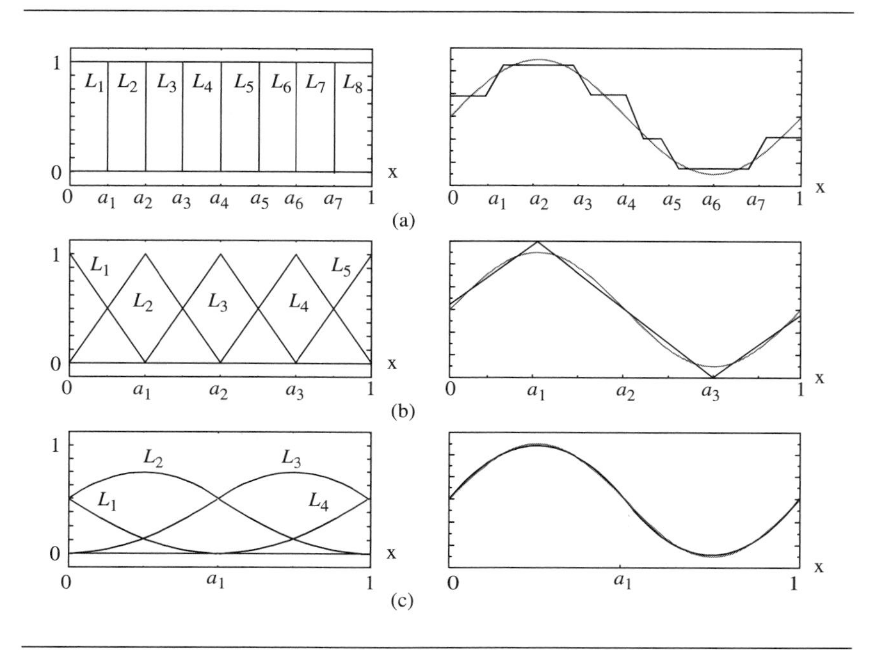

Figure 12.1. Approximation (in bold line) of (1+sin(2πx))/2 (in dashed line) with: (a) 8 rectangular labels; (b) 5 triangular fuzzy labels; and (c) 4 bell-shaped fuzzy labels.

where w_{pk} is the weight of one of the α^u active rules and f_{pk} is the parametric representation of its corresponding consequent.

The parameters that define this function $y_p(\bar{x})$ are a few of the global parameters that define the fuzzy system, $y(\bar{x})$. In this sense, a fuzzy system is viewed as a local piecewise interpolator that provides a piece, y_p, for each grid cell, GC_p. The interpolation provided can be piecewise constant, piecewise linear, or piecewise non-linear, in general, depending on whether y_p is constant, linear or non-linear in $\bar{x}$. These three cases are illustrated on the right of Figure 12.1 for the approximation of the function $(1+\sin(2\pi x))/2$.

The pieces y_p can be seen as the linear combination of the functions $w_{pk}/\Sigma_k w_{pk}$, which some authors have named "fuzzy basis functions" [Wang, 1992a]. This output is similar to that provided by the radial basis function networks, thus resulting in the analogy between them [Jang, 1993].

12.1.2 ANALYSIS OF SINGLETON FUZZY SYSTEMS

The simplest fuzzy systems among those that represent the consequents by several parameters are the singleton fuzzy systems. This sub-section focuses on analyzing the approximation capability of these systems, thus offering a basis for understanding the capability of more complex systems.

The linear or non-linear nature of the output pieces y_p that singleton fuzzy systems provide depends on the fuzzification (overlapping and shapes of the antecedent membership functions) and on the type of antecedent connectives. The other operators of a fuzzy inference engine: implication, aggregation, and defuzzification, merge into a weighted average operator in a singleton fuzzy system.

The overlapping degree of the antecedent membership functions is a factor that greatly influences the form of each output piece. As an example, Figure 12.2 illustrates the approximation of the function $(1+\cos(2\pi x))/2$ with a one-input one-output singleton fuzzy system which employs triangular membership functions to cover the input universe of discourse. If the overlapping degree is as shown in Figure 12.2a, the output pieces are linear, as can be seen in Figure 12.2b. Changing the overlapping degree to that illustrated in Figure 12.2c, the output pieces are non-linear, as can be seen in Figure 12.2d.

When considering the implementation of complex fuzzy systems with programmable integrated circuits, low overlapping degrees are preferred in order to reduce circuitry. Hence, let us analyze the approximation obtained with overlapping degrees of 1 (no overlapping) and 2.

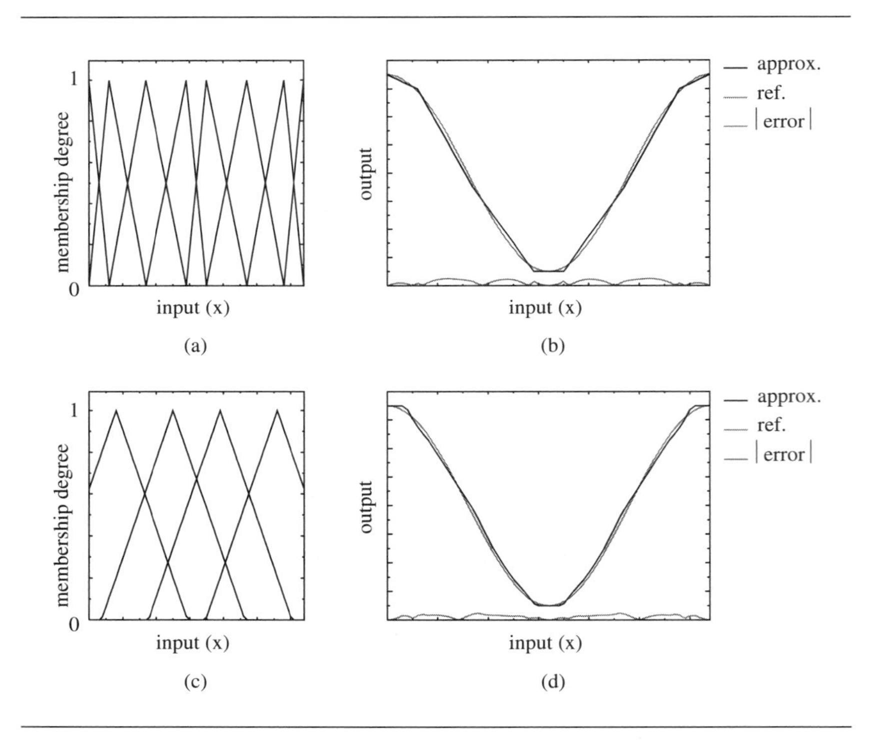

Figure 12.2. Illustration of how the overlapping degree of the antecedent membership functions influences the output.

Systems whose antecedent membership functions do not overlap (as was shown in Figure 12.1a) are not truly fuzzy. They are piecewise constant interpolators, that is, their output is constant for each grid cell (as can be seen in Figure 12.1a):

$$y_p(\bar{x}) = \frac{w_{p1}(\bar{x})f_{p1}}{w_{p1}(\bar{x})} = f_{p1} \tag{12.3}$$

To evaluate the error, $|f(\bar{x}) - y(\bar{x})|$, that this kind of system provides when approximating a differentiable function $f(\bar{x})$ within a given grid cell GC_p, let us define the ∞-norm for this error as [Powell, 1981]:

$$\|f(\bar{x}) - y(\bar{x})\|_\infty = sup_{\bar{x} \in GC_p} |f(\bar{x}) - y(\bar{x})| \tag{12.4}$$

$f(\bar{x})$ can be expressed by its Taylor expansion around the point $\bar{x}_o$ where $f(\bar{x}_o) = f_{p1}$, so that this error is bounded by:

$$\|f(\bar{x}) - y(\bar{x})\|_\infty = \left\| \sum_{i=1}^{u} \left.\frac{\partial f}{\partial x_i}\right|_\xi \cdot (x_i - x_{io}) \right\|_\infty \leq \sum_{i=1}^{u} \left\| \frac{\partial f}{\partial x_i} \right\|_\infty \cdot a \leq M_f \cdot a \tag{12.5}$$

where M_f is a constant that depends on the function f; $a = \max_i\{\max_{ji}(a_{i,ji+1} - a_{i,ji})\}$, with $i = 1, ..., u$, $j_i = 1, ..., N_i$; $a_{i,ji}$ are the partition points; and ξ verifies that:

$$|\xi_i - x_{io}| \leq |x_i - x_{io}| \quad \forall i = 1, \ldots, u \tag{12.6}$$

As a matter of fact, a look-up table is a particular case of these systems. In look-up tables with *p* bits of resolution, the input spaces are partitioned into 2^p intervals of the same width, and the output for each of the 2^{pu} grid cells is constant.

In the context of mathematical approximation theory, given a function $f(\bar{x}): I \subset R^u \to R$, a system $y(\bar{x})$ is said to be a k-th order accurate approximator for $f(\bar{x})$ if [Zeng, 1996]:

$$\|f(\bar{x}) - y(\bar{x})\|_\infty = sup_{\bar{x} \in I} |f(\bar{x}) - y(\bar{x})| < M_f \cdot a^k \tag{12.7}$$

Hence, systems with no overlapping are first-order accurate approximators. A feature of their output is discontinuity at the boundaries of the grid cells (Figure 12.1a).

Let us now analyze fuzzy systems with an overlapping degree of 2. As an example, let us consider a one-input one-output fuzzy system with triangular membership functions representing the antecedents. For a given input, two rules are activated:

If x is A_1 then y is c_1
If x is A_2 then y is c_2

where A_1 and A_2 are fuzzy sets described by triangular membership functions, $\mu(x)$, with slopes m_1 and m_2, respectively, in the activated interval. This is shown in Figure 12.3.

According to (12.2) and Figure 12.3, given an input *x*, output *y* is given by:

$$y = \frac{f_1 w_1 + f_2 w_2}{w_1 + w_2} = \frac{c_1[1 + m_1(a_1 - x)] + c_2[1 + m_2(x - a_2)]}{2 + m_1(a_1 - x) + m_2(x - a_2)} \tag{12.8}$$

Depending on the values of c_i and m_i, six cases can be distinguished, as illustrated in Figure 12.3 and summarized in Table 12.1. Among these cases, the situation of normalized membership functions (m_1=m_2 for all the overlapping regions) can simplify microelectronic implementations, as was noted in Chapters 7 and 9.

Having selected normalized membership functions with an overlapping degree of 2, the shapes of the functions also influence greatly the output. An example that compares triangular with Gaussian membership functions is illustrated in Figure 12.4.

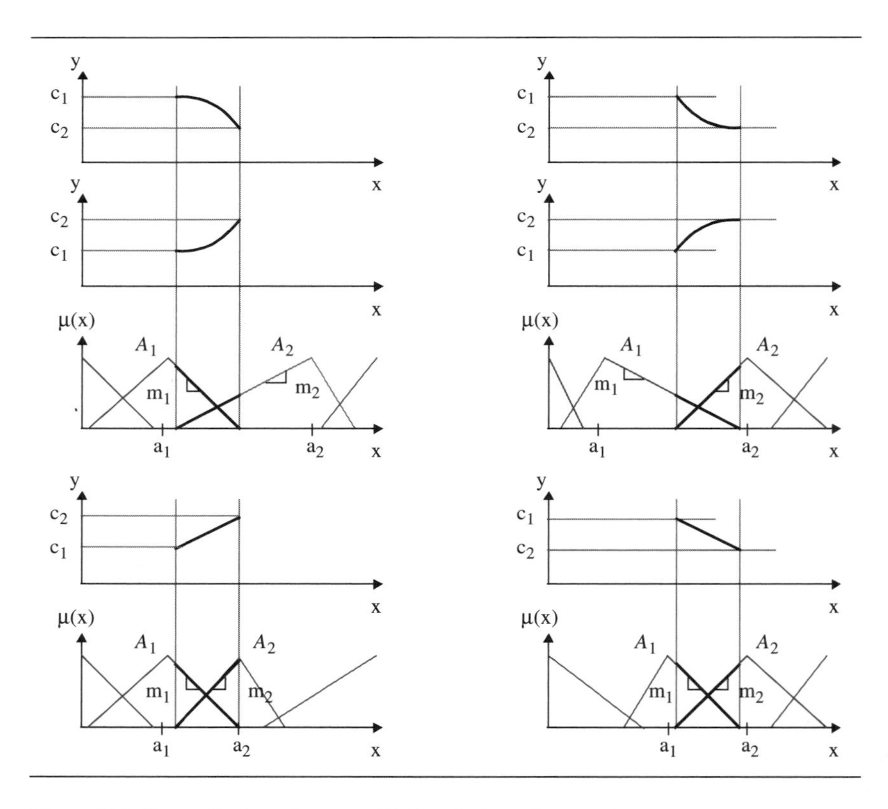

Figure 12.3. Types of output pieces provided by singleton fuzzy systems.

Table 12.1. Influence of the overlapping degree on the output.

	$c_1 > c_2$	$c_1 < c_2$
$m_1 > m_2$	convex and monotonously decreasing	concave and monotonously increasing
$m_1 < m_2$	concave and monotonously decreasing	convex and monotonously increasing
$m_1 = m_2$	linear and monotonously decreasing	linear and monotonously increasing

Given the same overlapping degree and shapes, the antecedent connective also varies the output. This is shown in Figure 12.5, where the minimum and product connectives are compared.

Fuzzy systems whose membership functions are normalized triangles and which connect them with the minimum operator provide first-order accurate approximation [Zeng, 1996], like look-up tables. The difference is that the fuzzy nature of the antecedents makes the output continuous at the boundaries of the grid cells (Figure 12.1b). Those that employ the product or the normalized conjunction instead of the minimum are piecewise multilinear interpolators that offer second-order accurate approximation [Zeng, 1996; Rovatti, 1998] so that:

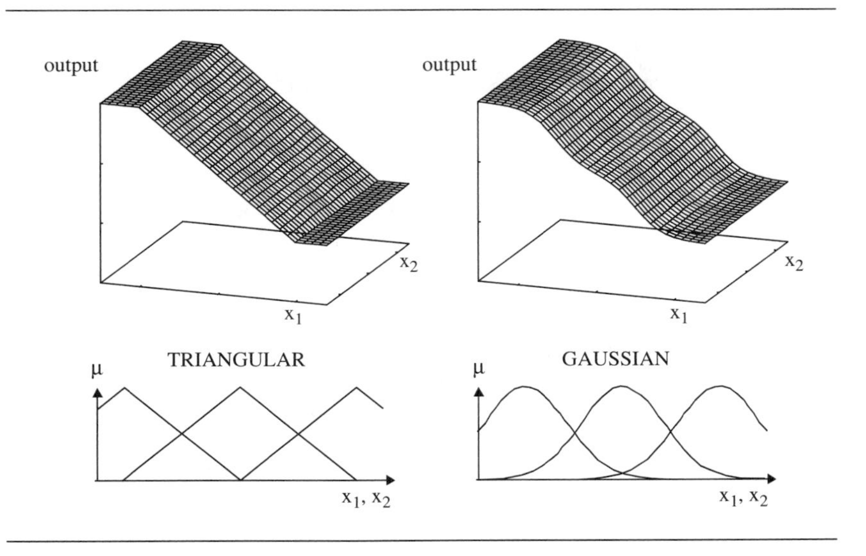

Figure 12.4. Influence of the shape of membership functions.

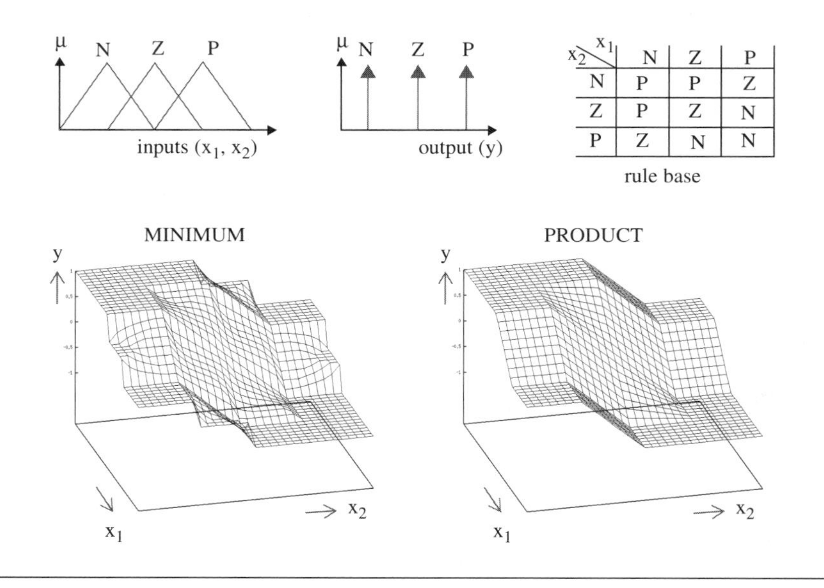

Figure 12.5. Influence of the antecedent connective on the output.

$$\|f(\bar{x})-y(\bar{x})\|_{\infty} \leq M \cdot \sum_{i,j} \left\| \frac{\partial^2 f}{\partial x_i \partial x_j} \right\|_{\infty} \cdot a^2 \tag{12.9}$$

where M is a constant.

Hence, they not only approximate the values of a desired function but also its first derivatives.

In general, singleton fuzzy systems with normalized triangles to represent the antecedent membership functions and the product as the antecedent connective are equivalent to piecewise multilinear Lagrange interpolators of degree 1 in x_i or equivalent to degree-1 B-spline interpolators [Zeng, 1996]. This means that if we consider, for instance, a two-input one-output fuzzy system, each piece of output is given by:

$$y_p(x_1,x_2) = Ax_1x_2 + Bx_2 + Cx_1 + D \tag{12.10}$$

where A, B, C, and D are constants related to the slopes of the membership functions and to the singleton values.

Fuzzy systems that use normalized membership functions for the antecedents but with an overlapping degree greater than 2 offer a higher order accurate approximation [Zeng, 1996] but they are more costly for hardware implementation, as discussed in [Baturone, 1999].

12.2 APPLICATIONS USING OFF-CHIP LEARNING

Let us consider applications that require a circuit to reproduce a desired input-output function. Such applications appear in problems like calibration of sensor non-idealities, waveform shaping or signal predistortion, linearizing non-ideal signal converters, or generating adequate control surfaces in the case of microelectronic implementations of non-linear controllers. Many authors have been concerned with the problem of designing and implementing these approximator circuits, known as *function synthesis circuits* [Gilbert, 1982; Fattaruso, 1987; Sánchez-Sinencio, 1989; Turchetti, 1993; Leme, 1993]. More recently, fuzzy circuits have also been described as approximator circuits [Manaresi, 1998].

The output of these circuits depends on the values of certain electrical parameters so that the designer must know these values to provide a given function. The usual approach is to calculate them with a learning algorithm running on a computer. Since this type of learning is performed outside the chip, it is usually called *off-chip learning*. If the circuit is conceived to be dedicated to a specific application, the values of these parameters can be fixed prior to the fabrication of the chip, thus resulting in a non-programmable device that is the least costly solution. However, the circuit must be programmable if it is going to be applied to different tasks. In the latter case, the programmable values suitable for a given task are calculated by the computer and downloaded to the circuit prior to its use. After this, the circuit operates autonomously.

The second approach mentioned above is illustrated in the following. Programmable circuits realized with FPGA implementation techniques are considered. Their programmable values suitable to reproduce a given function will be calculated off chip by a computer that runs the learning tool *xfbpa* available within the *Xfuzzy* environment.

12.2.1 MODELING OF NON-LINEAR STATIC SYSTEMS

System modeling has become a very important task in many fields of science and engineering. Conventional techniques are not efficient enough to model highly non-linear systems and that is why fuzzy systems have been recently employed. To illustrate the approximation capability of fuzzy systems, let us consider the modeling of the following non-linear two-input one-output systems:

$$F_1: \qquad z = \frac{1}{1 + e^{10(x-y)}} \qquad \text{and} \tag{12.11}$$

$$F_2: z = \frac{1}{2}(1 + \sin(2\pi x)\cos(2\pi y)) \tag{12.12}$$

Let us compare the approximation obtained by systems that employ the

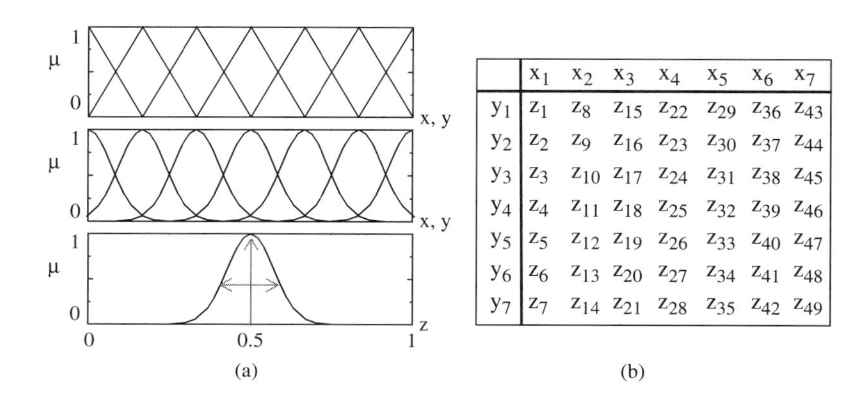

	x_1	x_2	x_3	x_4	x_5	x_6	x_7
y_1	z_1	z_8	z_{15}	z_{22}	z_{29}	z_{36}	z_{43}
y_2	z_2	z_9	z_{16}	z_{23}	z_{30}	z_{37}	z_{44}
y_3	z_3	z_{10}	z_{17}	z_{24}	z_{31}	z_{38}	z_{45}
y_4	z_4	z_{11}	z_{18}	z_{25}	z_{32}	z_{39}	z_{46}
y_5	z_5	z_{12}	z_{19}	z_{26}	z_{33}	z_{40}	z_{47}
y_6	z_6	z_{13}	z_{20}	z_{27}	z_{34}	z_{41}	z_{48}
y_7	z_7	z_{14}	z_{21}	z_{28}	z_{35}	z_{42}	z_{49}

Figure 12.6. Description of the inital fuzzy system to be adjusted. (a) Triangular or bell-shaped antecedent membership functions and consequent membership functions; (b) Rule base.

Fuzzy-Mean defuzzification method (singleton fuzzy systems) and systems that employ the Weighted-Fuzzy-Mean method, both of them using different shapes of membership functions and antecedent connectives. The knowledge bases capable of reproducing the input-output mappings in (12.11) and (12.12) are generated automatically by the learning tool *xfbpa*. In these examples, a set of 1000 patterns to train the system are obtained from a sweep of the input variables, x and y, for each function, F_1 and F_2.

At the beginning of the learning process, the fuzzy systems are described as shown in Figure 12.6. The input universes of discourse are covered by 7 membership functions equally spaced, which means 49 possible rules as a complete matrix rule base is considered. Two classes of membership functions are employed, triangular and bell-shaped functions, and two types of antecedent connectives are tested, the minimum and the product. The output universe of discourse is covered by 49 membership functions (one per rule) that are placed at the same point, 0.5 (Figure 12.6a). The fuzzy nature of the consequents is summarized into two parameters: the position of the function in the universe of discourse if Fuzzy-Mean (FM) defuzzification method is considered, and the position and width of the function if the Weighted-Fuzzy-Mean defuzzification method (WFM) is employed. The rule matrix of these starting systems is shown in Figure 12.6b. Since the consequents are all the same, the input-output surface provided by these systems is a horizontal plane defined by z=0.5, which in no way reproduces the intended behavior of (12.11) or (12.12).

The parameters that are adjusted by the learning tool are those defining the consequents: only the positions of the membership functions or both the positions and widths. Once those parameters have been adjusted, the learning tool allows the clustering of the resulting consequents labels in order to simplify the rule base. In the two examples discussed here, this step leads to a reduction of the number of consequent labels from 49 to 5. This is illustrated in Figure 12.7 for the two cases.

Table 12.2 shows the RMSE (root mean square error) and the XAE (maximum absolute error) obtained for the two modeling problems for each combination of membership function classes (triangles or bells), fuzzy connectives (minimum or product), and defuzzification methods (Fuzzy Mean, FM, or Weighted Fuzzy Mean, WFM) after clustering the consequents. Figure 12.8 provides a graphical representation of two of the approximations obtained, including the target surfaces, the learned surfaces, and the corresponding error surfaces. More details about the use of *xfbpa* to adjust fuzzy systems can be found in [Moreno, 1997].

12.2.2 ANALYSIS AND DISCUSSION OF FPGA IMPLEMENTATIONS

ASIC implementations of fuzzy systems that approximate given input/output mappings were illustrated in Chapters 7, 8, and 9. Let us now illustrate FPGA implementations. Among the different fuzzy systems adjusted previously with the off-chip learning, FPGA implementations will be discussed for systems that use triangular membership functions for the antecedents, the

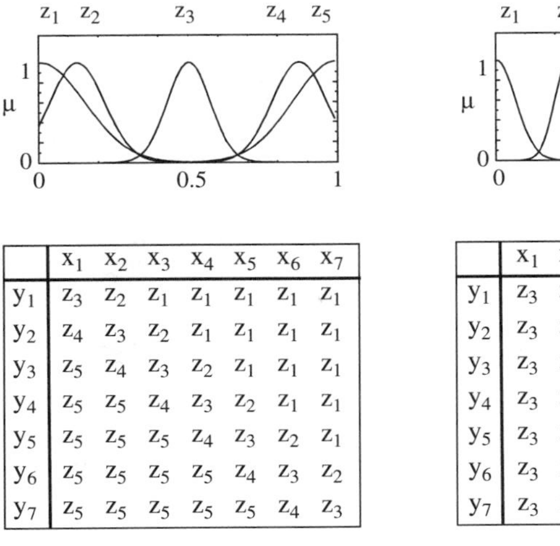

	x_1	x_2	x_3	x_4	x_5	x_6	x_7
y_1	z_3	z_2	z_1	z_1	z_1	z_1	z_1
y_2	z_4	z_3	z_2	z_1	z_1	z_1	z_1
y_3	z_5	z_4	z_3	z_2	z_1	z_1	z_1
y_4	z_5	z_5	z_4	z_3	z_2	z_1	z_1
y_5	z_5	z_5	z_5	z_4	z_3	z_2	z_1
y_6	z_5	z_5	z_5	z_5	z_4	z_3	z_2
y_7	z_5	z_5	z_5	z_5	z_5	z_4	z_3

F_1: Triangle-Minimum

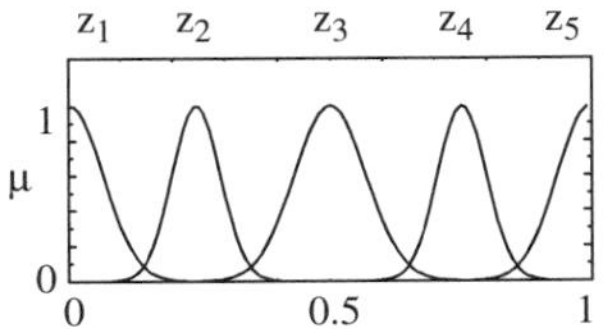

	x_1	x_2	x_3	x_4	x_5	x_6	x_7
y_1	z_3	z_5	z_5	z_3	z_1	z_1	z_3
y_2	z_3	z_4	z_4	z_3	z_2	z_2	z_3
y_3	z_3	z_2	z_2	z_3	z_4	z_4	z_3
y_4	z_3	z_1	z_1	z_3	z_5	z_5	z_3
y_5	z_3	z_2	z_2	z_3	z_4	z_4	z_3
y_6	z_3	z_4	z_4	z_3	z_2	z_2	z_3
y_7	z_3	z_5	z_5	z_3	z_1	z_1	z_3

F_2: Triangle-Minimum

Figure 12.7. Consequent membership functions and rule bases obtained after learning, considering minimum connective and triangular membership functions for the antecedents.

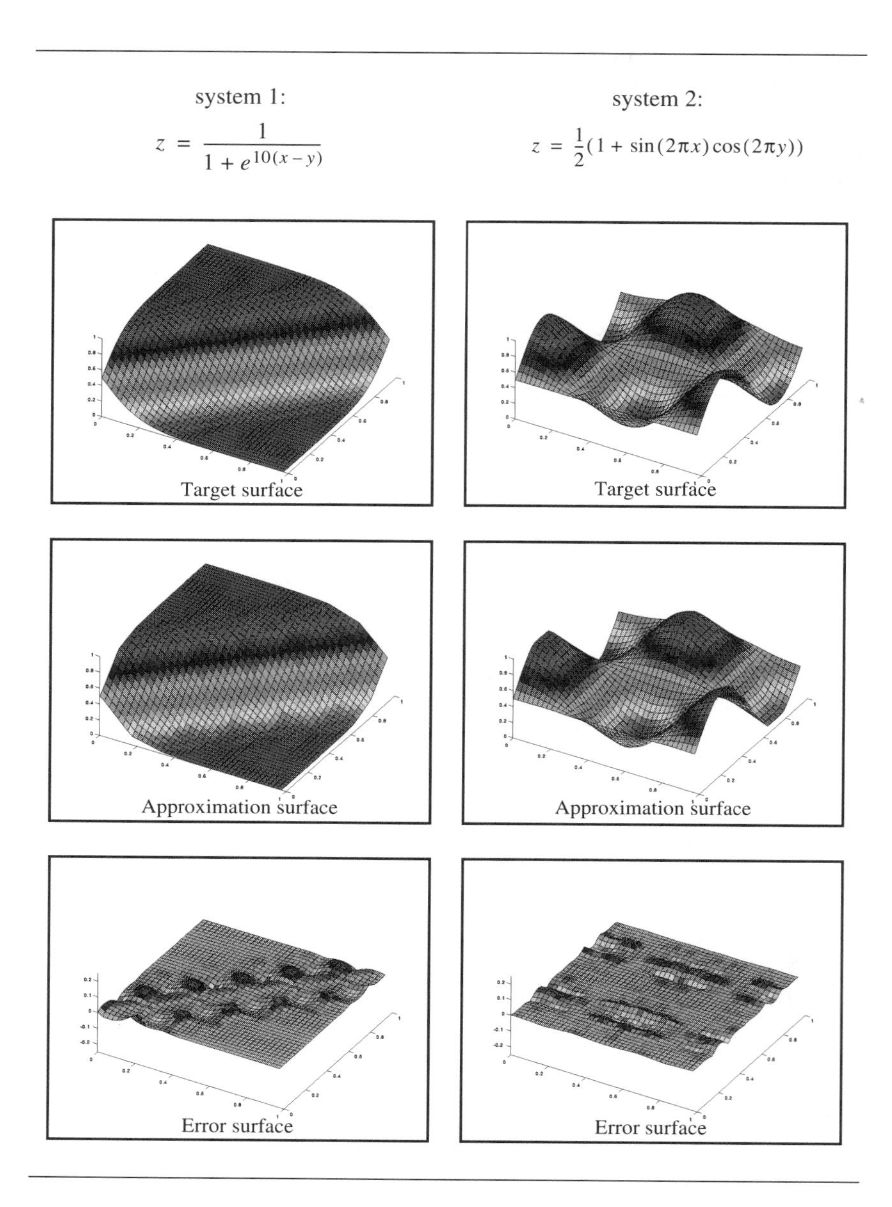

Figure 12.8. Examples of approximations with triangular membership functions for the antecedents, product connective, and Weighted-Fuzzy-Mean defuzzification method.

minimum connective and the Fuzzy Mean or Weighted Fuzzy Mean as defuzzification methods, in particular, systems whose consequent labels and rule bases are shown in Figure 12.7.

Table 12.2. RMSE // XAE obtained for the different fuzzy systems.

		system 1		system 2	
		FM	**WFM**	**FM**	**WFM**
min	**Bell**	1.54% // 6.01%	0.32% // 3.19%	2.54% // 6.27%	1.80% // 4.84%
	Triangle	1.60% // 5.74%	1.19% // 4.10%	2.46% // 7.48%	1.86% // 6.06%
prod	**Bell**	2.77% // 11.21%	1.81% // 5.41%	2.46% // 6.27%	2.18% // 4.43%
	Triangle	1.88% // 7.18%	1.00% // 3.21%	1.78% // 5.94%	1.45% // 5.80%

The XFL descriptions of these systems are introduced as input files for the hardware synthesis tools, *xftl* and *xfvhdl*, available within the *Xfuzzy* environment and described in Chapter 10. They allow the designer to explore different options within the design space: (1) implementations of look-up tables containing the output values the fuzzy system provides for each possible input combination, which means to follow an off-line strategy; (2) implementation of the fuzzy system with the active rule-driven architecture described in Chapter 9 using memory-based MFCs, which means to follow an on-line strategy; and (3) implementation of an on-line strategy as above but using the arithmetic-based MFCs described in Chapter 9. All the systems are designed using Xilinx XC4000 family FPGAs. To evaluate the performance of the different realizations, three figures will be considered: (1) implementation cost in terms of the number of configurable logic blocks, CLBs; (2) approximation accuracy measured by the root mean square error, RMSE; and (3) inference speed in terms of main clock cycles [Lago, 1997].

Figure 12.9a shows the evolution of the cost for the three implementation options as the number of bits increases. For low resolution implementations, the look-up table approach provides the best results. Conversely, as soon as the resolution grows (more than 5 bits), the on-line strategies address better results. In this sense, for higher resolution implementations, the arithmetic option seems to be more suitable than the memory option.

A similar analysis corresponding to the defuzzification methods is depicted in Figure 12.9b. Only the look-up table technique and the on-line strategy with arithmetic MFCs have been considered for simplicity. For look-up table techniques, the cost of WFM is slightly higher than that of FM. For on-line strategies, the cost of WFM is considerably higher than that of FM because specific hardware implementation of WFM requires one additional multiplier.

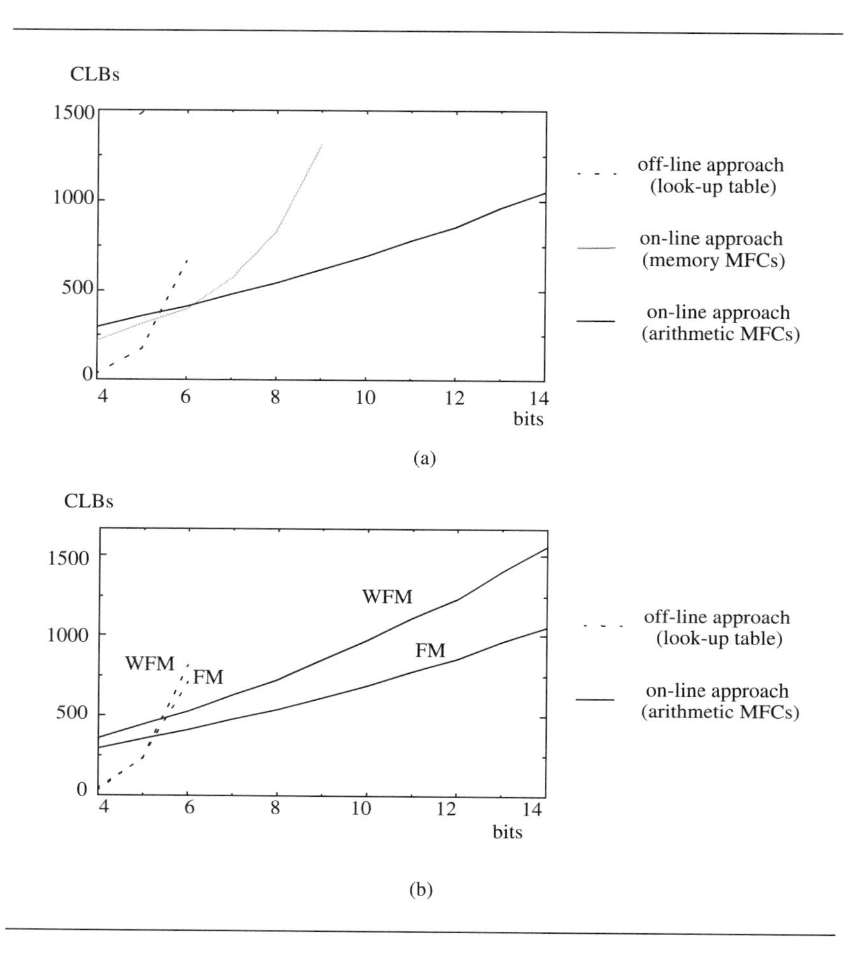

Figure 12.9. Implementation cost comparison. (a) System F_1 with FM; (b) System F_2 with FM and WFM.

Regarding approximation accuracy, three types of errors can be distinguished: (1) the inherent approximation error of the approximator system; (2) the truncation error due to the limited bus width; and (3) the error induced by the hardware implementation. The first error is inherent to any approximator system. This error can be minimized by increasing the number of membership functions for the antecedents and the number of singleton values for the consequents. To reduce the truncation error the bus width should be increased. Finally, the implementation error is a consequence of the fixed-point arithmetic used in the fuzzy system. In certain parts of the circuit this error is accumulative, as happens at the defuzzification stage.

Figure 12.10a represents the RMSE as a function of the number of bits. The behavior is similar for the three implementation options. For low resolutions (up to 6 bits) the truncation error is the dominant factor. When resolution is increased, approximation and hardware errors become the main error sources. Comparing the three techniques, implementation errors have higher influence when on-line strategies are used.

The results obtained for the system F_2 permit analysis of how the defuzzification method influences the RMSE. As shown in Figure 12.10b, in general WFM provides better results than FM. The difference between both defuzzification methods is more significant for the look-up table approach.

Finally, another aspect worth considering is the inference speed provided by the different implementation options. Implementations based on look-up tables are combinational circuits, so that inference speed is limited by the propagation delay of the FPGA critical path. The system operation is slower when an on-line strategy is used because several clock cycles are needed to carry out

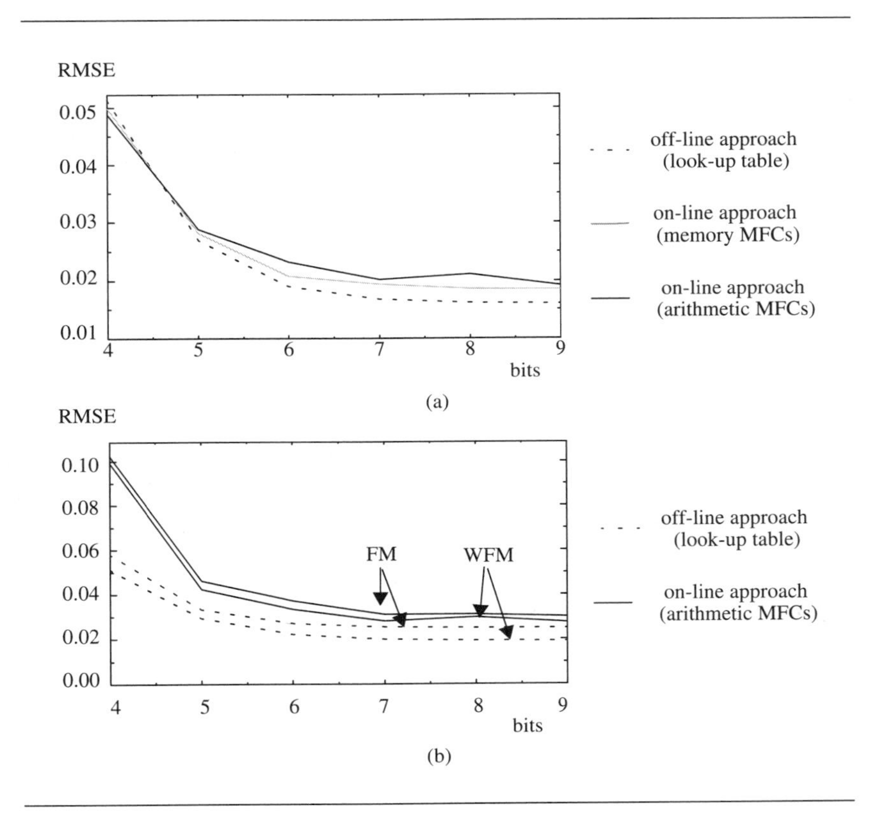

Figure 12.10. RMSE comparison. (a) System F_1 with FM; (b) System F_2 with FM and WFM.

an inference. On the one hand, the system throughput depends on the bus width. On the other hand, latency is also fixed by the number of pipeline stages (two for memory-based MFCs and three for arithmetic-based MFCs). Experimental results for systems with a resolution of five and six bits implemented using the Xilinx 4013PQ160-5 FPGA prove that a clock frequency of 5.5 MHz. is sufficient to ensure correct operation. This means an inference speed above 5 MFLIPS for the look-up table strategy, and close to 1 MFLIPS for the on-line strategies.

As a summary of these results, a trade-off among cost, accuracy, and speed must be considered when using FPGAs to implement a fuzzy system, as also happens with ASIC implementations. Although the look-up table approach, which is the most straightforward approach, is the best choice in terms of accuracy and speed, cost criteria make this solution impracticable when the resolution grows. As a consequence, techniques based on an on-line strategy implementing a fuzzy system are preferred for higher resolution systems, in particular techniques that implement arithmetic MFCs.

12.3 APPLICATIONS USING ON-CHIP LEARNING

Static applications like those described in the previous section employ fuzzy systems whose knowledge base does not change with time. Hence, it can be precalculated by an off-chip learning and then fixed or downloaded to the chip. However, there are complex applications for which a dynamically changing knowledge base is needed. Examples of this type of application are identification and modeling of non-linear dynamic systems, adaptive prediction of time series, adaptive noise cancellation, adaptive control, etc. In these cases, the fuzzy systems employed are called *adaptive fuzzy systems* because they adapt their knowledge base to match the new situation, thus resorting to the use of the learning mechanism as they are operating.

12.3.1 ANALYSIS AND DISCUSSION OF ASIC IMPLEMENTATIONS

The block diagram of an adaptive fuzzy system is shown in Figure 12.11. The fuzzy inference engine implements the fuzzification, rule processing, and defuzzification stages. The adaptive loop contains a performance monitor that measures the quality of the performance and sends this information to the tuning/learning block. This block implements the learning algorithm, which updates the knowledge base to match the new situation.

Concerning ASIC implementation of adaptive fuzzy systems, two situations can be distinguished. One situation, known as *chip-in-the-loop training*, relies on a host computer to train the chip. The computer acts as a performance monitor, which provides the input training vectors to the chip and compares the desired output vectors with the output generated by the chip. The fuzzy inference engine is implemented by a programmable device that is adequately pro-

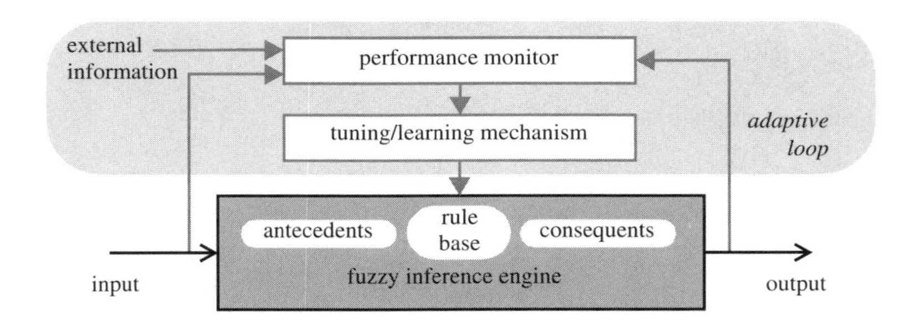

Figure 12.11. Block diagram of a typical adaptive fuzzy system.

grammed and reprogrammed by using an off-chip learning mechanism implemented by the host computer. The other situation is that the learning mechanism is implemented on chip together with the inference mechanism. The latter is the feasible solution when requirements of low power consumption, embedded operation (low area occupation), and real-time performance apply not only to the inference engine but also to the learning loop.

On-chip learning is a problem that is currently the object of much research. It began in the field of neural systems and currently covers also the field of adaptive fuzzy systems [Vidal-Verdú, 1994; Baturone, 1998; Miki, 1999].

An adaptive fuzzy chip with on-chip learning should contain circuitry to implement the fuzzy inference method and to carry out the learning algorithm. Selection of an efficient architecture is crucial to simplify the total circuitry and to provide a high inference and learning speed. In this sense, an active rule-driven architecture is advantageous to implement not only the inference method but also the learning algorithm because given an input, only the parameters associated with the active rules take part in the learning phase, similar to what happens during the inference phase.

Since the output of a singleton fuzzy system is a linear function of the consequent singleton values, a gradient-descent method is very suitable to optimize them because there is only one (global) minimum. As was shown in Chapter 5, the generic gradient-descent updating equation for a singleton value, c_i, is:

$$c_i\big|_{new} = c_i\big|_{old} - \eta \cdot \frac{\partial E}{\partial c_i} \tag{12.13}$$

where η, the learning rate, is a suitably chosen constant and E, the error, is a performance index of how well the desired function, y_d, is approximated by the fuzzy system output, y. E is usually defined as the mean square error over a finite set of training data or a finite time interval [k+1–T, k]:

$$E = \frac{1}{T} \cdot \sum_{p=k+1-T}^{k} (y - y_d)^2 \Big|_p \tag{12.14}$$

Taken for y the expression of a singleton fuzzy system:

$$y = \sum_{r=1}^{R} h_r \cdot c_r / \sum_{r=1}^{R} h_r \tag{12.15}$$

where h_r is the activation degree of the r-th rule, the parameter c_i is adjusted as:

$$c_i(k+1) = c_i(k+1-T) - \zeta \cdot \sum_{p=k+1-T}^{k} (y - y_d) \frac{h_i}{\Sigma h_r} \Big|_p \tag{12.16}$$

where $\zeta = 2\eta / T$.

Adjustment of the antecedent parameters, a_i, is much more difficult than consequent tuning. One reason is that the dependence of the fuzzy system output on the antecedent parameters is non-linear so that the optimization process based on gradient-descent techniques can be trapped at local minima. Another reason that makes antecedent tuning difficult is that the expressions of the partial derivatives of the error function are much more complex than for the consequents. To avoid this problem, a weight-perturbation learning algorithm is quite adequate. As was explained in Chapter 5, this is an steepest-descent method with a very simple updating equation:

$$a_i(k+1) = a_i(k+1-T) - \beta \cdot \sum_{p=k+1-T}^{k} \Delta E|_p \tag{12.17}$$

where β is a suitably chosen constant and $\Delta E|_p$ is the error variation for the pattern p and for a small perturbation in parameter a_i. Variation $\Delta E|_p$ is calculated as:

$$\Delta E|_p = |y_{pert} - y_d| - |y - y_d| \tag{12.18}$$

where y_{pert} is the output of the fuzzy system when parameter a_i is perturbed.

Equations (12.16) and (12.17) represent an off-line learning algorithm where each value is updated after the presentation of T patterns. On-line learning, where the values are updated after the presentation of 1 pattern (T=1), is more adequate for hardware implementation because no accumulator is required, although it is less effective in general. The on-line tuning of the consequents is enough for many applications as will be shown in the examples of Sections 12.3.2 and 12.3.3. In other applications like adaptive noise cancellation (Section 12.3.4), the desired output is contaminated by an additional signal that should be averaged out during the training process, so that off-line learning can be preferred.

To carry out the learning process of the singletons, the parameter change $\zeta \cdot (y - y_d) \cdot h_i / \Sigma h_r$ has to be computed and the new value has to be written in the consequent memory. Since the values of y_d and ζ depend on the particular problem and even y can be an external signal related to the fuzzy system output, signal $\zeta \cdot (y - y_d)$ can be obtained from outside the chip with a suitable conditioning circuitry which allows a continuous variation of ζ. Hence, the parameter change can be computed by multiplying the external signal $\zeta \cdot (y - y_d)$ with the value $h_i / \Sigma h_r$ which can be provided by the inference circuitry. Concerning the on-chip learning of antecedents and supposing that signal $\beta \cdot (y - y_d)$ is also provided from outside the chip, only a scaler (to scale by β) and adders are required to implement the learning algorithm.

Analog or digital circuits can be employed to implement these operations (addition, scaling, and multiplication). As occurs with the inference mechanism, analog circuitry is feasible because there are many applications that do not require a high precision. In the examples commented on in the following, resolution below 7 bits is enough. One of the factors that greatly influences the final error obtained after training is the quantization of the tuning parameters if they are stored into digital memories, and the quantization of the learning process operations if this is implemented by digital or mixed-signal circuitry. For instance, when off-line learning is considered, an analog integrator instead of a digital accumulator may be preferred to calculate the parameter change.

Several application examples that illustrate the use of adaptive fuzzy chips are shown in the following. The chips considered in these examples employ current-mode continuous-time analog circuitry (described in Chapter 7) for implementing both the inference and the learning processes. They make use of quantizers (basically a combination of A/D and D/A converters, similar to the successive-approximation divider circuits described in Chapter 7) in order to store the discrete level closest to the parameter value learned by the tuning process [Baturone, 1997].

12.3.2 IDENTIFICATION OF A DYNAMIC PLANT

In this application, an adaptive fuzzy system is employed as a building block of a dynamic identification model that reproduces the behavior of a plant. As an example, let us consider a non-linear plant usually reported in the literature [Narendra, 1991; Wang, 1992b], governed by the following difference equation:

$$y(k+1) = 0.5y(k) + 0.6y(k-1) + g[u(k)] \tag{12.19}$$

where $g(\cdot)$ is an unknown function.

For this application, the flow diagram of the configuration is shown in Figure 12.12. The following series-parallel model is used to identify the plant:

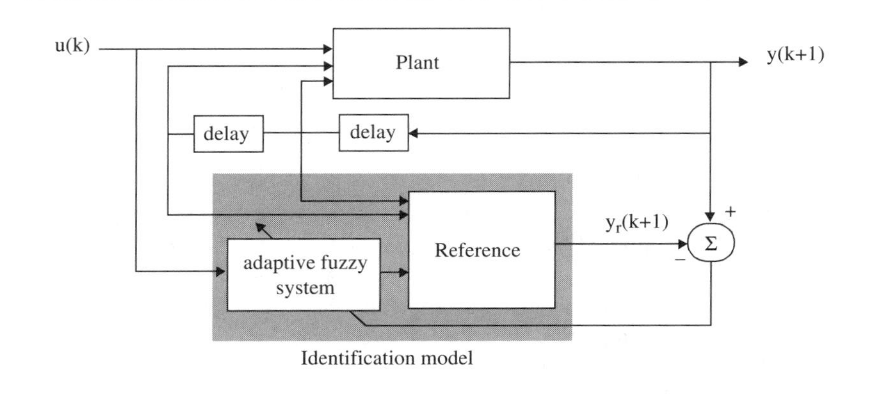

Figure 12.12. Scheme of the identification of a dynamic plant.

$$y_r(k+1) = 0.3y(k) + 0.6y(k-1) + F[u(k)] \tag{12.20}$$

where F[u] is the response of an adaptive fuzzy system that approximates the unknown function $g(\cdot)$.

This configuration has been simulated considering a singleton fuzzy chip with 18 rules whose antecedent fuzzy sets are represented by normalized triangular membership functions covering the input space uniformly and whose singleton values are learned by using a generic gradient descent learning mechanism. Contrary to the examples in the previous section, the learning is now performed on-line after each time step, k.

During training, the input to the plant and to the model was a sinusoid $u(k)=\sin(2\pi k/40)$ and the desired output patterns of the fuzzy system are taken as $y(k+1) - y_r(k+1)$. To simulate the plant, function $g(\cdot)$ has been considered as:

$$g(u) = 0.6\sin(\pi u) + 0.3\sin(3\pi u) + 0.1\sin(5\pi u) \tag{12.21}$$

Starting from a situation in which the 18 singleton values were zero and considering a resolution of 4 bits for the quantizers of the chip, the model follows the output of the plant after 70 time steps of training, as illustrated in Figure 12.13. This figure shows the output of the identification model and the plant for the input $u(k)=\sin(2\pi k/250)$ for $70 \le k \le 320$ and $571 \le k \le 770$ and $u(k)=0.5\sin(2\pi k/250)+0.5\sin(2\pi k/25)$ for $321 \le k \le 570$.

12.3.3 PREDICTION OF TIME SERIES

Adaptive prediction of time series are applications where the fuzzy system has to adjust its knowledge base to predict the output provided by the time series. As an example, let us consider the two-input one-output time series defined by:

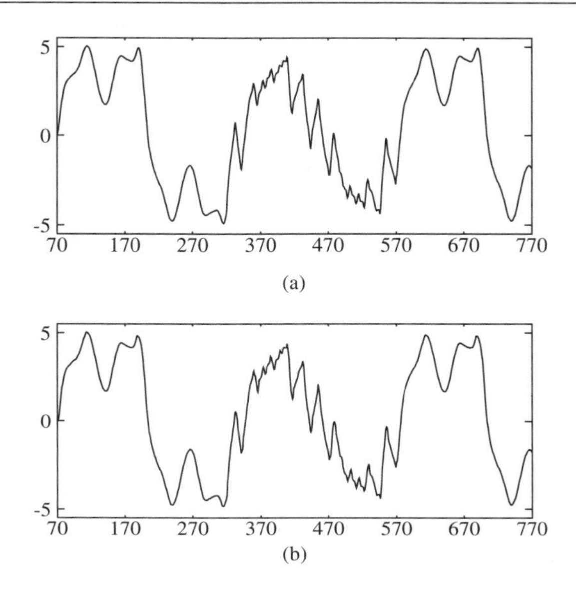

Figure 12.13. Output of (a) the plant and (b) the model after the training process.

$$y(t) = \left(0.8 - 0.5e^{-y(t-1)^2}\right) \cdot y(t-1) - \left(0.3 + 0.9e^{-y(t-1)^2}\right) \cdot y(t-2) + 0.1\sin(\pi y(t-1)) \tag{12.22}$$

which has been also analyzed in [Brown, 1994]. This series has an unstable equilibrium at the origin and a globally attracting limit cycle, as shown in Figure 12.14. This figure illustrates the evolution of the time series starting from the initial conditions y(t–1) = 0.1 and y(t–2) = 0.1.

The block diagram of the adaptive fuzzy system for this application is shown in Figure 12.15. A tapped delay line that consists of two delay units is used to provide a history of the dynamic operation of the series to the fuzzy system (the signals y(t–1) and y(t–2)).

A singleton fuzzy chip employing normalized and triangular membership functions to represent the antecedents has been simulated. The input spaces of y(t–2) and y(t–1) are covered by 2 and 4 membership functions, respectively, so that the fuzzy system contains 8 rules. Its 8 singleton values, which are initially zero, are adapted after each time step, that is, after the presentation of

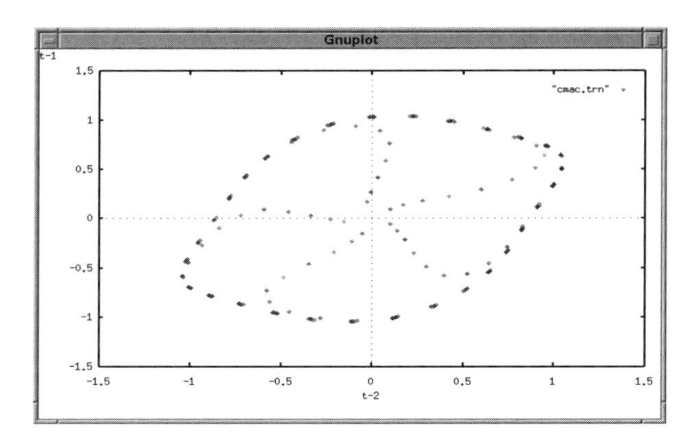

Figure 12.14. Time series evolution from [0.1,0.1] as initial conditions.

each training pattern by using a generic gradient descent method. The resolution of the quantizers is 6 bits.

Figure 12.16 illustrates how the fuzzy system learns to reproduce the behavior of the time series. This evolution corresponds to that shown in Figure 12.14 of y(t–2)=0.1 and y(t–1)=0.1 as initial conditions. The fuzzy system is trained during the first 100 time steps, so that the approximation error decreases as more training patterns are considered. The next time steps (from t=100) are used to validate the learned knowledge base.

12.3.4 ADAPTIVE NOISE CANCELLATION SYSTEM

The basic concept of adaptive noise cancellation systems is shown in Figure 12.17. A signal *s* is transmitted over a channel to a sensor which also receives a

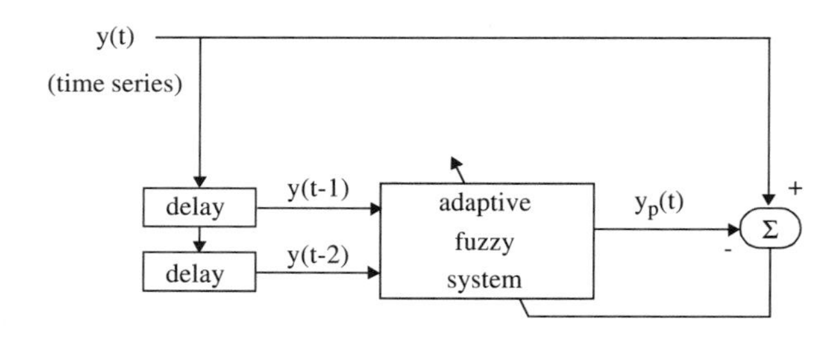

Figure 12.15. Scheme of adaptive prediction of time series.

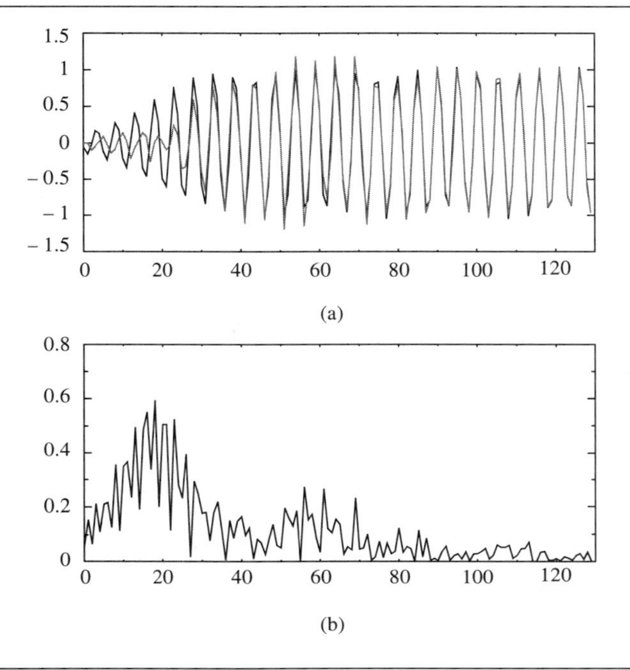

Figure 12.16. (a) Output of the fuzzy system (gray line) and the time series (black line) during the learning process; (b) Evolution of the error $|y(t) - y_p(t)|$.

noise n_0, uncorrelated with the signal, thus transmitting a corrupted signal x. The noise n_0 is a non-linear function of a noise source n_1 due to the existence of a non-linear channel. A second sensor receives the noise n_1 which is uncorrelated with the signal but correlated in some unknown way with the noise n_0.

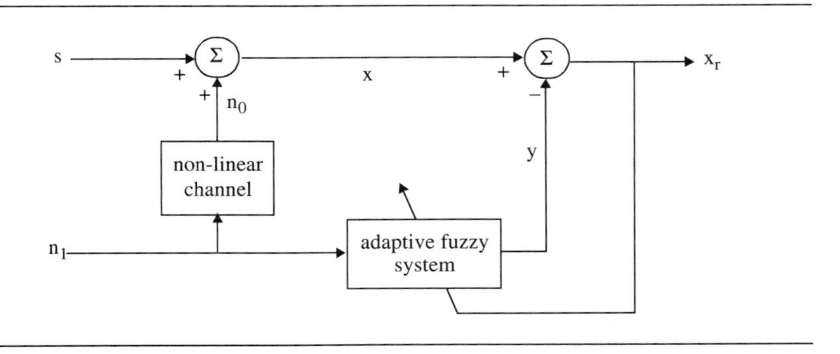

Figure 12.17. Scheme of adaptive noise cancellation application.

In order to generate a replica of n_0, the fuzzy system is trained with pairs of data (n_1, $s+n_0$). The example reported in [Lin, 1997] has been considered, where:

$$s(k) = \sin(0.06k) \cdot \cos(0.01k) \tag{12.23}$$

$$n_0(k) = 0.6 \cdot [n_1(k)]^3 \tag{12.24}$$

The fuzzy chip simulated contains 9 rules whose antecedent membership functions are normalized triangles that cover the input space uniformly. The 9 singleton values are updated after each epoch of 628 training data, which means an off-line learning. A generic gradient descent method is again used and the quantizers of the chip are taken with 6 bits of resolution. The good performance obtained is illustrated in Figure 12.18. The bottom part of this figure shows the recovered signal, x_r, after 16 epochs of training.

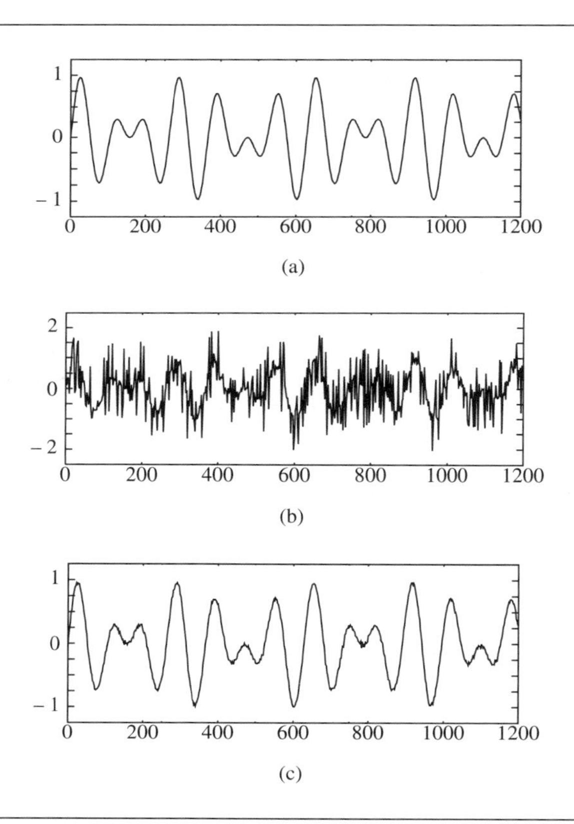

Figure 12.18. (a) The information signal, s; (b) The corrupted signal, x; (c) The recovered signal, x_r.

REFERENCES

1. **Baturone, I., Sánchez-Solano, S., Huertas, J. L.**, Self-checking current-mode analogue memory, *Electronics Letters*, Vol. 33, N. 16, pp. 1349-1350, 1997.
2. **Baturone, I., Sánchez-Solano, S., Huertas, J. L.**, Towards the IC implementation of adaptive fuzzy systems, *IEICE Transactions on Fundamentals*, Vol. E81-A, N. 9, pp. 1877-1885, 1998.
3. **Baturone, I., Sánchez-Solano, S., Barriga, A., Huertas, J. L.**, Design issues for the VLSI implementation of universal approximator fuzzy systems, in *Proc. World MultiConference on Circuits, Systems, Communications and Computers*, pp. 6471-6476, Athens, 1999.
4. **Bossley, K. M.**, Neurofuzzy modelling approaches in system identification, Ph.D. dissertation, University of Southampton, 1997.
5. **Brown, M., Harris, C.**, *Neurofuzzy Adaptive Modelling and Control*, Prentice-Hall, 1994.
6. **Buckley, J. J.**, Universal fuzzy controllers, *Automatica*, Vol. 28, pp. 1245-1248, 1992.
7. **Castro, J. L.**, Fuzzy logic controllers are universal approximators, *IEEE Transactions on Systems, Man, and Cybernetics*, Vol. 25, N. 4, pp. 629-635, 1995.
8. **Fattaruso, J., Meyer, R.**, MOS analog function synthesis, *IEEE Journal of Solid-State Circuits*, Vol. 22, N. 6, pp. 1056-1063, 1987.
9. **Gilbert, B.**, A monolithic microsystem for analog synthesis of trigonometric functions and their inverses, *IEEE Journal of Solid-State Circuits*, Vol. 17, N. 6, pp. 1179-1191, 1982.
10. **Jang, J-S. R., Sun, C-T.**, Functional equivalence between radial basis function networks and fuzzy inference systems, *IEEE Transactions on Neural Networks*, Vol. 4, N. 1, pp. 156-159, 1993.
11. **Kosko, B.**, Fuzzy systems as universal approximators, *IEEE Transactions on Computers*, Vol. 43, N. 11, pp. 1329-1333, 1994.
12. **Lago, E., Hinojosa, M. A., Jiménez, C. J., Barriga, A., Sánchez-Solano, S.**, FPGA implementation of fuzzy controllers, in *Proc. XII Conference on Design of Circuits and Integrated Systems*, pp. 715-720, Seville, 1997.
13. **Leme, C. A., Baltes, H.**, Programmable analog function synthesis, in *Proc. IEEE Int. Symposium on Circuits and Systems*, pp. 1397-400, Chicago, 1993.
14. **Lin, C-T., Juang, C-F.**, An adaptive neural fuzzy filter and its applications, *IEEE Transactions on Systems, Man, and Cybernetics*, Vol. 27, N. 4, pp. 635-656, 1997.
15. **Manaresi, N., Rovatti, R., Franchi, E., Baccarani, G.**, A current-mode piecewise-linear function approximation circuit based on fuzzy logic, in *Proc. IEEE Int. Symposium on Circuits and Systems*, pp. 127-30, Monterey, 1998.
16. **Miki, T., Yamakawa, T.**, Analog implementation of neo-fuzzy neuron and its on-board learning, in *Computational Intelligence and Applications*, pp. 144-149, World Scientific and Engineering Society Press, 1999.
17. **Moreno, F. J., López, D. R., Barriga, A., Sánchez-Solano, S.**, XFBPA: A tool for automatic fuzzy system learning, in *Proc. 5th European Congress on Intelligent Techniques and Soft Computing*, pp. 1084-1088, Aachen, 1997.
18. **Narendra, K. S., Parthasarathy, K.**, Gradient methods for the optimization of dynamical systems containing neural networks, *IEEE Transactions on Neural Networks*, Vol. 2, N. 2, pp. 252-262, 1991.
19. **Park, J., Sandberg, I. W.**, Universal approximation using radial-basis-function networks, *Neural Computation*, Vol. 3, pp. 246-257, 1991.
20. **Powell, M. J. D.**, *Approximation Theory and Methods*, Cambridge University Press, 1981.

21. **Rovatti, R.**, Fuzzy piecewise multilinear and piecewise linear systems as universal approximators in Sobolev norms, *IEEE Transactions on Fuzzy Systems*, Vol. 6, N. 2, pp. 235-249, 1998.

22. **Sánchez-Sinencio, E., Ramírez-Angulo, J., Linares-Barranco, B., Rodríguez-Vázquez, A.**, Operational transconductance amplifier-based nonlinear function synthesis, *IEEE Journal of Solid-State Circuits*, Vol. 24, N. 6, pp. 1576-1586, 1989.

23. **Turchetti, C., Conti, M.**, A general approach to nonlinear synthesis with MOS analog circuits, *IEEE Transactions on Circuits and Systems*, Vol. 40, N. 9, pp. 608-612, 1993.

24. **Vidal-Verdú, F., Rodríguez-Vázquez, A.**, Circuits and algorithms for adaptive neuro-fuzzy analog chips, in *Proc. MicroNeuro*, pp. 331-338, 1994.

25. **Wang, L. X., Mendel, J. M.**, Fuzzy basis functions, universal approximation, and orthogonal least-squares learning, *IEEE Transactions on Neural Networks*, Vol. 3, N. 5, pp. 807-814, 1992a.

26. **Wang, L. X., Mendel, J. M.**, Back-propagation fuzzy systems as nonlinear system identifiers, *Proc. Int. Conference on Fuzzy Systems*, pp. 1409-1418, San Diego, 1992b.

27. **Ying, H.**, Sufficient conditions on general fuzzy systems as function approximators, *Automatica*, Vol. 30, pp. 521-525, 1994.

28. **Zeng, X-J., Singh, M. G.**, Approximation theory of fuzzy systems-SISO case, *IEEE Transactions on Fuzzy Systems*, Vol. 2, N. 2, pp. 162-176, 1994.

29. **Zeng, X-J., Singh, M. G.**, Approximation theory of fuzzy systems-MIMO case, *IEEE Transactions on Fuzzy Systems*, Vol. 3, N. 2, pp. 219-235, 1995.

30. **Zeng, X-J., Singh, M. G.**, Approximation accuracy analysis of fuzzy systems as function approximators, *IEEE Transactions on Fuzzy Systems*, Vol. 4, N. 1, pp. 44-63, 1996.

Index